OP AMP NETWORK DESIGN

OP AMP

NETWORK DESIGN

John R. Hufault

A Wiley-Interscience Publication
JOHN WILEY & SONS
New York • Chichester • Brisbane • Toronto • Singapore

Library of Congress Cataloging in Publication Data:

Hufault, John R.
Op amp network design.

"A Wiley-Interscience publication."
Includes index.
1. Operational amplifiers. 2. Electronic
circuit design. I. Title.
TK7871.58.06H82 1985 621.381'735 85-9495

ISBN 0-471-81327-3
Printed in the United States of America

10 9 8 7 6 5 4 3 2 1

This book is dedicated to all the frustrated, down-trodden, over-worked, and unloved circuit designers of the world.

PREFACE

Since the advent of the semiconductor integrated circuit (IC), the operational amplifier has become a standard "component" of circuit design. Recent breakthroughs in IC metal-oxide-silicon (MOS) technology have now incorporated the Op Amp "building block" into very-large-scale-integration (VLSI) chips. These events have led to a phenomenal increase in the application of Op Amp networks in computer interfacing and signal processing applications.

The emphasis throughout this book concentrates on the industrial application of today's existing operational amplifier technology. The material presented is based on proven network techniques that have been successfully used in the field. No time is spent on experimental ideas that would involve extensive research and development to perfect. The circuits and techniques shown are a collection of the best known methods available today for achieving a desired design goal in the least amount of design time.

Section I discusses network parameters. First, a brief description of the general purpose operational amplifier is covered. The network is then examined with respect to each specific parameter of the Op Amp. By using this approach the designer is given a first-hand knowledge of the inherent trade-offs associated with network design. Furthermore, the technical discussions attempt to emphasize the "professional feel" for proper design rather than just a mathematical approach. However, several pertinent mathematical derivations are made available to the reader in the appendices for reference.

Section II illustrates useful Op Amp networks. This portion of the text presents over 300 practical networks that are designed with component values included. They are concisely explained in a manner that concentrates on the "end result". From this vantage point, remembering the trade-offs and limitations from Section I, the reader can use the simplified design equations to easily modify component values for specific requirements without spending time on tedious mathematical derivations.

External circuit components that do not contribute directly to the basic networks concept have been purposely omitted from the circuit diagrams for clarity. These components include external phase compensation circuits, current and voltage offset compensation, and various incidental components that are usually explicitly shown on the particular Op Amp data sheet. Many of the more popular Op Amps require little, if any, extra components in most applications.

It is also assumed that the designer will incorporate proper power supply decoupling, signal shielding, and grounding techniques with all the networks presented in this book.

John R. Hufault

Tucson, Arizona
October 1985

ACKNOWLEDGMENTS

I wish to thank the following individuals for their generous efforts and contributions toward completing this book: Mrs. Linda Coltrin and her young son Brian (who contributed his fingerprints) for the layout of the original manuscript and illustrations, Ken Lipson for much of the schematic artwork, and Mrs. Barbara Bickel for the excellent typing format and her uncanny skills in deciphering my illegible "hen scratchings."

I would also like to give thanks to my good friends Vic Borg and Jim Self for the encouragement they've given me over the years to "hang in there."

And last, but not least, I owe a very special thanks to my loving wife, Cathy, for her endless patience during numerous tough times while seeing this endeavor through.

J.R.H.

CONTENTS

Page

APPENDIX

TEXT LEGEND LIST

TEXT LEGEND LIST

APPENDIX LEGEND LIST

APPENDIX LEGEND LIST

OP AMP NETWORK DESIGN

SECTION I

NETWORK PARAMETERS

NETWORK PARAMETERS

INTRODUCTION

Although the Op Amp is extremely versatile in many network variations, the basic design concepts can be easily understood by thoroughly understanding the simple inverting and non-inverting circuit configurations.

This section is devoted to developing an effective and concise approach to learning the fundamental techniques used in designing practical Op Amp networks. Electron current flow is used throughout.

It will be seen that the parameters of an Op Amp are defined in a manner that generally makes them each independent variables. With a foundation of the parameter concepts in SECTION I, a designer can then proceed to SECTION II and quickly select the most advantageous configurations for solutions to specific design problems with a useful knowledge of the engineering tradeoffs.

NETWORK PARAMETERS

Op Amp Network Sample Design

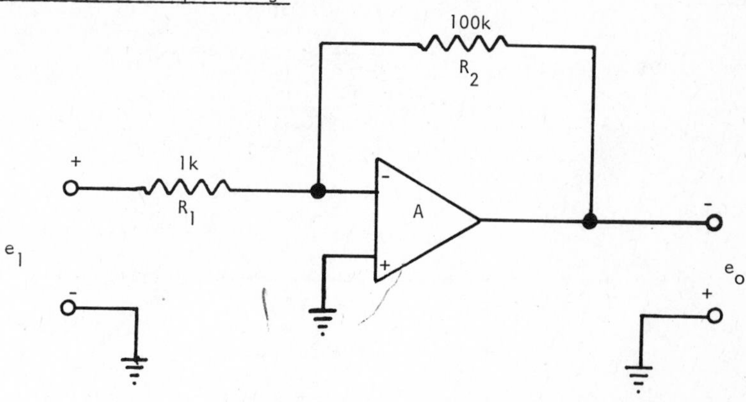

Sample Op Amp Specifications

parameter title	value	unit	remarks
Open loop gain	100,000	V/V	
	100	dB	
Unity gain bandwidth	1	MHz	
Slew rate	0.2	V/μsec	
Input voltage offset	1	mV	
Input voltage drift	50	μV/°C	
Power supply rejection	100	μV/V	
Input bias current	100	nA	
Input offset current	10	nA	
Input current drift	1	nA/°C	
Output impedance	1k	ohms	
Common mode rejection ratio	80	dB (input 10 V_{max})	
Input voltage noise	5	$\mu V_{(rms)}$	10 Hz to 1 kHz
Input current noise	100	$pA_{(rms)}$	10 Hz to 1 kHz
Settling time	5	μsec	@ 1% full scale
Capacitive loading	1000	pf	
Input impedance	1	MΩ	

Figure 1. Op Amp Network Sample Design

Sample Design Op Amp Network (continued)

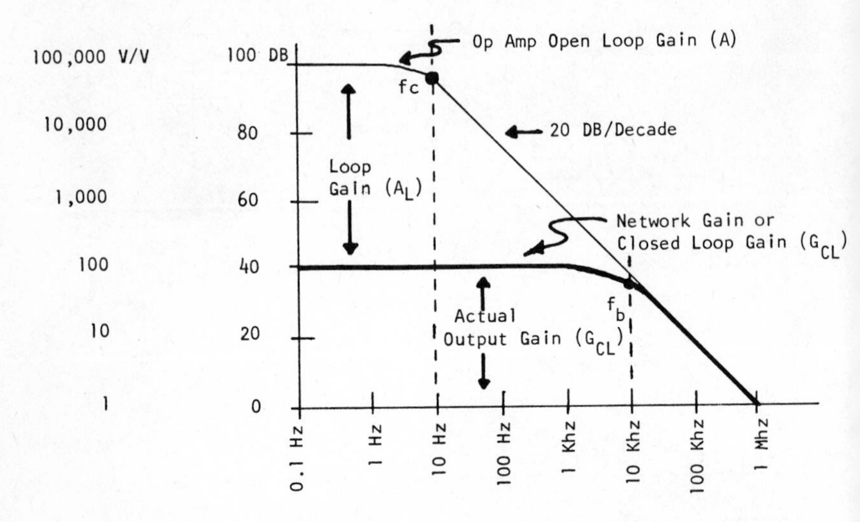

Figure 2. Op Amp Gain-Bandwidth Curves for Sample Design

The network shown in Figure 1 will be repeatedly used as an example circuit throughout Section 1 of the text. The network is a typical Op Amp inverting configuration that is designed to produce an output voltage gain of 100.

In Figure 2 the frequency characteristics of the network are shown. The open loop gain (A) of the Op Amp and the closed loop gain (G_{CL}) of the network are superimposed on the same graph for convenience. It is the closed loop gain (G_{CL}) that actually amplifies the input signal e_1 and produces the appropriate voltage at the output terminals e_o in the sample design Op Amp network of Figure 1.

NETWORK PARAMETERS

Gain

Although the gain of the Op Amp itself is 100,000 (or 100 DB), the actual output gain of the sample network shown in Figure 1 is constrained to 100. This output gain is commonly referred to as the "closed loop gain" G_{CL}.

$$G_{CL} = \text{OUTPUT GAIN} = \frac{R_2}{R_1} = \frac{100k}{1k} = 100$$

or expressed in decibels,

$$DB = 20 \log (G_{CL}) = 20 \log (100) = 40 \text{ DB}.$$

The "left over" gain is called the loop gain of the network A_L.

$$\text{DB LOOP GAIN} = \text{OP AMP GAIN} - \text{OUTPUT GAIN}$$

$$= (100 \text{ DB}) - (40 \text{ DB}) = 60 \text{ DB (from DC to fc)}.$$

Loop gain is essentially what enables an Op Amp network to be a very precise circuit. As seen from the frequency response curve in Figure 2, the loop gain begins to decrease as the frequency increases. The affect of this characteristic on all Op Amps is that the output accuracy will begin to deteriorate. This condition is physically seen by observing the output voltage begin to decrease. The output decrease is not in direct proportion to the loop gain decrease, however. In Figure 2 the loop gain begins decreasing rapidly after 10 Hz (20 DB/Decade). Yet, the actual output gain does not show any significant affect until the loop gain almost disappears near 10 Khz. This effect can be shown in the more complex output gain equation which is derived in Appendix P.

Gain (continued)

$$G_{CL} = \text{OUTPUT GAIN} = \underbrace{\frac{R_2}{R_1}}_{\text{Ideal Gain}} \underbrace{\left(\frac{1}{1 + \frac{1}{A_L}}\right)}_{\text{Gain Error}}$$

See Appendix P

A_L = Loop Gain

The percent gain error is $\left(1 - \frac{1}{1 + \frac{1}{A_L}}\right) \times 100$

or, $\% \text{ Error} = \frac{1}{1 + A_L} \times 100$

Example:

The gain error at 1 Hz, where loop gain is

60 DB (or 1000)

$$1 \text{ Hz gain error} = \frac{100}{1 + 1000} \simeq 0.1\%$$

The output voltage will therefore be slightly lower than the ideal gain calculation of 100 — that is, the output gain will be 99.9. This is well below the tolerance of standard precision 1% metal film resistors.

The gain error at 100 Hz where loop gain has now decreased by 20 DB leaving it at 40 DB (or 100).

$$100 \text{ Hz gain error} = \frac{100}{1 + 100} \simeq 1\%$$

Gain (continued)

It now can be seen that the output voltage will drop about 1% from its original DC value as a result of the loop gain decreasing with frequency. It should be noted that phase shift enters into these calculations. A more rigorous analysis is required at higher frequency. This is covered in Appendix A.

At 10 Khz the loop gain is essentially 0 DB or unity. This is the point where the output is 70.7% of its original value. This is the frequency cutoff point just as in the case of the common RC network. It is the point where the phase shift reaches 45°, and for all practical purposes the Op Amp network can no longer be called a precision circuit. This phenomenon is discussed in more detail in the following sections on bandwidth and phase shift.

For the non-inverting case the above error analysis also applies, except the actual output gain equation is:

$$\text{Non-inverting Output Gain} = \left(1 + \frac{R_2}{R_1}\right)\left(\frac{1}{1 + \frac{1}{A_L}}\right) .$$

Open loop gain (A) can be measured by applying a very small signal (10 μV peak) directly to the inputs of the Op Amp (Figure 3).

Gain (continued)

(a)

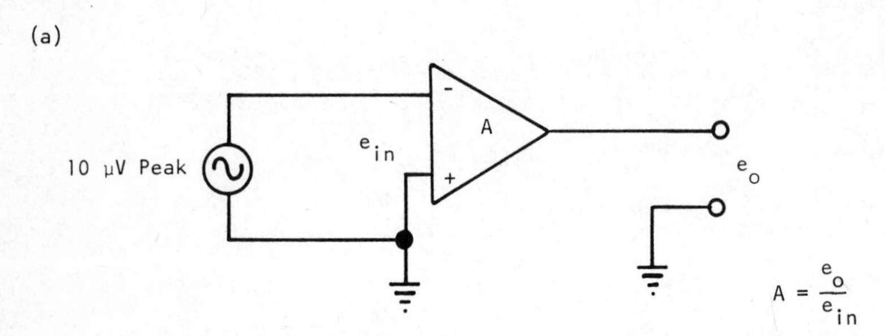

Figure 3. Open Loop Gain Measurement, Direct Method

(b)

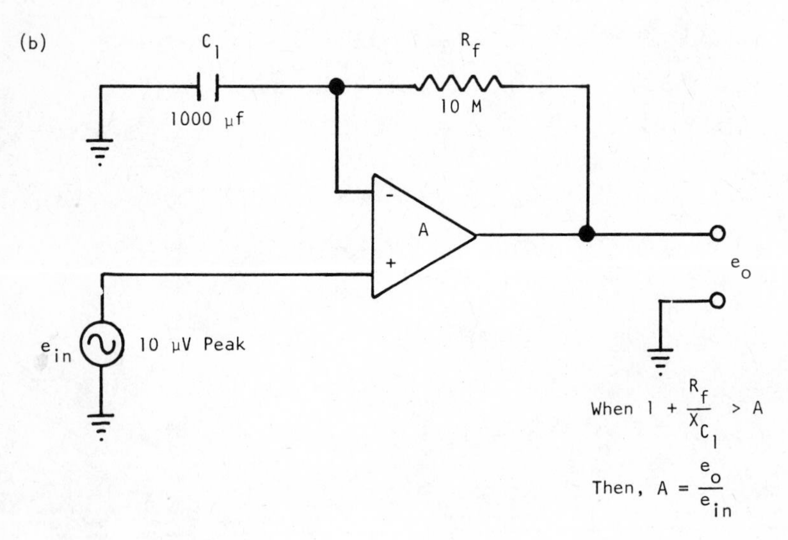

Figure 4. Open Loop Gain Measurement, DC Offset Bypassed

Gain (continued)

However, the input DC drift and offset voltages normally will make the output of the Op Amp saturate, or lock-up at one of the power supply voltages when trying to measure open loop gain directly. To avoid this problem the circuit shown in Figure 4 is used. The capacitor C_1 stores and cancels out the DC offsets while allowing the AC signal to pass to the output. The frequency must be below the Op Amp open loop gain breakpoint f_c. See Figure 2.

The gain of the R_f C_1 combination at this frequency (10 Hz) must be greater than the open loop DC gain (A).

$$1 + \frac{R_f}{X_{C_1}} > A$$

For an Op Amp with 100 DB DC gain to be measured by this circuit at 10 Hz, the R_f C_1 values should be 100 MΩ and 1000 μf, respectively.

$$\text{AC Gain at 10 Hz} = 1 + \frac{R_f}{X_{C_1}} = 2\pi f\ C_1 R_f$$

$$= 2\pi(10)(1000\mu f)(10M)$$

$$= 6.28\ (10^5)\ V/V$$

or, approximately

$$DB = 20 \log[6.28(10^5)] \simeq 116\ DB$$

This result shows that the circuit could measure Op Amp open loop gains as high as 116 DB.

Gain (continued)

However, if the DC response of the Op Amp was flat only to 1 Hz rather than 10 Hz, then the maximum gain that could be measured with the above values could be only 96 DB.

$$\text{AC Gain at 1 Hz} = 2\pi f C_1 R_f$$

$$= 6.28\ (10^4)\ \text{V/V}$$

or, approximately

$$\text{DB} = 20 \log [6.28\ (10^4)] = 96\ \text{DB}$$

Most Op Amp open loop responses are flat out to 100 Hz. Therefore, the R_fC_1 values shown are generally quite practical.

Sometimes the output may lock-up when the amplifier is initially plugged in. This is due to the inability of the large capacitor (C_1) to quickly compensate for the voltage offset (slow time constant). The problem can usually be overcome by initially reducing the value of R_f. Often, just temporarily placing a smaller resistor across the resistor (R_f) will bring the circuit out of saturation within a few seconds.

NETWORK PARAMETERS

Bandwidth

The bandwidth of the amplifier network in the sample design is 10 Khz (f_b). See Figure 2.

Network Bandwidth = 10 Khz

At this point, the loop gain is essentially 0 DB (or unity) and the output begins rolling off at 20 DB/Decade. For a more rigorous analysis see Appendix A.

A frequency response, or Bode plot, of the Op Amp is not always available on all data sheets. It is, however, possible to derive this curve by referring to the "Unity Gain-Bandwidth Product" specification (sometimes called Gain-Bandwidth Product).

On the sample design shown in Figure 1 and Figure 2, the Unity Gain-Bandwidth is 1 Mhz. To derive the frequency response, plot a point at unity gain (0 DB) and 1 Mhz. Move back one decade in frequency (100 Khz) and plot another point at 20 DB. For every decade in frequency the Op Amp gain will increase 20 DB until the open loop DC gain of the Op Amp is finally reached. In the case of Figure 2, the frequency where the DC open loop gain intersects with the 20 DB roll-off line is 10 Hz (f_c).

A more detailed look at this gain roll-off and the phase shift effects is covered in the following section on phase shift.

NETWORK PARAMETERS

Phase Shift

The phase shift is controlled by the DC loop gain as well as the frequency (see Appendix A for derivation).

$$\tan \phi = \frac{f}{A_L f_c}$$

ϕ = Phase Shift Angle

f_c = Cutoff Frequency of Open Loop Op Amp

A_L = Loop Gain at DC or Before Open Loop Cutoff Frequency.

In the sample design of Figure 1 the DC loop gain (A_L) is 60 DB, or 1000. The open loop cut off frequency (f_c) is 10 Hz.

Therefore at f = fc = 10 Hz (see Figure 2)

$$\tan \phi = \frac{10}{(1000)\ (10)} : \phi = 0.057^{\circ}$$

and at $f = f_b$ = 10 Khz

$$\tan \phi = \frac{10}{(1000)\ (10)} = 1 : \phi = 45^{\circ}$$

It can be seen that the output phase shift increases with frequency and is reduced by loop gain. It should also be noted that the closed loop cutoff frequency f_b of the network is extended out to the intersection of the open loop response curve and the closed loop output gain line. At that point the phase shift is $\phi = 45^{\circ}$, and the output voltage is 3 DB down or 70.7% of its original value.

Phase Shift (continued)

The analysis is identical for the non-inverting configuration.

A more rigorous analysis of bandwidth and phase shift error is covered in Appendix A. However, it is usually not necessary to use this rigorous approach in most design applications.

Slew Rate

This parameter is often the cause of confusion.

The slew rate limitation of an Op Amp will limit the output frequency response of the Op Amp for large signal outputs.

The large signal output is referred to as Full Power Frequency response. This is the frequency limitation of the Op Amp when the output is operating at its maximum specified voltage swing. Beyond this frequency the output sine wave will begin to triangulate, or "slew rate limit" (Figure 5).

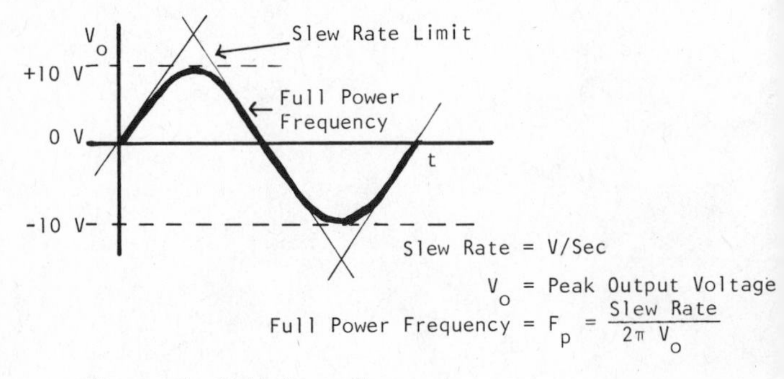

Figure 5. Full Power Frequency

In the sample design the slew rate = 0.2 V/μsec and the specified maximum output voltage is 10 volts peak (see Figure 1 specifications).

$$F_p = \frac{(0.2\ \text{V}/\mu\text{sec})\ (10^6)}{2\pi\ (10\text{V})} = 3.2\ \text{Khz} \quad \text{(Full Power Frequency)}$$

Although the actual bandwidth of this network extends out to 10 Khz as seen in Figure 2, an output 10 volt signal will begin to triangulate or rate limit itself at 3.2 Khz.

Slew Rate (continued)

It should be noted that by lowering the output voltage swing (this is done by reducing the input signal), this frequency can be extended. If the output swing is reduced to 5 volts, the frequency can be increased to 6.4 Khz before it is limited by the slew rate (see Figure 6).

Carrying this logic further, as the output voltage swing is continually lowered, the output frequency response will approach the actual output gain response curve shown in Figure 2, of f_b = 10 Khz.

Conversely, to find the maximum voltage swing possible to reach 10 Khz without triangulating (output cutoff frequency) in the sample design of Figure 1.

$$V = \frac{\text{Slew Rate}}{2\pi\ F_{max}}$$

$$= \frac{(0.2\ \text{V/ sec})\ (10^6)}{2\pi\ (10)\ (10^3)} = 3.2 \text{ volts (peak)}$$

This means if the Op Amp network shown in Figure 1 is operated with an input voltage of 32 mV peak, the output will be 3.2 volts peak from DC to 10 Khz with no triangulation of the output signal.

If, however, the input voltage is increased to 100 mV peak, the output will be 10 volts peak from DC to only 3.2 Khz. Beyond 3.2 Khz the output signal will begin to triangulate or slew rate limit (See Figure 6).

Slew Rate (continued)

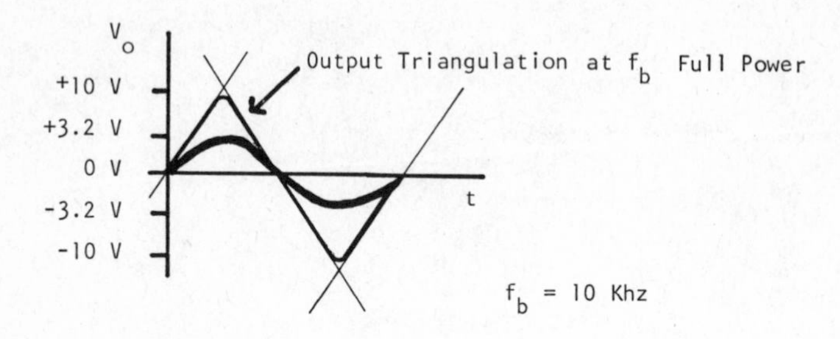

Figure 6. Triangulation Eliminated by Reducing Output Voltage Swing

Voltage Offset

The output voltage offset is calculated by first letting e_1 equal zero. The input offset can be simulated by inserting a voltage source at the non-inverting input to the Op Amp as shown in Figure 7.

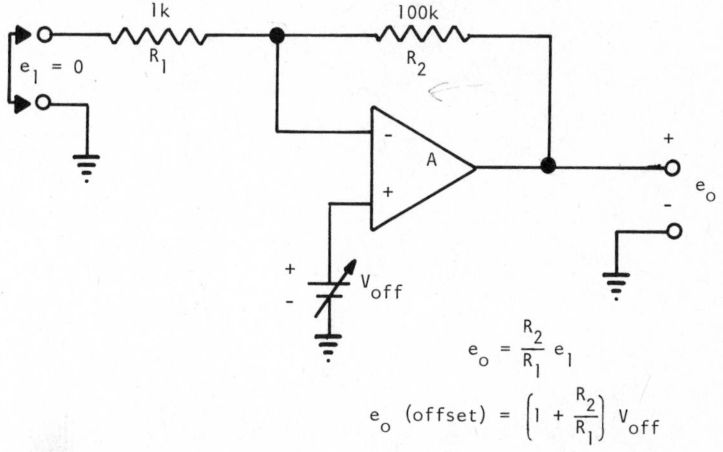

Figure 7. Voltage Offset Equivalent Network

Notice that the output offset is calculated from the non-inverting configuration regardless of whether the Op Amp is actually being used in the inverting configuration.

$$\text{OUTPUT OFFSET VOLTAGE} = \left(1 + \frac{R_2}{R_1}\right) \times (\text{INPUT OFFSET})$$

$$e_o \text{ (offset)} = \left(1 + \frac{100k}{1k}\right) (1 \text{ mV}) \simeq 100 \text{ mV}$$

Voltage Offset (continued)

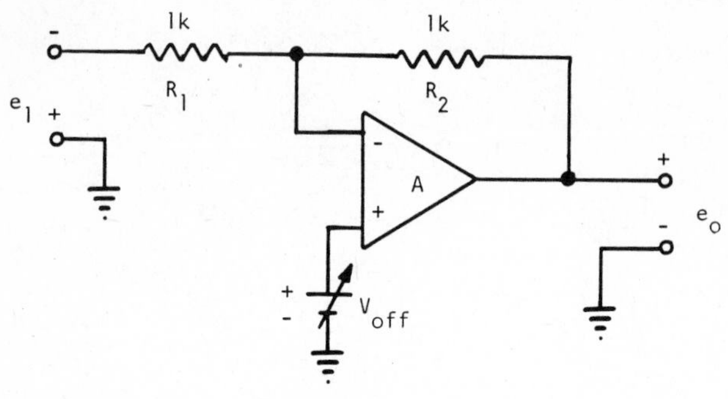

$$\text{Output Signal} \quad e_o = \frac{R_2}{R_1} e_1 \; : \; e_o = e_1$$

$$\text{Output Offset } e_o \text{ (offset)} = \left(1 + \frac{R_2}{R_1}\right) V_{off} \; : \; e_o \text{ (offset)} = 2\, V_{off}$$

Figure 8. Unity Gain Configuration

It should be noted that when an Op Amp is used in the unity gain inverting configuration, the offset output gain will be two (2), _not unity_ (see Figure 8).

Most Op Amps have the option of inserting an external trimming resistor on certain specified terminals to null out the voltage offset internally. This is the most effective method of minimizing the voltage offset error.

If the external trim option is not available, the offset voltage can be cancelled out by feeding an equal and opposite error signal into the network. The method is different depending on the Op Amp configuration, whether inverting or non-inverting (see Figures 9 and 10).

Voltage Offset (continued)

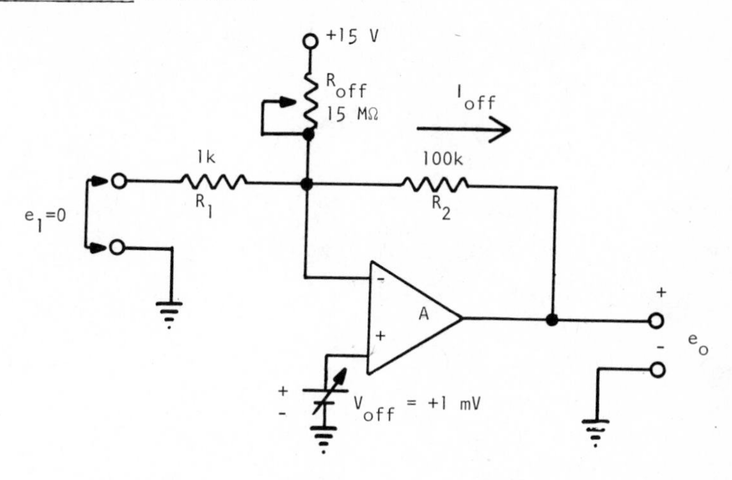

Figure 9. Output Voltage Offset Cancellation for Inverting Configuration

In Figure 9 the value of R_{off} is determined by the magnitude of the output voltage offset caused by V_{off}.

$$e_o \text{ (offset)} = \left(1 + \frac{100k}{1k}\right) \text{ (1 mV)} \approx 100 \text{ mV}$$

To cancel this offset voltage, an offset current must be fed into the network equal and opposite to I_{off} (Figure 9).

$$I_{off} = \frac{e_o}{R_2} = \frac{100 \text{ mV}}{100k} = 1 \text{ µA}$$

Voltage Offset (continued)

Therefore,

$$R_{off} = \frac{V}{I_{off}} = \frac{+15\ V}{1\ \mu A} = 15\ M\Omega$$

Note that the polarity of the voltage driving R_{off} must be chosen to cancel the polarity of the output error. This error can have either polarity and therefore must be measured before connecting R_{off} to the proper voltage source.

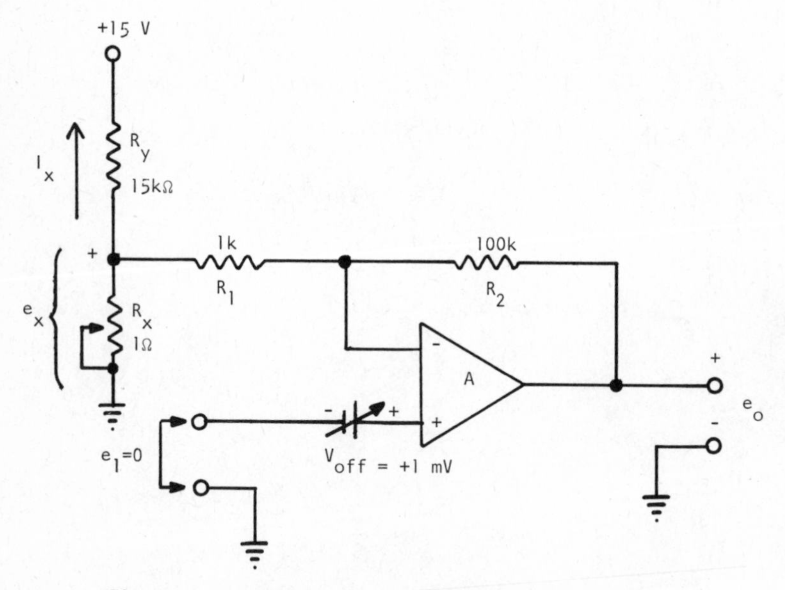

Figure 10. Output Voltage Offset Cancellation for Non-Inverting Configuration

Voltage Offset (continued)

For the non-inverting configuration in Figure 10, the voltage e_x is used to cancel the output voltage offset caused by V_{off}.

$$e_o \text{ (offset)} = \left(1 + \frac{100k}{1k}\right) (1 \text{ mV}) \simeq 100 \text{ mV}$$

since,

$$\frac{e_o}{e_x} = \frac{R_2}{R_1}$$

then

$$e_x = \frac{R_1}{R_2} e_o = \frac{1k}{100k} (100 \text{ mV}) \simeq 1 \text{ mV} .$$

The voltage e_x is generated from the external divider R_x, R_y as shown in Figure 10. It should be noted that the value of R_x must be small compared to R_1 to avoid excessive "loading" of the divider circuit.

Making

$$R_x = 1\Omega$$

then,

$$I_x = \frac{e_x}{R_x} = \frac{1 \text{ mV}}{1\Omega} = 1 \text{ mA}$$

and

$$R_y = \frac{V}{I_x} = \frac{+15 \text{ V}}{1 \text{ mA}} = 15 \text{ k}\Omega$$

Voltage Offset (continued)

Note the polarity of e_x must be chosen to cancel the output voltage offset polarity.

The voltage offset discussed in this section is measured with the Op Amp at room temperature and at a specified supply voltage.

The voltage offset changes with temperature (voltage drift), supply voltage (supply rejection), and with time (stability). All these specifications are on the data sheets and will be discussed in detail in later sections.

Voltage Drift

The data sheet voltage drift specification is always given as referred to the input. The output drift can then be calculated based on the external feedback gain circuitry for the specific engineering application.

In general, the polarity of the voltage drift temperature coefficient is unpredictable. The voltage drift of an Op Amp is the change in voltage offset with temperature. The output voltage error due to drift is determined in the same manner as the voltage offset error discussed in the previous section. However, unlike the voltage offset, the drift cannot be nulled out easily. This is usually the most troublesome parameter to contend with in precision Op Amp designs.

The input voltage drift can be simulated by inserting a voltage source at the non-inverting input. The output voltage drift is then

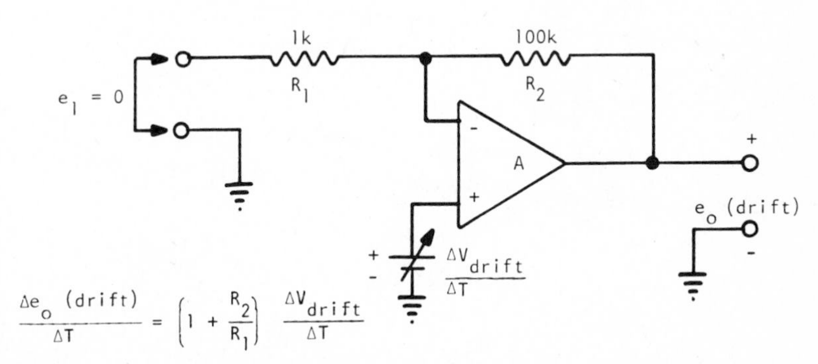

$$\frac{\Delta e_o\ (\text{drift})}{\Delta T} = \left(1 + \frac{R_2}{R_1}\right) \frac{\Delta V_{drift}}{\Delta T}$$

Figure 11. Voltage Drift Equivalent Network

Voltage Drift (continued)

calculated by letting the input signal e_1 equal zero.

In the case of the sample design (Input Voltage Drift = 50 $\mu V/^{\circ}C$ from Figure 1).

$$\frac{\Delta e_o \text{ (drift)}}{\Delta T} = \left(1 + \frac{100k}{1k}\right) (50\ \mu V/^{\circ}C)$$

$$\approx \pm 5\ mV/^{\circ}C \quad \text{output voltage drift}$$

If the Op Amp is heated from room temperature $+25^{\circ}C$ to $+55^{\circ}C$.

$$\Delta T = 55^{\circ}C - 25^{\circ}C = 30^{\circ}C$$

and

$$\Delta e_o \text{ (drift)} = \frac{\Delta e_o}{\Delta T} \text{ (drift) } \Delta T$$

$$= (5\ mV/^{\circ}C)(30^{\circ}C)$$

$$= \pm 150\ mV \quad \text{output voltage change}$$

The output will change by 150 mV from where it was at room temperature. The polarity of the change is unpredictable and very dependent on different manufacturer designs.

Voltage Drift (continued)

The total output voltage error is the algebraic sum of the initial room temperature output voltage offset and the output voltage drift.

$$e_o = e_o \text{ (offset)} + \dot{e_o} \text{ (drift)}$$

For the sample design in Figure 11

$$e_o = \left(1 + \frac{100k}{1k}\right) (\pm 1 \text{ mV}) + \left(1 + \frac{100k}{1k}\right) (50 \text{ } \mu V/^{\circ}C)(30^{\circ}C)$$

$$= (\pm 100 \text{ mV}) + (\pm 150 \text{ mV})$$

$$e_o = \pm 250 \text{ mV} \quad \text{(worse case output error)}$$

Although the voltage drift is specified as a linear parameter, this is not actually the case, except for chopper stabilized Op Amps.

For bi-polar and FET Op Amps, the drift range from 20 $\mu V/^{\circ}C$ and above is essentially linear. However, for drift ranges below 10 $\mu V/^{\circ}C$ the straight line approximation can sometimes be misleading.

Voltage Drift (continued)

Figure 12 shows the various combinations of drift curves that can occur for a given bi-polar or FET Op Amp family. Different manufacturers specify these maximum limits in various ways. The user must refer to the data sheet for the exact method of interpretation.

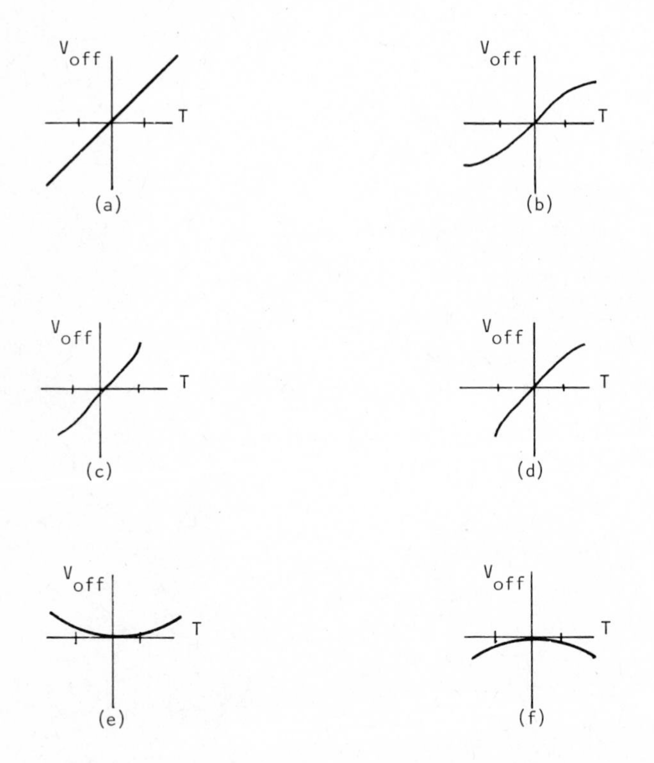

Figure 12. Voltage Drift Family of Curves for Bi-Polar and FET Op Amps

Voltage Drift (continued)

It should also be noted that the best drift linearity occurs between $0^{\circ}C$ and $+60^{\circ}C$. Beyond these two limits internal second order effects begin to increase significantly.

Generally, the bi-polar Op Amp is more linear with temperature than the FET Op Amp. High-quality bi-polar Op Amps are available with drifts as low as 0.1 $\mu V/^{\circ}C$ while the FET Op Amps are usually limited to the 1 $\mu V/^{\circ}C$ range.

The chopper stabilized Op Amp is a very linear low-drift Op Amp which essentially overcomes the non-linear problems mentioned above. The chopper Op Amps are available with drifts in the 0.01 $\mu V/^{\circ}C$ range.

The common disadvantage of the chopper type Op Amp is that it must be operated in the inverting configuration only. This can be a serious problem when trying to instrument high impedance transducers.

There are a few "differential" (both inverting and non-inverting) chopper Op Amps available, but parameters such as noise and drift are sacrificed in order to achieve operation in the non-inverting configuration. The high-quality bi-polar Op Amps mentioned above often have good performance in the low drift non-inverting applications. They are also priced more competitively, however, this will change as FET technology improves.

Supply Rejection

As with the other offset specifications, the supply rejection is also referred to the input of the Op Amp.

This parameter reflects the change in input offset voltage for a given change in power supply voltage. It is normally specified in microvolts input offset per one volt change in power supply (μV/V).

The output voltage error is calculated in the same manner as voltage offset and drift.

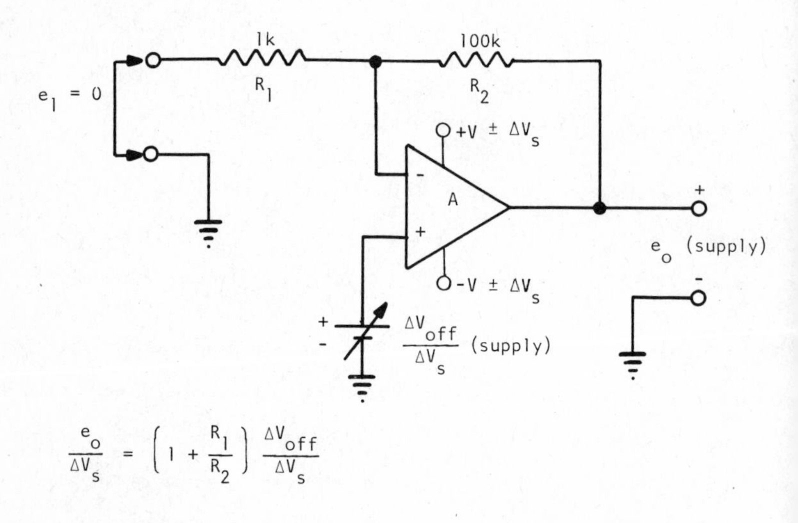

$$\frac{e_o}{\Delta V_s} = \left(1 + \frac{R_1}{R_2}\right) \frac{\Delta V_{off}}{\Delta V_s}$$

Figure 13. Supply Rejection Equivalent Circuit

Supply Rejection (continued)

In the sample design the supply rejection (Figure 1),

$$\frac{\Delta V_{off}}{\Delta V_{cc}} \text{ (supply)} = 100 \ \mu V/V$$

Therefore the output error for a one volt supply change is

$$e_o \text{ (supply)} = \left(1 + \frac{100k}{1k}\right) (100 \ \mu V/V)$$
$$= 10 \text{ mV}$$

This error is caused by an internal fluctuation in the main biasing network of the Op Amp. If the internal biasing technique used is a resistor divider network, the supply rejection will be in the order of 100 to 1000 μV/V. If, however, internal constant current sources are used, as in the case of high-quality Op Amps, the supply rejection is typically less than 20 μV/V.

The supply pin polarity of this error varies from unit to unit. Depending on the internal design, one supply pin will usually show a considerably greater error than the other. The larger of the two errors is usually the specification limit found on the data sheet.

External power supply regulation directly translates to supply rejection output errors of an Op Amp network.

Bias Current

The bias current error as seen at the output is determined by the value of the external feedback components (Figure 14),

$$e_o \text{ (bias)} = I_{bias} R_2$$
$$= (100 \text{ nA})(100\text{k})$$
$$= 10 \text{ mV}$$

It should be noted that this actual output value will occur only if the input voltage offset is at zero. Otherwise, the output will be the algebraic sum of both voltage and current offsets.

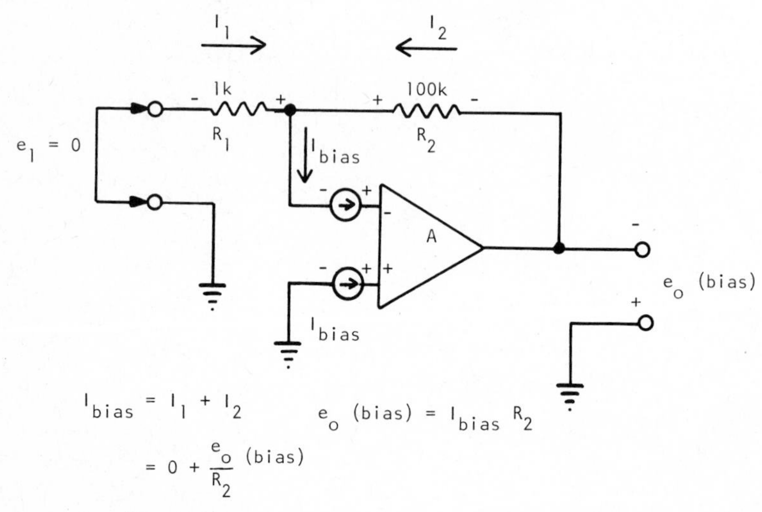

$$I_{bias} = I_1 + I_2 \qquad e_o \text{ (bias)} = I_{bias} R_2$$
$$= 0 + \frac{e_o \text{ (bias)}}{R_2}$$

Figure 14. Bias Current Equivalent Circuit

Bias Current (continued)

In a bi-polar Op Amp, the bias current is the base current (I_B) drawn by the input transistor. The magnitude of this current is determined by the gain (β) of the input transistor and the quiescent current of the input stage. The Integrated Circuit (IC) Op Amps generally have high quiescent current due to inherent processing limitations, therefore, their bias currents are high, relative to somewhat larger size discrete component modular Op Amps.

The typical low cost IC Op Amp bias current is in the range of 100 nA to 1 μA. The discrete component modular Op Amps are usually 1 nA to 10 nA. With the advent of the super-beta transistor, some discrete bi-polar Op Amps require less than 500 pA of input bias current.

In the FET Op Amp, the bias current is essentially the leakage current (I_G) of the input Field Effect Transistor (FET). It is not dependent on gain or quiescent of the input stage as in the bi-polar transistor. However, the cost is usually higher.

The typical low cost FET Op Amp bias current ranges from 10 pA to 200 pA. High quality FET Op Amps are available with bias currents as low as 0.05 pA (50 fA).

Bias Current (continued)

The bias current error can be cancelled (Figure 15) by utilizing the bias current from the non-inverting (+) input of the Op Amp. Typically, the bias currents of each input are very close to being equal, therefore, with the proper technique the non-inverting input bias current can be used to cancel the inverting input bias current error.

$$e_o \text{ (Bias, Inv.)} = e_o \text{ (Bias, Non-Inv.)}$$

Referring to Figure 15, resistor R_B can be used to generate a voltage ($I_{bias} R_B$) to oppose the inverting bias current error.

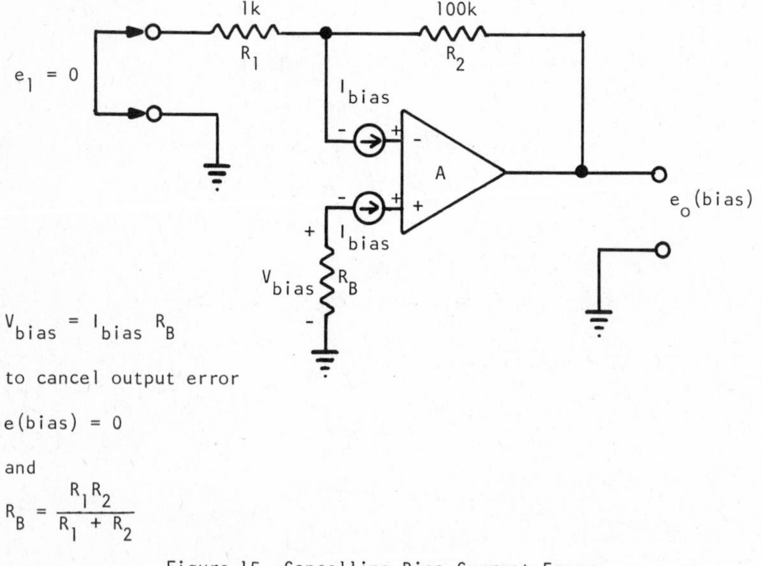

Figure 15. Cancelling Bias Current Error

Bias Current (continued)

The simulated input bias voltage is multiplied by the non-inverting gain to give the output voltage.

$$e_o \text{ (bias, non-inv)} = I_{bias} R_B \left(1 + \frac{R_2}{R_1}\right)$$

Equating the two bias errors

$$e_o \text{ (bias, inv)} = e_o \text{ (bias, non-inv)}$$

$$I_{bias} R_2 = I_{bias} R_B \left(1 + \frac{R_2}{R_1}\right)$$

and

$$R_B = \frac{R_2}{1 + \frac{R_2}{R_1}} = \frac{1}{\frac{1}{R_1} + \frac{1}{R_2}}$$

Therefore, the value of R_B that will exactly cancel the bias current output error is the parallel combination of the two feedback components.

$$R_B = R_1 \,||\, R_2$$

$$= \frac{(1k)(100k)}{(1k + 100k)} \simeq 1k$$

From this analysis it can be seen that the feedback resistor R_2 must be kept to a minimum, particularly with IC Op Amps, where input bias currents can be relatively high. Also, when using the Op Amp in the non-inverting configuration, the source impedance should be kept to a minimum for the same reason.

Bias Current (continued)

To overcome these disadvantages, the FET Op Amp is usually the most economical alternative. For extremely low bias characteristics, the varactor-type Op Amp is sometimes recommended.

It should be noted that when the value of R_B approaches the megohm range, resistor thermal noise becomes a noticeable factor. It is usually advisable to by-pass R_B with a good quality capacitor (polystyrene or polycarbonate). The section on noise will discuss this in more detail.

Current Offset

Input current offset is a commonly misunderstood specification. It is often confused with input bias current. However, these are not identical parameters.

Input current offset is the difference between the two input bias currents. In other words, it is a measure of the "match" of the input stage transistor pair.

$$I_{offset} = I_{bias}^{(+)} - I_{bias}^{(-)}$$

As mentioned earlier in the bias current section, these two input bias currents are usually in the same direction. Therefore the input current offset specification is usually 10 to 100 times less than the individual Op Amp input bias currents.

In general, the input current offset specification is not an extremely useful parameter. Unfortunately, it seems to have been over-emphasized in advertising, particularly with the IC Op Amps, since the input bias currents are sometimes embarrassingly high.

Using the techniques discussed in the previous section (Bias Current), the offset current error is easily eliminated by slightly adjusting R_B (Figure 15) in a direction to null the mismatch in bias currents.

Current Drift

The current drift from each of the Op Amp inputs is the change of the input bias current with temperature. The technique for determining the output error is the same as that for calculating the output bias error (Figure 16).

$$I_{drift} = \frac{\Delta I_{bias}}{\Delta T} \qquad \text{Input Current Drift}$$

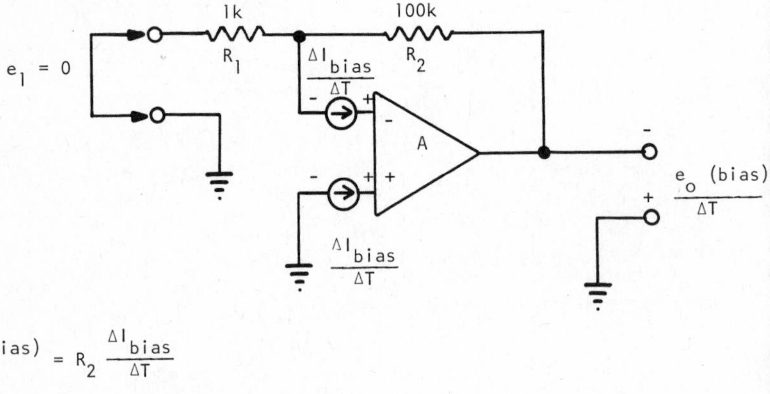

$$\frac{e_o(bias)}{\Delta T} = R_2 \frac{\Delta I_{bias}}{\Delta T}$$

Figure 16. Current Drift Equivalent Circuit

The output drift voltage due to input current drift for the sample design of Figure 1 is

$$\frac{e_o(bias)}{\Delta T} = R_2 \frac{\Delta I_{bias}}{\Delta T}$$

$$= (100k)\ (1\ nA/^{o}C)$$

$$= 100\ \mu V/^{o}C$$

Current Drift (continued)

For a temperature change from room environment $+25^{\circ}C$ down to $-5^{\circ}C$

$$\Delta T = (25^{\circ}C) - (-5^{\circ}C) = 30^{\circ}C$$

The output voltage error due to current drift at $-5\,^{\circ}C$ is

$$e_o(\text{drift}) = \frac{e_{o(\text{bias})}}{\Delta T}\,\Delta T$$

$$= (100\ \mu V/^{\circ}C)(30^{\circ}C)$$

$$= 3\ mV$$

The total error due to current bias is the sum of the output bias error at room temperature and the output current drift error.

At room temperature the output bias error is

$$e_o(\text{bias}) = R_2\ I_{\text{bias}}$$

$$= (100k)(10\ nA)$$

$$= 10\ mV$$

Current Drift (continued)

The total output error caused by current bias and drift at -5°C is

$$e_o(\text{total}) = e_o(\text{bias}) + e_o(\text{drift})$$
$$= 10\ \text{mV} + 3\ \text{mV}$$
$$= 13\ \text{mV}$$

Generally, the bias current at each input will drift in the same direction. This means the same technique can be used to cancel the current drift error as was used to cancel bias current error. Referring to Figure 17, resistor R_B can be used to generate a voltage to oppose the inverting drift error.

$$\frac{\Delta V_{bias}}{\Delta T} = \frac{\Delta I_{bias}}{\Delta T} R_B$$

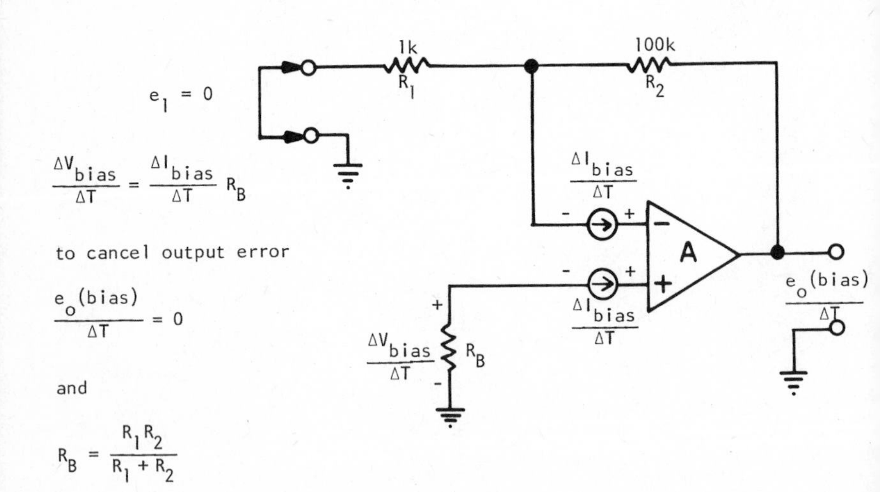

Figure 17. Cancelling Current Drift Error

Current Drift (continued)

The simulated input bias drift is multiplied by the non-inverting gain to give the output drift voltage.

$$\frac{\Delta e_o(\text{bias, non-inv})}{\Delta T} = \frac{\Delta I_{bias}}{\Delta T} R_B \left(1 + \frac{R_2}{R_1}\right)$$

Equating the two drift errors

$$\frac{\Delta e_o(\text{bias, inv})}{\Delta T} = \frac{\Delta e_o(\text{bias, non-inv})}{\Delta T}$$

$$\frac{\Delta I_{bias}}{\Delta T} R_2 = \frac{\Delta I_{bias}}{\Delta T} R_B \left(1 + \frac{R_2}{R_1}\right)$$

and

$$R_B = \frac{R_1}{1 + \frac{R_2}{R_1}} = \frac{1}{\frac{1}{R_1} + \frac{1}{R_2}}$$

The value of R_B that will exactly cancel the drift current output error is the parallel combination of the two feedback components.

$$R_B = R_1 \,||\, R_2$$

Using bi-polar Op Amps, the input current drift has a predictable direction. As the temperature is increased, the input bias current will decrease. The reason being that gain (β) of the input stage transistors increases with rising temperature, therefore less current is required to bias them. Conversely, at lower temperatures bias current increases.

Current Drift (continued)

In higher quality bi-polar Op Amps the input stage current source is deliberately designed to increase collector current as temperature rises. This compensates for the change in gain (β) and results in a lower overall current drift.

With the FET Op Amp, the current drift is a non-linear parameter. The bias current doubles every $10^{\circ}C$. It is essentially the same temperature characteristic as a reversed biased diode. As in the case with the bi-polar Op Amp, the bias current drifts track very closely and can therefore be cancelled in the same manner as discussed.

When the value of R_B approaches the megohm resistor range, resistor thermal noise becomes a noticeable factor. It is usually advisable to by-pass R_B with a good quality capacitor. The section on noise will discuss this in more detail.

Output Impedance

The actual output impedance (Z_o) of the Op Amp network is very dependent on the loop gain of the Op Amp circuit. (Note that feedback resistors are in parallel. This factor is usually negligible if the resistors R_1, R_2 are large compared to Z_o.

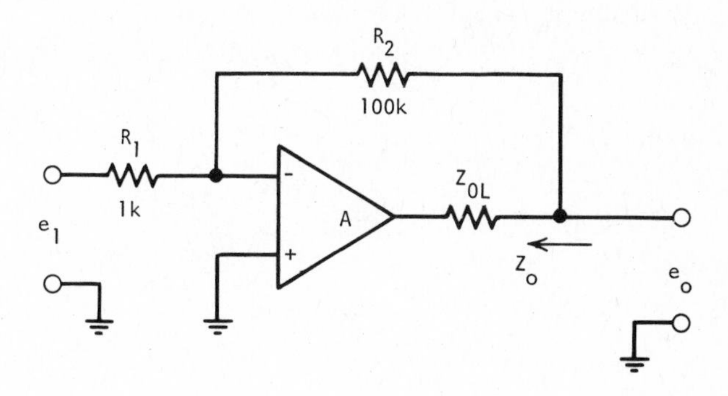

$A = 100{,}000$ (or 100 DB)

$G_{CL} = 100$ (or 40 DB)

$A_L = 100$ (or 60 DB)

Figure 18. Output Impedance for Closed Loop Network

$$\text{OUTPUT IMPEDANCE } Z_o = Z_{OL} \frac{1}{1 + A_L} \; || \; (R_1 + R_2)$$

Z_{OL} = Op Amp open loop output impedance

A_L = Loop Gain

Output Impedance (continued)

or, approximately,

$$Z_o \approx \frac{Z_{OL}}{1 + A_L}$$

In the sample design (see Figure 18), for DC to 10 Hz as shown in Figure 2 the output impedance is reduced by the loop gain (A_L = 1000) and the feedback network resistors R_1, R_2 are negligible.

$$Z_o = \frac{(1k\Omega)}{(1 + 1000)} \;||\; (1k + 100k)$$

$$= \frac{1000}{1 + 1000} \;||\; 101k\Omega \approx \frac{1000}{1 + 1000}$$

$$\approx 1\Omega \quad \text{(DC to 10 Hz)}$$

For frequencies above 10 Hz, the loop gain progressively decreases and therefore the output impedance increases. For example, at 10 Khz the loop gain is unity (ODB) and the output impedance becomes one-half of the open loop value.

$$Z_o = \frac{(1k\Omega)}{(1 + 1)} \;||\; 101k = 500\Omega \text{ at 10 Khz}$$

Beyond this point the output impedance rapidly approaches the open loop value, and the Op Amp feedback network no longer is effective reducing the output impedance.

As a quick rule of thumb for most DC and mid-frequency applications the closed loop output impedance calculation can be reduced to simply dividing the open loop output impedance by the network loop gain.

$$Z_o = \frac{Z_{OL}}{A_L} = \frac{1k}{1000} = 1\Omega$$

Common Mode Rejection

The common mode rejection is another Op Amp parameter that is often misunderstood. Common mode error is the error signal that occurs when both input terminals of the Op Amp are varied together.

It should be pointed out that a common mode error does not occur for the case of the inverting network configuration. When the Op Amp is used in this type network the two input terminals do not experience a common voltage swing. They both are essentially locked to ground because the non-inverting (+) input terminal is directly connected to ground.

Always keep in mind that with linear Op Amp networks, one Op Amp input terminal will always follow the other. This case will be clearly shown by the discussion that will follow (see Figures 19 and 20).

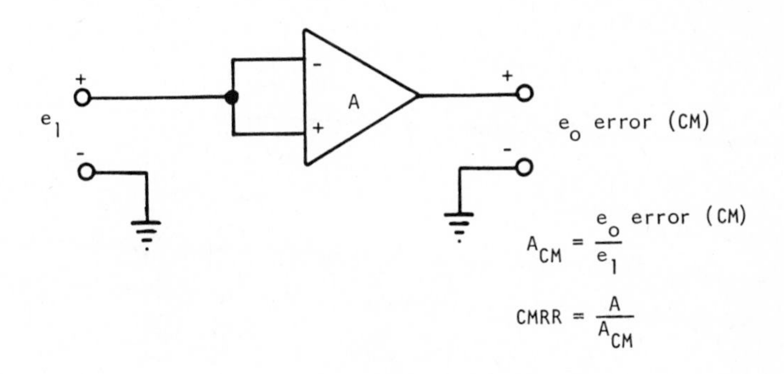

Figure 19. Common Mode Gain

Common Mode Rejection (continued)

As in the case with most Op Amp error parameters, the error signal is referred back to the input. This reference specification is called the Common Mode Rejection Ratio (CMRR). CMRR is defined as the ratio of the Op Amp differential gain (open loop gain, A) to the Common Mode Gain, A_{CM}.

$$CMRR = \frac{A}{A_{CM}}$$

When identical signals are applied to both input terminals simultaneously, ideally, the Op Amp output should be zero (see Figure 19). However, due to the "dynamic mismatch" of the input transistors, the gain on one side of the differential stage will always be slightly different than the gain on the other side. The difference of these two gains is known as the Common Mode Gain A_{CM} and will show up as an output error signal (Figure 19).

$$e_o = \text{error (CM)} = e_1 \, A_{CM}$$

The error voltage referred to the input is then

$$V_{CM} = \frac{e_1 \, A_{CM}}{A}$$

$$= \frac{e_1}{CMRR}$$

Common Mode Rejection (continued)

For the Op Amp used in the network shown in Figure 20, the Common Mode Rejection Ration (CMRR) is 80 DB, or 10,000:1. This means that if an input voltage of 100 mV was applied at e_1 (Figure 20), the equivalent input error caused by this common mode signal would be e_1 reduced by 10,000.

$$V_{CM} = \frac{e_1}{CMRR}$$

$$= \frac{100 \text{ mV}}{10,000}$$

$$= 10 \ \mu V \text{ (input error voltage)}$$

This error must now be multiplied by the closed loop gain of the Op Amp network to determine the actual output error signal. For the network shown in Figure 20,

$$e_o \text{ error (CM)} = V_{CM} \left(1 + \frac{R_2}{R_1}\right)$$

$$= (10 \ \mu V) \left(1 + \frac{100k}{1k}\right)$$

$$= 1 \text{ mV (output error voltage)}$$

In other words, if an input voltage of 100 mV is applied at e_1, the output will be 10.1 volts. However, an error of ± 1 mV will be introduced at the output due to the mismatch of the Op Amp input circuitry. This error is generally referred to as the "common mode error."

Common Mode Rejection (continued)

Notice that a common mode voltage swing will occur at both of the Op Amp input terminals [(+) and (-)]. This "common mode voltage swing" will be the voltage e_1.

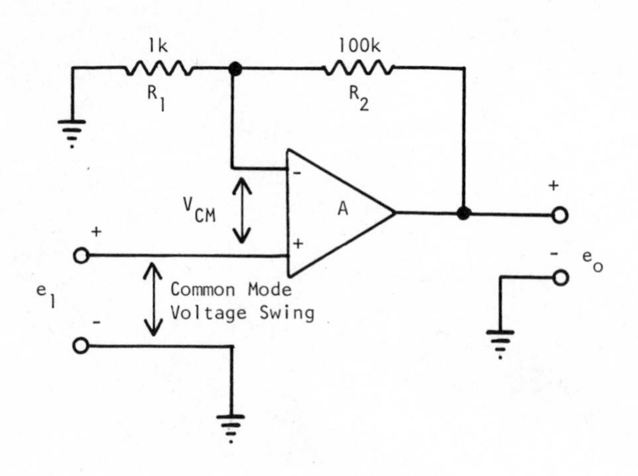

$$V_{CM} = \frac{1}{CMRR}$$

$$\%e_o \text{ (CM) error} = \frac{1}{CMRR} \times 100\%$$

Open Loop Gain = 100 DB

CMRR = 80 DB or 10,000:1

Figure 20. Common Mode Error is Present Only in the Non-Inverting Type Network

Common Mode Rejection (continued)

The output error due to common mode swing is

$$\%\ \text{error(CM)} = \frac{1}{\text{CMRR}} \times 100\%$$

The derivation of this equation is shown in Appendix D. For the example shown in Figure 20.

$$\%\ \text{error(CM)} = \frac{1}{10{,}000} = 100\%$$

$$= 0.01\%$$

This means for a full-scale output signal (e_o) of 10 V, the output error due to common mode voltage swing will be

$$e_o\ \text{error(CM)} = e_o \times \%\ \text{error(CM)}$$

$$e_o\ \text{error(CM)} = (10\ \text{V})(0.01\%)$$

$$= 1\ \text{mV}$$

Since the Common Mode Rejection Ratio is derived directly from the difference ratio of the inverting and non-inverting open loop gains of the Op Amp, it follows that the frequency response of the CMRR parameter will be similar to that of the Open Loop Gain frequency response (see Figure 21). Therefore, the Common Mode output error will increase as frequency increases.

Common Mode Rejection (continued)

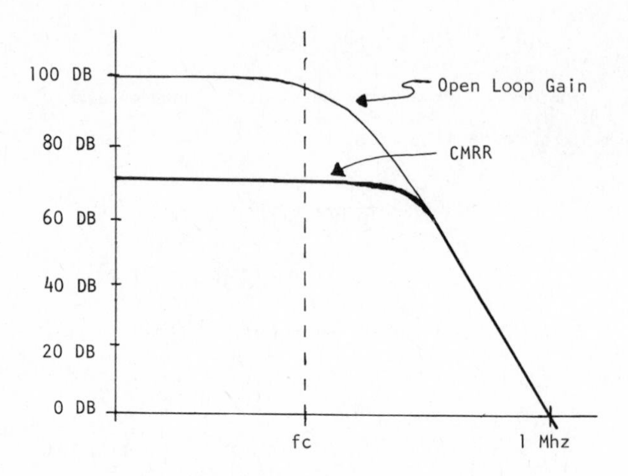

Figure 20. CMRR Response

Another point that should be mentioned is the magnitude of the common mode voltage. Some Op Amp manufacturers specify the CMRR only for small common mode voltages (less than 1 volt).

However, if the Op Amp is to be used as a large signal unity gain buffer amplifier as shown in Figure 22, the common mode voltage can be 10 volts (using $\pm$15 V supply voltage).

Common Mode Rejection (continued)

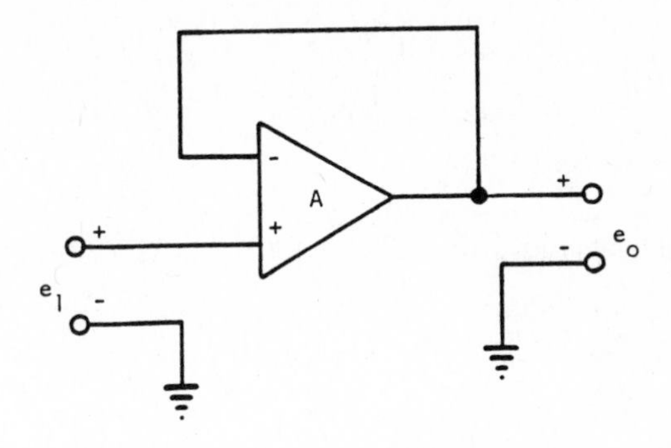

Figure 22. Unity Gain Buffer

A problem can occur when input signals approach the power supply voltage rails. This is due to the input transistors beginning to saturate. Common mode error will increase drastically if saturation occurs.

This effect is more prominent in FET Op Amps because of the inherent non-linear characteristics of the input FET transistors. They are more difficult to match and therefore the CMRR specifications for FET Op Amps are generally lower than for bipolar Op Amps.

A correctly specified Op Amp should stipulate the common mode voltage range for which the CMRR specification applies.

NETWORK PARAMETERS

Noise

There are three categories of noise commonly referred to in Op Amps: (1) thermal noise, or wideband white noise, (2) flicker noise, or 1/f noise, and (3) shot noise, or popcorn noise.

Thermal noise is the wideband white noise due to the molecular vibration and agitation associated with temperature effects. In short, how hot is it.

Flicker noise is a low frequency effect due to the generation of carrier charges at the boundaries of the semiconductor material. A good analogy for demonstrating this effect is that of boiling water in a frying pan. Individual bubbles will be generated from random locations at the bottom of the pan. This noise is usually measured at the frequencies from 1 Hz to 100 Hz and its amplitude characteristically decreases as an inverse proportion to the frequency (1/f). The absolute peak value usually is higher than the wideband thermal noise mentioned above.

The shot noise is the lowest frequency noise present in the Op Amp. This effect shows itself in a very similar manner to that of popcorn being heated over an open fire. The effect is seen as a few high amplitude noise spikes randomly occurring over a time period of seconds.

Rigorous mathematical analysis of these noise effects is beyond the scope of this text. Practical values for these parameters are found on the Op Amp data sheets. This is all that is necessary for proper design.

Noise (continued)

Another noise that should briefly be mentioned is that which is characteristically found in the chopper stabilized Op Amps. It is known as chopper noise and is due to the switching circuitry associated with the chopper section of the Op Amp. In more recently designed chopper Op Amps, this noise has been virtually eliminated due to improved shielding techniques. However, in the earlier chopper Op Amp produced, this noise is quite significant. The noise is seen in the form of sharp spikes repeating periodically, usually from 100 Hz to 500 Hz, corresponding to the frequency of the particular chopper oscillator.

Op Amp specifications usually show thermal noise over the bandwidth from 0 to 1 Khz and measured in RMS values. The flicker noise (1/f) is usually specified from 0 Hz to 10 Hz and measured in peak-to-peak units.

The shot noise (popcorn) is usually specified over a bandwidth from 0 to 1 Hz and measured in peak-to-peak units. The noise is usually monitored for about 20 seconds on an oscilloscope in order to obtain a good confidence level for the maximum peak values of the spikes.

One of the best ways to get a feel for Op Amp noise is by using the analogy of a pan of boiling water.

The thermal wideband noise can be thought of as the actual temperature of water itself. It is the overall background noise.

Noise (continued)

The flicker (1/f) noise can be looked upon as the bubbles present in boiling. Notice they are large and can be heard. This can be construed as being a low frequency and somewhat higher amplitude noise signal.

Finally, the shot noise (popcorn) can be simulated by putting the cover on the boiling pan of water. As the pressure builds up under the cover, it will lift and puff steam causing sporadic rattles. These are very low frequency noise transients that only occur a few times per second.

Remember that all these categories of noise are present in both the form of current as well as voltage noise.

The voltage noise categories are the most commonly specified in Op Amps. In some cases, manufacturers do specify both the current as well as the voltage noises.

The voltage noise is the more important of the two, since its value gets directly amplified by the non-inverting closed loop gain.

The current noise, on the other hand, is seen only if it is dropped across a fairly large input resistor (1MΩ). Since typically input resistance is usually less than 10kΩ, this noise is usually negligible.

Note that it is the parallel combination of the two resistors (input and feedback) that constitute the equivalent input noise current resistor at the input terminal.

Noise (continued)

Noise Figure has historically always seemed to cause confusion. The reason is because of the basic definition. Remember that Noise Factor is the ratio of the signal-to-noise ratios of input to output.

$$\text{Noise Factor} \quad F = \frac{\left(\frac{S_i}{N_i}\right)}{\left(\frac{S_o}{N_o}\right)}$$

S_o = Signal Output Power

N_o = Noise Output Power

S_i = Signal Input Power

N_i = Noise Input Power

The Noise Figure is the logarithm of the noise factor in decibels.

$$N.F. = 10 \log F$$

It is a well known fact that a minimum Noise Figure is achieved by using an optimum value of input source Resistance R(opt). The theoretical value of this resistor is calculated from the ratio of the input noise voltage (e_n) and input noise current (i_n).

$$R_s(\text{opt}) = \sqrt{\frac{\overline{e_n^2}}{\overline{i_n^2}}}$$

Noise (continued)

The optimum source resistance is different for bi-polar input Op Amps than for FET input Op Amps. The major reason is because the noise current of a FET transistor is much less than that of a bi-polar transistor; thus the optimum source resistance for FETs is higher.

Figure 23 shows a noise figure plot of a typical bi-polar Op Amp and a FET Op Amp. Both have the same input noise voltage. Notice that the low noise current has extended the FET optimum source resistance point.

Usually a bi-polar input Op Amp, R(opt) will fall between 10kΩ and 100kΩ while the FET input Op Amp will range from 100kΩ to 1MΩ (Figure 23).

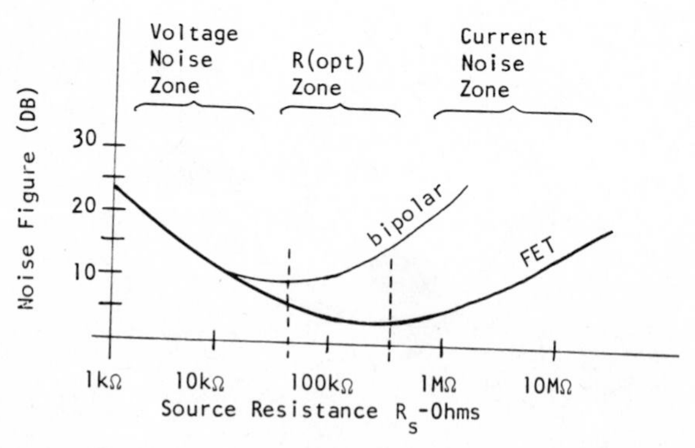

Figure 23. Typical Op Amp Noise Figures vs Source Resistance

Noise (continued)

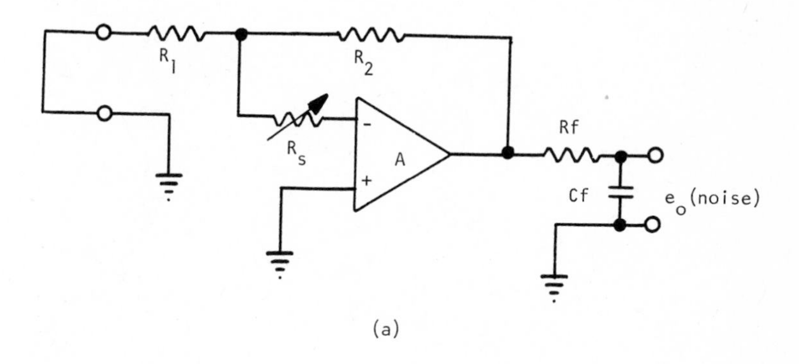

(a)

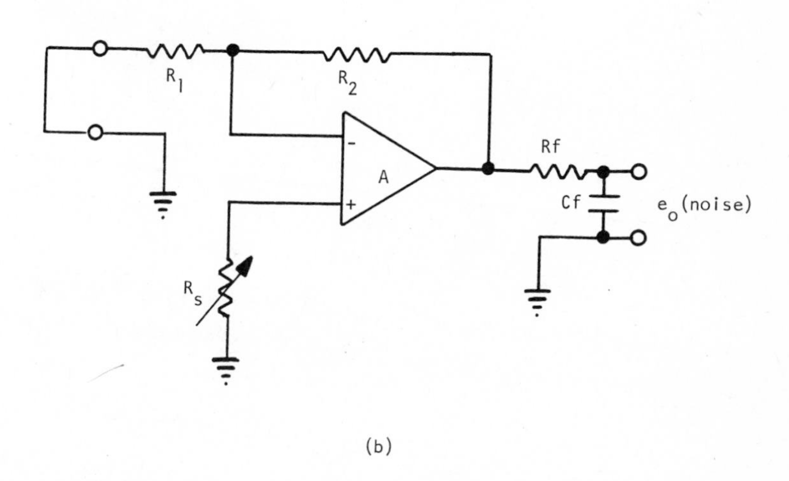

(b)

Figure 24. Noise Figure Measurement for R_s(opt)

Noise (continued)

The output filter is used to standardize the noise bandwidth for consistency in specifying the Op Amp noise.

The output filter noise bandwidth is calculated from the following formula (see Figure 25).

$$f_n = \frac{1}{4\, R_f C_f} \quad , \qquad f_n = \text{Noise Bandwidth}$$

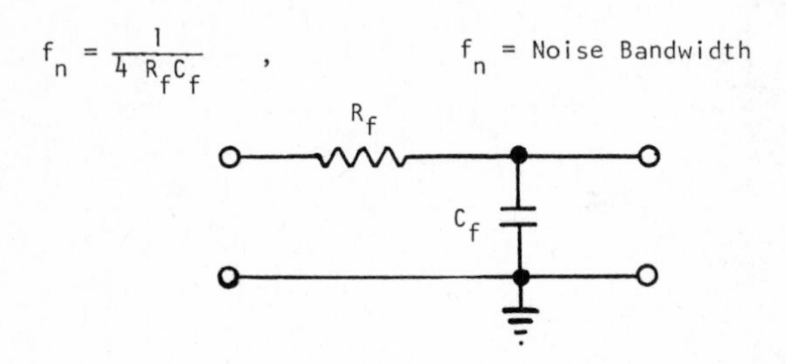

Figure 25. Output Filter Circuit

Another important point that causes considerable confusion is the actual output noise as compared to the noise figure.

First of all, there are two input noise sources in the Op Amp: (1) voltage noise, and (2) current noise.

The easiest way to distinguish between these two parameters is by studying the different test methods. The voltage noise is measured by using a low source impedance high gain circuit as shown in Figure 26. The current noise is measured by using a high source impedance high gain circuit as shown in Figure 27.

Noise (continued)

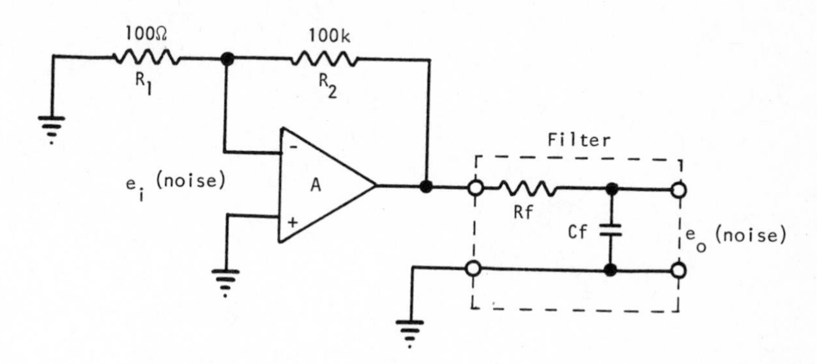

$$e_i(\text{noise}) = \frac{e_o(\text{noise})}{\left[1 + \frac{R_2}{R_1}\right]}$$

Figure 26. Voltage Noise Test Circuit

In Figure 26 the output voltage noise signal will be

$$e_o(\text{noise}) = e_i(\text{noise}) \left[1 + \frac{R_2}{R_1}\right]$$

if,

$e_i(\text{noise}) = 5\ \mu V$ (rms), from Sample Op Amp Specification shown with Figure 1.

then,

$e_o(\text{noise}) \approx 5$ mV , output noise due to Input Voltage Noise.

However, current noise will swamp out the voltage noise if source impedance is high. In Figure 27 the output noise signal will become large as the input current noise is multiplied by the large source

Noise (continued)

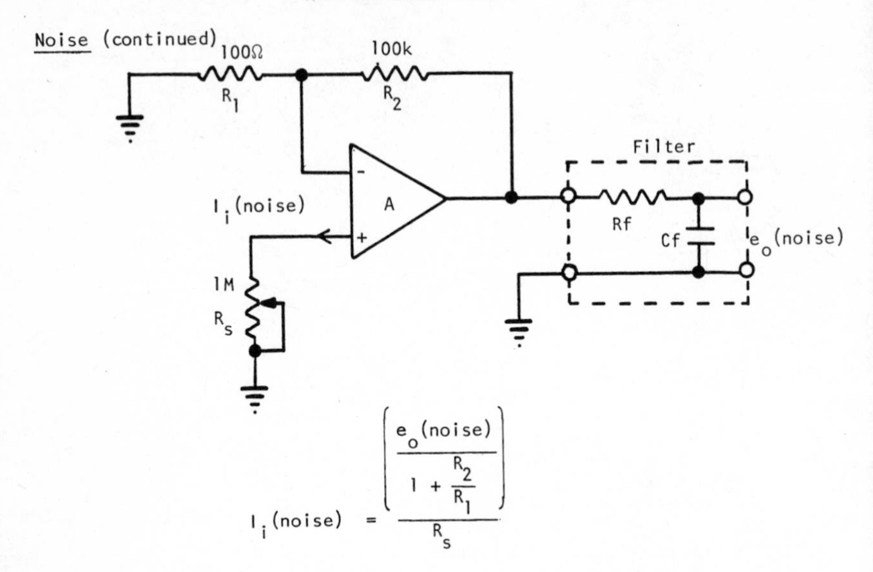

$$I_i(\text{noise}) = \frac{\left(\dfrac{e_o(\text{noise})}{1 + \dfrac{R_2}{R_1}}\right)}{R_s}$$

Figure 27. Current Noise Test Circuit

resistor R_s. The output noise signal will be the sum of both, but it will be primarily due to current input noise.

$$e_o(\text{noise}) = [I_i(\text{noise})\ R_s + e_i(\text{noise})]\left[1 + \frac{R_2}{R_1}\right]$$

or, $$e_o(\text{noise}) \approx I_i(\text{noise})\ R_s \left[1 + \frac{R_2}{R_1}\right]$$

if, $I_i(\text{noise}) = 100$ pA (rms), from Sample Op Amp Specification, Figure 1

then, $$e_o(\text{noise}) \approx (100\ \text{pA})\ (1\text{M}) \left[1 + \frac{100\text{k}}{100}\right]$$

≈ 100 mV output noise due to Input Current Noise.

Noise (continued)

It can be seen that output noise can become quite substantial when high impedances are used at the inputs of the Op Amp.

Generally speaking, it pays to keep input impedances as low as possible when dealing with Op Amp instrumentation designs. This may seem to conflict with the noise figure results shown in Figure 23, but remember, the noise figure is the ratio of two ratios and therefore, it is not an absolute number. In other words, the ratio of the two signal-to-noise ratios of input to output may be optimum, but the absolute magnitude of the output noise could be unnecessarily high. This is simply because the input source resistor itself can cause excessive thermal noise that could actually over-ride the Op Amp noise.

For instrumentation design, do not increase input source resistors in order to optimize noise figure.

Even if the signal-to-noise ratio of the input to output is optimized, this does not mean that the signal-to-noise of the output by itself is optimum. And it is the output signal-to-noise ratio that is the major concern in instrumentation design.

When dealing with instrumentation problems it is not necessary to optimize "power transfers" from input to output. One is concerned with signal reproduction--clean and simple--regardless of the power it takes to do it.

Noise (continued

Therefore, do not bother optimizing the noise figure. Optimize the output signal-to-noise ratio only, because that is what is needed.

Optimization of this ratio is done by minimizing input current noise effects. Use small input resistors and bypass the source resistor with a capacitor C_s to reduce the noise bandwidth (Figure 28). This technique also reduces thermal noise generated from the resistor itself, as discussed later in this section.

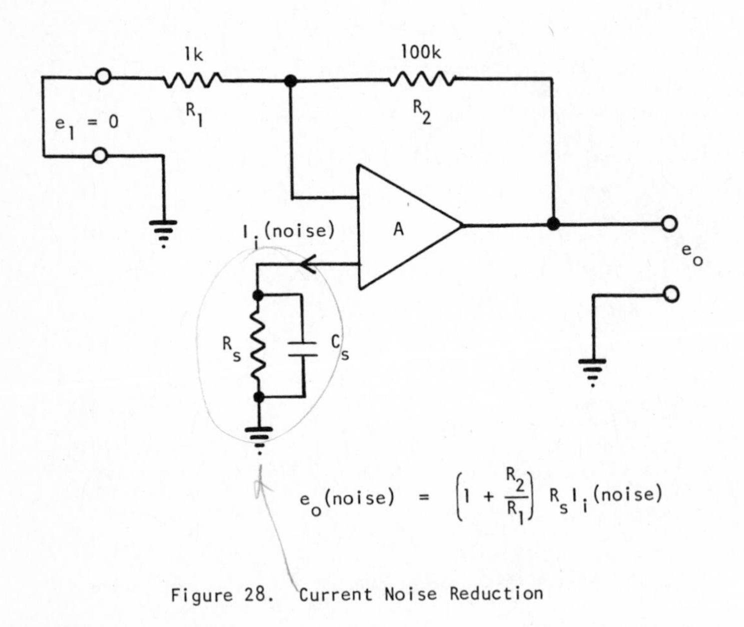

Figure 28. Current Noise Reduction

Noise (continued)

The output voltage noise is determined by the non-inverting amplification of the input voltage noise specification.

$$e_o(\text{noise}) = \left(1 + \frac{R_2}{R_1}\right) e_i(\text{noise})$$

In the sample design shown in Figure 28, for e_i (noise) = 5 μVrms,

$$e_o(\text{noise}) = \left(1 + \frac{100k}{1k}\right) (5\ \mu\text{Vrms})$$

$$= 500\ \mu\text{Vrms}$$

for 1Khz bandwidth

This value can be reduced by reducing the bandwidth of the output. It can be easily accomplished by connecting a capacitor across R_2 (see Figure 29). However, it must be remembered that the frequency response of the output signal will also be decreased. Therefore, choose a cutoff frequency above the highest frequency of the input signal. The value of the capacitor is determined directly from the feedback network. (See Figure 29).

$$f_c = \frac{1}{2\pi R_2 C}, \qquad f_c = \text{Cutoff Frequency of Network}$$

$$f_n = \frac{1}{4R_2 C}, \qquad f_n = \text{Equivalent Noise Bandwidth}$$

Noise (continued)

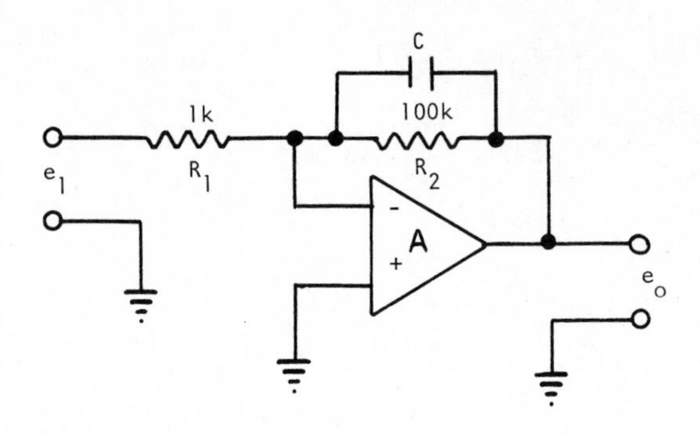

Figure 29. Minimizing Output Thermal Noise Using Feedback By-Pass Capacitance

When a source resistor is used at the input to cancel bias current effects (see section on bias current) it is desirable to bypass this resistor to eliminate the wideband thermal noise caused by the resistor.

The magnitude of the resistor thermal noise is determined by the following equation:

$$e_{noise\ (rms)} = \sqrt{4kT\ f_n R_s}$$

k = Boltzman constant
T = Temperature in Kelvin
f_n = Noise Bandwidth (Hz)
R_s = Source Resistance (MΩ)

at room temperature $+25^{\circ}C$,

$$e_{noise(rms)} \simeq \frac{1}{8}\sqrt{f_n R_s} \text{ in microvolts } (\mu V)$$

Noise (continued)

For a source resistor of 1 megohm and a bandwidth of 10 Khz

$$e_{noise\ (rms)} = \frac{1}{8} \sqrt{(10\ \text{Khz})(1\text{M}\Omega)}$$

$$= 12.5\ \mu\text{V(rms)} \quad \text{for 10 Khz bandwidth}\ .$$

It can be seen that this voltage noise, when multiplied times the gain of the network, could be substantial at the output.

$$e_o(\text{noise}) = \left(1 + \frac{R_2}{R_1}\right) e_1(\text{noise})$$

$$= \left(1 + \frac{100}{1\text{k}}\right) 12.5\ \mu\text{Vrms}$$

$= 1250\ \mu$Vrms Output Noise Due to Source Resistor (1MΩ), from 0 - 10 Khz Bandwidth.

If a feedback capacitor is not used, the noise bandwidth will be determined by the closed loop roll-off of the network f_c.

$$f_n = \frac{\pi}{2} f_c$$

f_c = Closed Loop Roll-Off Point (3 DB)

f_n = Equivalent Noise Bandwidth

See Appendix E for Derivation of this equation.

Noise (continued)

A practical circuit for testing the noise voltage of an Op Amp is shown in Figure 30. The filter components R_f and C_f are selected to roll-off the noise at the desired bandwidth.

$$f_c = \frac{1}{2\pi R_f C_f}$$

$$= 636 \text{ Hz}$$

Since the roll-off of this filter is 20 DB/decade, the effective noise bandwidth is

$$f_n = \frac{\pi}{2} f_c = \frac{1}{4R_f C_f}$$

$$= 1 \text{ kHz}$$

Dividing the measured output voltage by the network gain gives the equivalent input noise voltage,

$$e_i(\text{noise}) = \frac{e_o(\text{noise})}{1 + \frac{R_2}{R_1}}$$

$$\simeq \frac{e_o(\text{noise})}{1000}$$

In other words, reading the output voltage in volts is equivalent to the input in millivolts for a 1 kHz noise bandwidth.

Noise (continued)

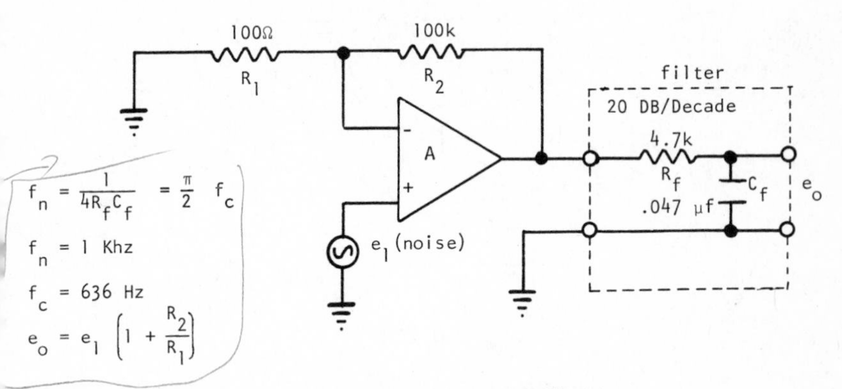

Figure 30. Noise Measurement Circuit

Extreme care should be taken in shielding noise test circuitry in order to avoid hum pickup and extraneous RF signals.

Note that the output will carry the DC component caused by the input voltage offset. When measuring the RMS output noise be sure the output instrument is eliminating this DC level from its reading. Most audio output meters are inherently AC coupled so this is usually not a problem. If this is not the case, insert a large capacitor C_1 (5000 μf) in series with R_1 to ground (see Figure 31).

Noise (continued)

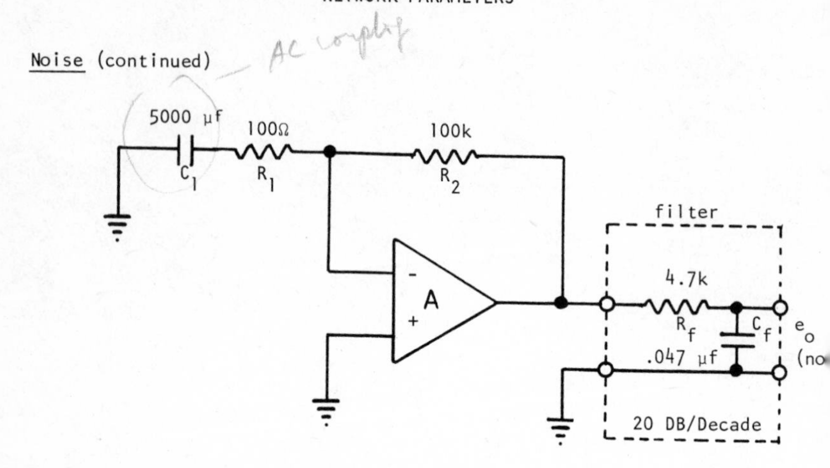

Figure 31. Noise Test Circuit Eliminating Offset Voltage

Settling Time

Settling time has been specified in many ways. So, first a word of caution to the user: Be sure to observe the definition as well as the test conditions for which this parameter has been measured.

This specification is slowly becoming somewhat standardized. However, the Integrated Circuit Op Amp manufacturers have evolved into a different standard than the Modular Op Amp manufacturers.

Usually, the Modular Op Amps are completely self-contained devices and use no external phase compensation components. Therefore, these products will generally define settling time in terms of the total time it takes for a unity gain inverting Op Amp network to respond to a full-scale input square wave, or pulse signal (see Figure 33). The output of this device will first react with a slight "delay"; then it will ramp, or "slew" toward full-scale. As the output reaches full-scale, the waveform will usually "overshoot" and then return back to the final full-scale output value. Slight ringing is often present during this return, particularly with wide bandwidth Op Amps.

The Integrated Circuit Op Amps are not always completely self-contained devices. Some devices must be phase compensated externally with a small capacitor, or combination of resistors and capacitors.

Settling Time (continued)

The values of these external components drastically affect slew rate and settling time. For this reason the Integrated Circuit manufacturers usually specify settling time by excluding the slew time. Unfortunately, this slew time is usually quite long, particularly for unity gain stability, in IC Op Amps, and therefore should not be neglected in the overall design calculations.

The particular parameter that is most directly affected by phase compensation is the "slew". The smaller the phase compensation capacitor, the faster the output will slew toward final value. However, if the capacitor is made too small, the output will overshoot and ring excessively, causing the Op Amp output settling time to increase. If, on the other hand, the capacitor value is increased too much, the slew time will become very long, thus causing the output to take an excessive time to reach the final value level. Therefore the total time for the output to settle is impaired either by oscillations, (or ringing, sometimes referred to as conditional stability) or slow rise time (due to slow slew rate).

Every IC Op Amp has optimum values for external components to achieve the best settling time. These values, or range of values, are generally specified on the data sheets. If it is not shown on the

Settling Time (continued)

data sheet, the user must either run experimental tests with the Op Amp or consult the manufacturer.

It should be pointed out that the processing methods used in manufacturing IC Op Amps are somewhat critical. This means that the optimum external component values for one individual Op Amp will not necessarliy be exactly the same as for the other Op Amp right next to it. This range may vary as wide as 400%, and more in some cases.

Figure 32 shows the most generally accepted circuit for which settling time is specified. Resistors R_1 and R_2 are usually kept as small as possible without overloading the output of the Op Amp. This is done to minimize the effect of stray capacitance across the resistors.

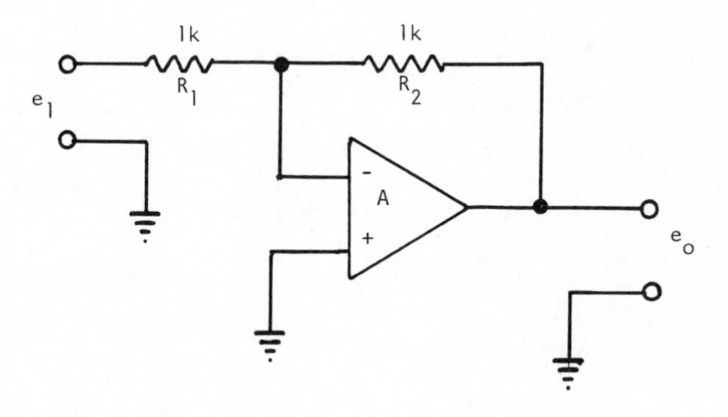

Figure 32. Typical Configuration for Specifying Settling Time

Settling Time (continued)

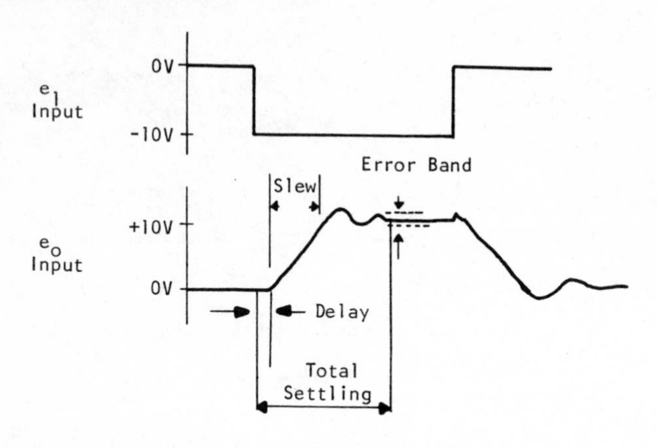

Figure 33. Settling Time Definition for Modular Op Amps

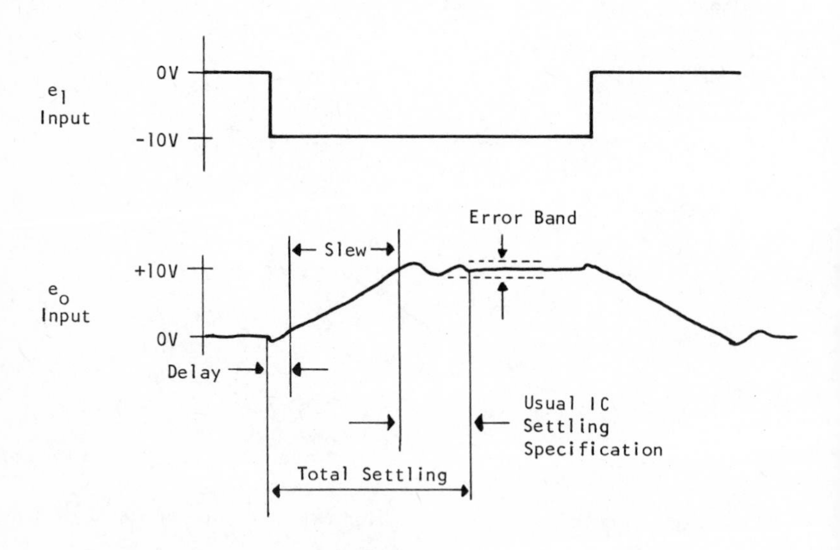

Figure 34. Settling Time Definition for IC Op Amps

Settling Time (continued)

When specifying settling time it must also be stated as to what "error band" the measurement is made. As an example, it may take 1 μsec to settle to within 1% of full scale, while it may take 50 μsec for the same signal to settle to within a 0.1% error band of full scale.

Settling time is a difficult parameter to measure accurately using the circuit shown in Figure 32. When measuring error bands in the vicinity of 0.1% and 0.01% another approach must be used. This is due to the large output signal saturating a sensitive high gain measuring instrument (i.e., the vertical amplifier of the oscilloscope).

Another method for measuring settling time is to monitor the summing junction at the (-) input terminal of the Op Amp. Once the

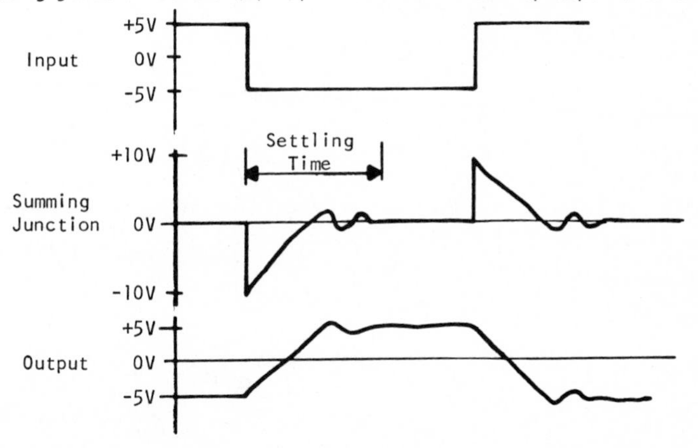

Figure 35. Summing Junction Error During Settling Time

Settling Time (continued)

Op Amp output has settled, the summing junction error will be essentially null (Figures 35 and 36).

For the unity gain circuit in Figure 32, the summing junction is essentially the difference signal between input and output. Therefore, settling time can be measured directly at the summing junction, provided the large 10 volt spikes are eliminated. This can be done with a pair of clipping diodes D_1, D_2, as shown in Figure 36.

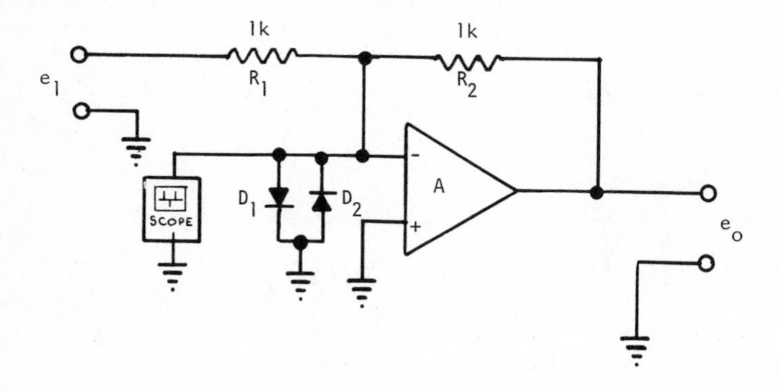

Figure 36. Diodes Eliminating Large Spike at Summing Junction Allowing Settling Time to be Measured Accurately

Now the high-gain scope input amplifier will not be overdriven, and the settling time measurement will be a valid reproduction of the error signal at low levels.

Settling Time (continued)

This is a valid way for measuring settling time except for one thing--the summing junction is a very sensitive point. Usually when an instrument is connected to the summing junction of an Op Amp the entire network will oscillate because of the input impedance of the instrument itself. For this reason, this technique is generally an undesirable method for measuring settling time.

The most accepted way of measuring the true settling time is by employing an artificial, or "false", summing junction. Figure 37 shows such a circuit.

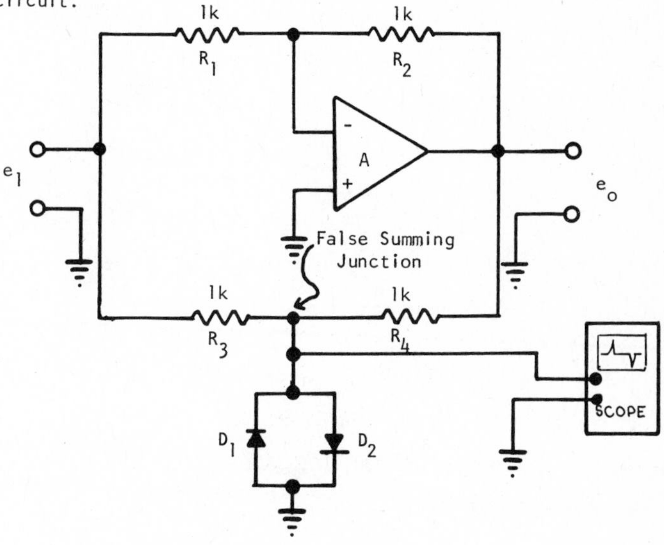

Figure 37. Settling Time Management Using False Summing Junction

Settling Time (continued)

If the Op Amp were ideally fast, resistors R_3 and R_4 would act like a voltage "see-saw" where the false summing junction would be the pivot point and would always be at zero volts. This is because e_1 and e_o would always be equal and opposite voltages.

Since the Op Amp is not ideal during the settling time, the false summing junction will jump away from zero volts when the output fails to respond to the speed of the input pulse. Therefore, this circuit simulates the same error as the true summing junction. The diodes serve the purpose of eliminating (clipping) the initial high voltage transient ($\pm$10V). This point is now isolated from the input terminals of the Op Amp, and therefore causes no inherent measuring error or network oscillations.

For ultra-high speed Op Amps (200 V/μs, 500ns) it is recommended that hot carrier diodes be used for the clipping network rather than the conventional silicon switching diodes.

High quality non-inductive resistors, such as metal film type, should be used in this test circuit. Also, R_3 or R_4 may have to be trimmed slightly in order to obtain zero volts at the false summing junction during steady state conditions (null condition).

Capacitance Loading

Adding external capacitance (C-loading) at the output of an Op Amp network has an internal affect on the open loop response of the Op Amp itself.

A correctly compensated Op Amp has a smooth continual 20 DB/Decade roll-off all the way down to unity gain (zero DB). This roll-off insures frequency stability at all gain settings. However, somewhere below unit gain in every Op Amp there exists the internal (output) driver circuitry frequency breakpoint. Therefore, connecting a capacitive load to the output of the Op Amp will shift the driver circuitry breakpoint (see Figure 38). As long as this breakpoint remains substantially below unity gain, the Op Amp network will remain stable. But if the external capacitive load is sufficiently large (usually greater than 1000 pf), the breakpoint will move toward the unity gain cross-over point f_2 of the open loop response, and "conditional stability" will begin to occur. In typical closed loop applications, this will result in "peaking" at the outer edge of the network frequency range. The corresponding square wave response will show output "overshoot" and dampened ringing after each transition edge.

In network configurations involving feedback resistors, it is often possible to easily eliminate the peaking and ringing effects of capacitive loading. Placing a capacitor in parallel with the output feedback resistor to the summing junction will accomplish this objective.

C-Loading (continued)

Usually, the value of this capacitor will range from 100 pf to 0.01 μf, depending on the particular application. However, since this technique directly reduces the frequency response (bandwidth) of the network, additional care must be taken for high-frequency applications.

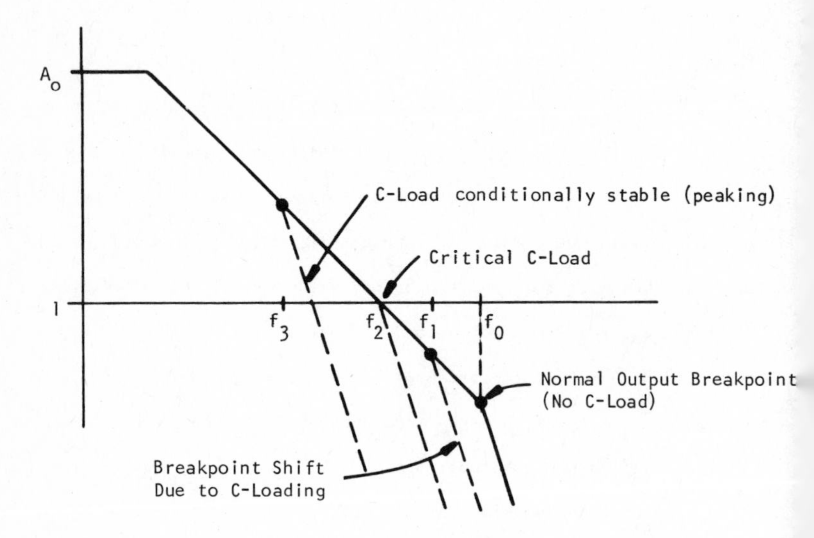

Figure 38. Breakpoint Op Amp Output Circuitry f_0

Figure 38 shows the effect of C-loading on a typical Op Amp open loop response. As the capacitance increases, f_0 shifts to f_1, f_2 and f_3 respectively.

C-Loading (continued)

Conditional stability occurs if the C-load causes the output circuitry to break before f_2; the Op Amp then tries to oscillate when used in a closed loop feedback configuration.

When large capacitive loads are an application requirement (greater than 1000pF), then it is advisable to use a buffer amplifier (current booster, or driver as they may sometimes be called). This technique helps isolate the capacitive load from the Op Amp output (see Figure 39).

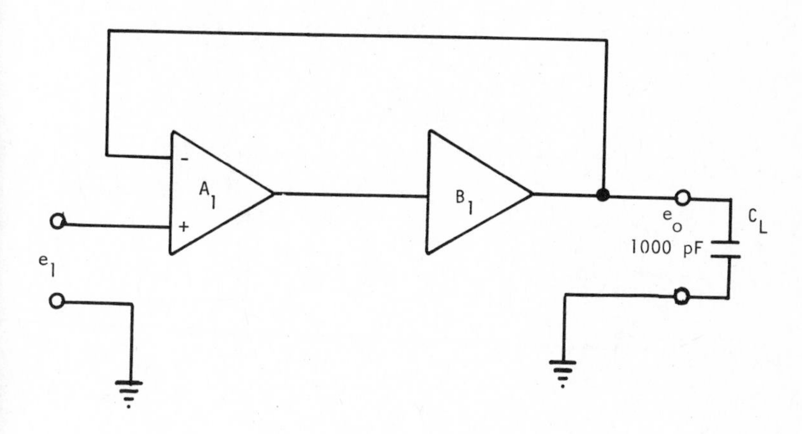

Figure 39. Buffer Amplifier Used to Drive Capacitive Load

The buffer amplifier B_1 is not an Op Amp; it is an emitter-follower type circuit (approximately unity gain) used for driving large output currents. See Appendix M for a typical buffer circuit schematic.

C-Loading (continued)

In special high frequency pulse work the buffer circuit may require additional isolation to improve settling time. Figure 40 shows a useful technique for achieving this requirement. The resister R_1 and capacitor C_1 combination form an integrator circuit that is used to remove the overshoot from the output signal. In addition, resistor R_2 is used to isolate any capacitance coming from the buffer input. The component values shown in Figure 40 are imperical and can vary considerably based on stray capacitance and other extraneous high-frequency effects. High-quality grounding and power supply bypassing techniques are essential in these type applications.

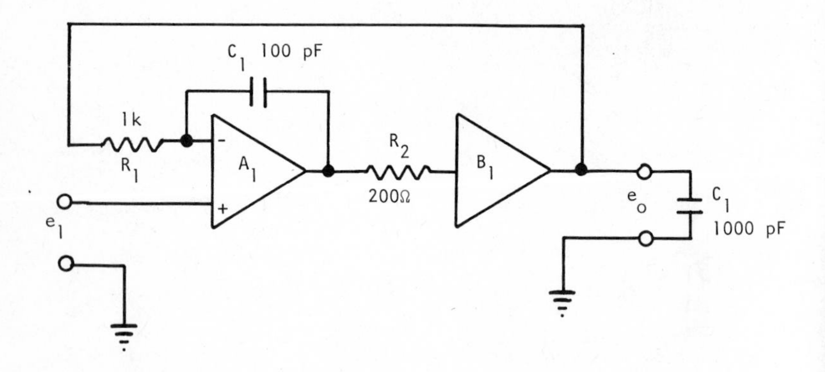

Figure 40. Additional Isolation Used to Optimize Settling Time

Input Impedance - Inverting Configuration

The input impedance to the inverting configuration is, for all practical purposes, equal to the value of the input summing resistor R_1 because the summing junction is a virtual ground (see Figure 41).

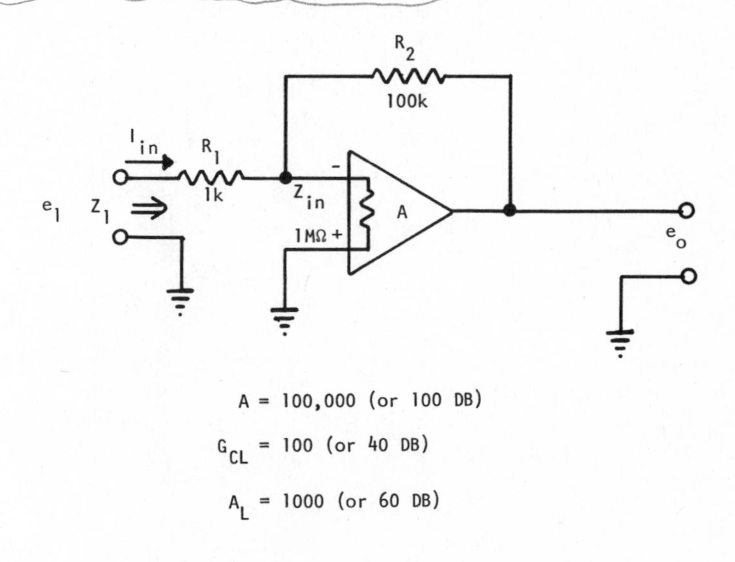

$$A = 100{,}000 \text{ (or 100 DB)}$$

$$G_{CL} = 100 \text{ (or 40 DB)}$$

$$A_L = 1000 \text{ (or 60 DB)}$$

Figure 41. Input Impedance for Inverting Configuration

$$Z_1 = R_1$$

$$= 1\text{k}\Omega$$

Input Impedance - Non-Inverting Configuration

The input impedance to the non-inverting configuration is very high.

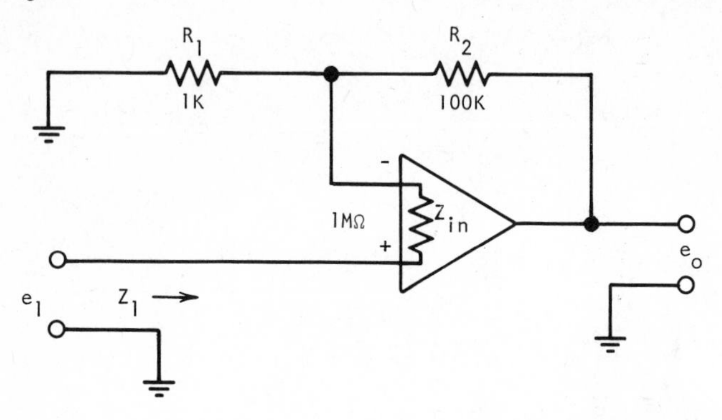

$$A = 100{,}000 \text{ (or 100 DB)}$$

$$G_{CL} = 100 \text{ (or 40 DB)}$$

$$A_L = 1000 \text{ (or 60 DB)}$$

Figure 42. Input Impedance for Non-Inverting Configuration

The actual value of the input impedance Z_1 is the input impedance of the Op Amp itself (Z_{in}) multiplied by the loop gain (A_L).

$$Z_1 = Z_{in} A_L$$

Input Impedance - Non-Inverting Configuration (continued)

In Figure 42, the typical low-cost bi-polar Op Amp with a specified differential input impedance (Z_{in}) of 1 megohm will have a non-inverting network input impedance (Z_1) of,

$$Z_1 = (10^6)A_L$$

$$Z_1 = 1000\ M\Omega, \text{ where } A_L = 1000 \text{ (or 60 DB)}$$

Generally, the input impedance of most good quality Op Amps is considerably higher. Typical FET (Field Effect Transistor) input Op Amps have differential input impedances in the 10^{12} ohm range.

Using the FET Op Amp in the non-inverting configuration offers the designer excellent isolation circuit for high impedance input sources (i.e., pH probes).

A word of caution: Sometimes one might be tempted to use high input impedance bi-polar IC Op Amps for this purpose. This is not usually advisable. The input bias currents normally make this prohibitive. Remember, the input current parameters are independent from input impedance parameters. Therefore, both these parameters must be carefully considered when designing for high impedance sources. See Sections on bias current and offset current.

SECTION II

USEFUL OP AMP NETWORKS

SECTION II

General

This section serves as a takeoff point for the designer. With an understanding of the Op Amp parameters from section I, the designer will be able to apply these concepts to the circuits shown in this section.

Each network can be modified to fit the particular application. Actual component values are shown in each circuit so as to give the designer a feel for the relative relationships between components.

A brief description will sometimes follow a network in order to direct the designer into a better understanding of the particular advantage or disadvantage of the approach.

DC VOLTAGE OUTPUT

Introduction

The networks shown in this category are generally referred to as DC Voltage Amplifiers. The inputs to the networks may be either voltages or current sources. The outputs of the circuits, however, are those which deliver an output voltage only.

Networks illustrated include:

Inverting Amplifiers

Non-inverting Amplifiers

T Network Amplifiers

Buffer Amplifiers

Current-to-Voltage Converters

Voltage Reference Networks

Variable Gain Control Network Techniques

Input Capacitance Neutralizing

Output Current Boosters

Push-Pull Outputs

Null Detection Amplifiers

Adders

Averagers

Subtractors

Differential Output Op Amps

Differential Instrumentation Amplifiers

DC VOLTAGE OUTPUT

Inverting Amplifier

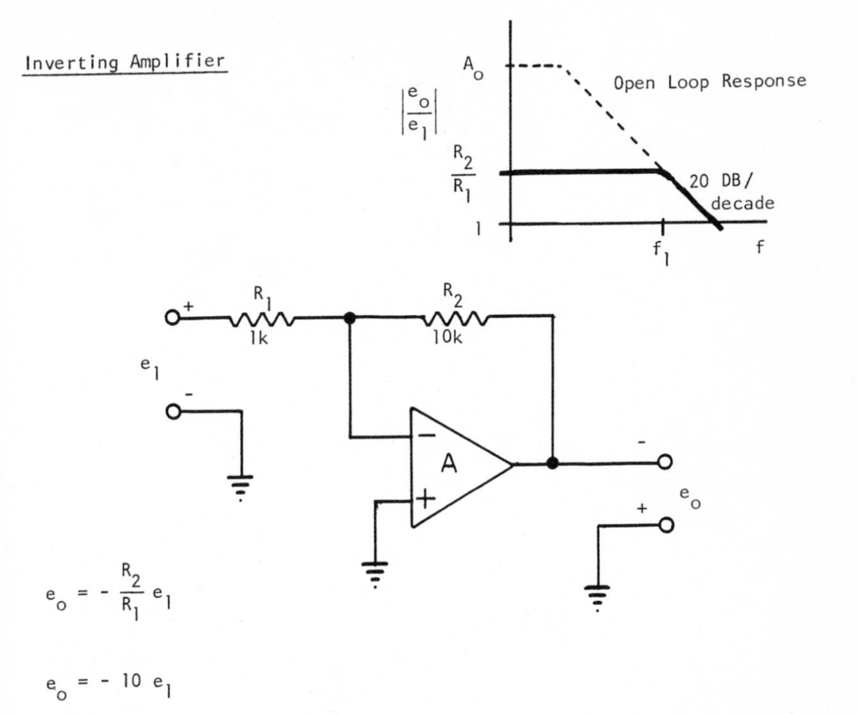

$$e_o = -\frac{R_2}{R_1} e_1$$

$$e_o = -10\, e_1$$

This is the most popular Op Amp circuit. Miniature Op Amp modules are available to yield output voltage swings of more than ± 100 volts.

The basic disadvantage of this circuit is that the input impedance is low (the value of R_1). Frequency f_1 is the overall small signal bandwidth. The large signal (full power response) frequency limit is determined by the maximum slew rate of the particular Op Amp.

$$F_{(max)} = \frac{\text{Slew Rate}}{2\pi E_{(max)}}$$

Slew rate - volts/sec

$E_{(max)}$ - full scale output voltage (peak)

$F_{(max)}$ - full power frequency

High Gain Wide Band Amplifier

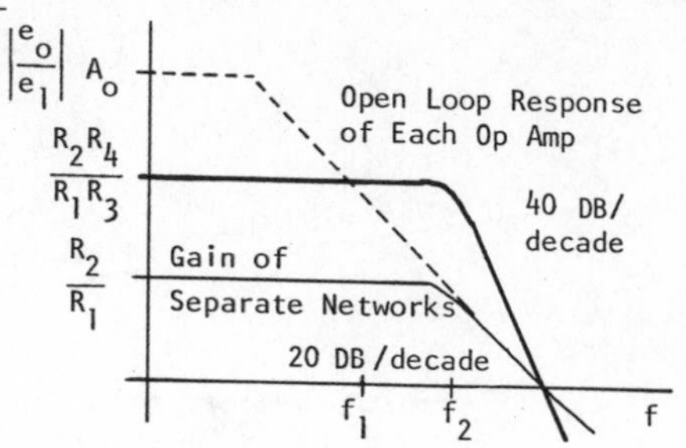

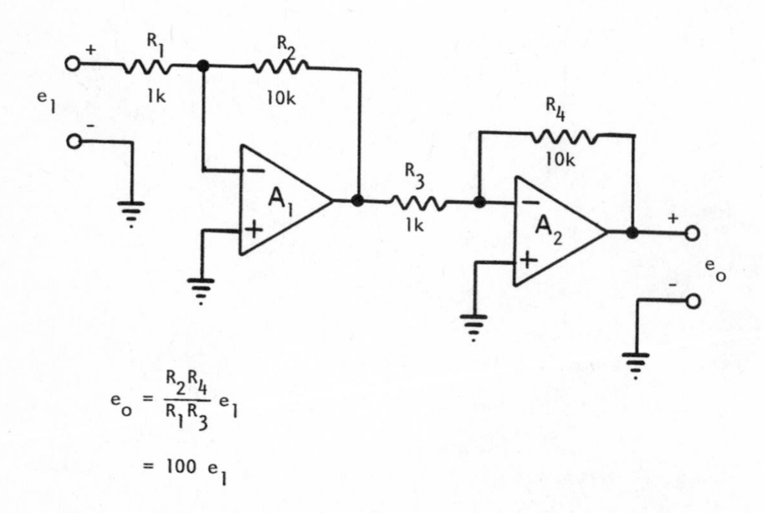

$$e_o = \frac{R_2R_4}{R_1R_3} e_1$$

$$= 100\ e_1$$

Frequency f_2 is the overall network small signal bandwidth. If only one Op Amp were used instead of two, the bandwidth would be frequency f_1. Notice that the roll-off for the two amplifiers combined is 40 DB/decade as compared to 20 DB/decade for the single amplifier. This results in a greater phase shift at the output at high frequencies.

Non-inverting Amplifier

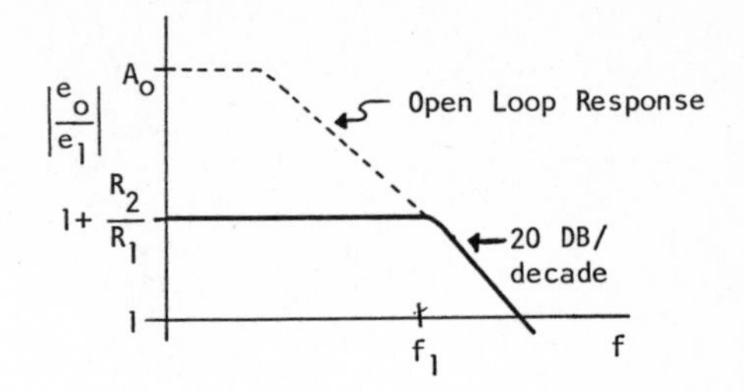

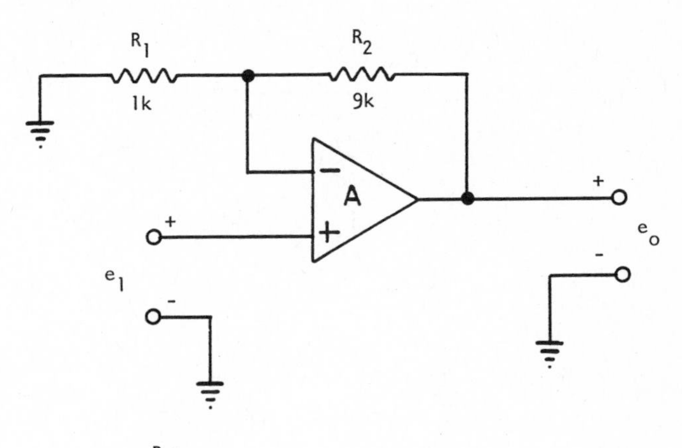

$$e_o = \left(1 + \frac{R_2}{R_1}\right) e_1$$

$$= \left(1 + \frac{9k}{1k}\right) e_1 = 10\ e_1$$

This is the second most popular Op Amp Circuit. It offers an advantage over the inverting configurations in that the input impedance is that of the Op Amp itself rather than R_1.

A disadvantage is the input signal swing is limited to the common mode voltage swing of the Op Amp.

Unity Gain Non-inverting Amplifier

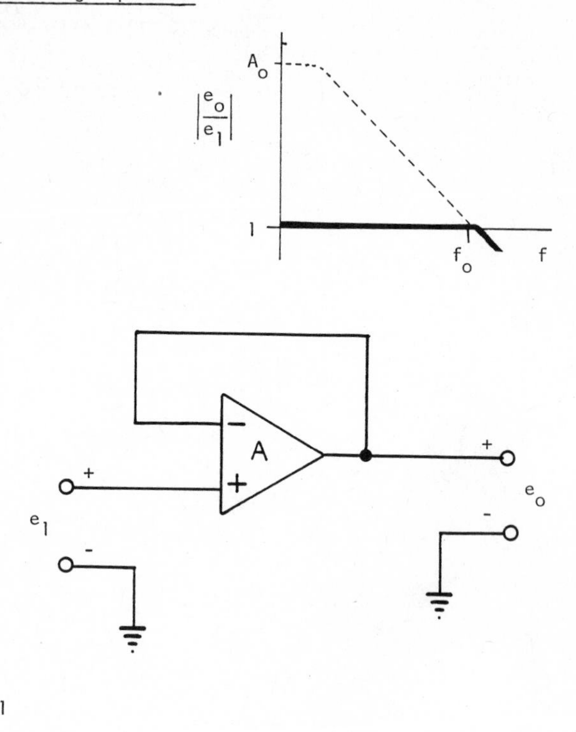

$e_o = e_1$

This is an extremely effective isolation technique. It is commonly called a unity gain buffer or a voltage follower circuit. Care should be taken to by-pass all power supply leads close in to the Op Amp pins. This is a critical circuit for oscillations and is usually quite sensitive to capacitive loading.

DC Amplifier T Network

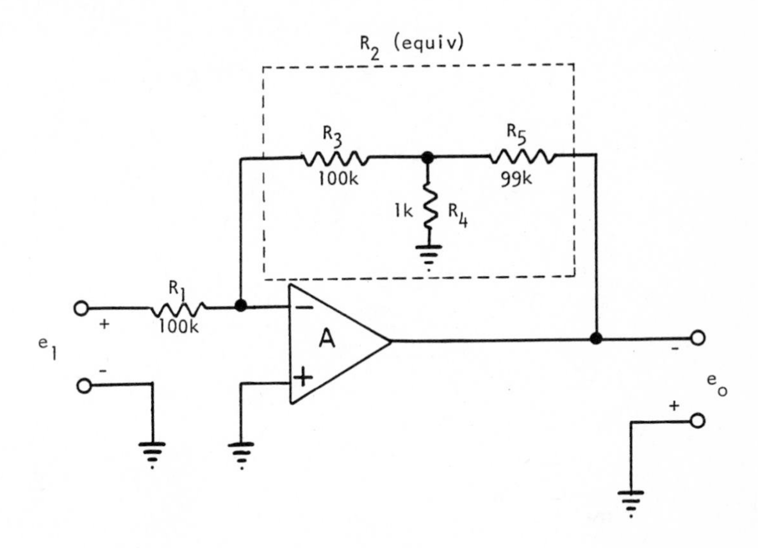

$$e_o = -\frac{R_2 \text{ (equiv)}}{R_1} e_1 , \qquad \text{where } R_2\text{(equiv)} = R_3\left(1 + \frac{R_5}{R_4}\right),$$

$$= -100\, e_1 \qquad\qquad R_4 < R_3/10$$

$$= 10\ \text{M}\Omega$$

High output gain can be achieved while using relatively small resistors in the feedback circuit. This technique offers the advantage of allowing higher input impedance (R_1) without trade-off in gain due to the unavailability of precision metal film resistors.

The value of R_4 should be at least 10 times less than R_3 for the equations to hold. (See Appendix F for derivation.)

DC Amplifier T Network (continued)

If this same circuit is redrawn it is sometimes easier to understand as a combination of both inverting and non-inverting configurations.

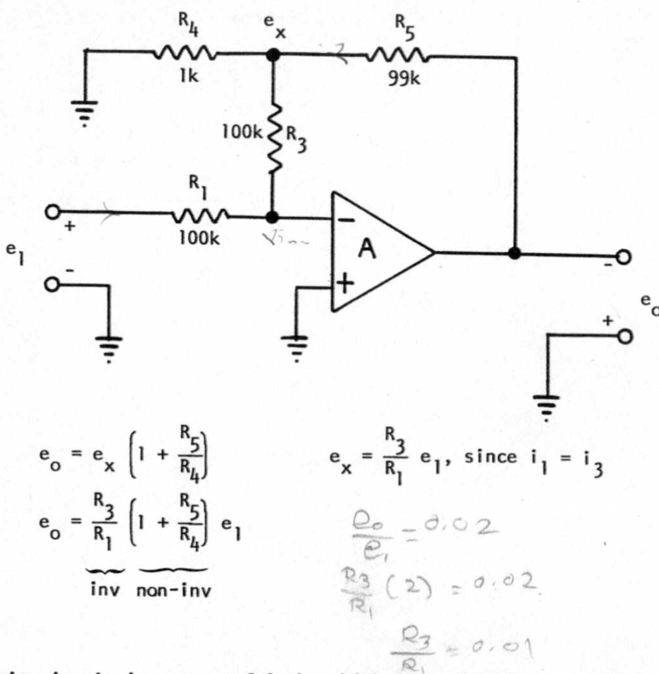

$$e_o = e_x \left(1 + \frac{R_5}{R_4}\right) \qquad e_x = \frac{R_3}{R_1}\, e_1, \text{ since } i_1 = i_3$$

$$e_o = \underbrace{\frac{R_3}{R_1}}_{\text{inv}} \underbrace{\left(1 + \frac{R_5}{R_4}\right)}_{\text{non-inv}} e_1$$

This circuit is very useful when high input impedance is needed with low drift. Chopper Stabilized Op Amps (low drift) usually can only be used in the inverting mode.

Buffer Amplifier

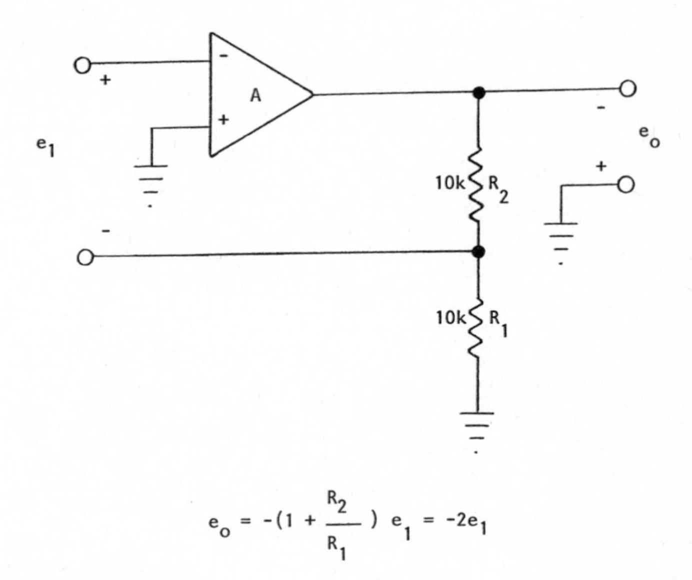

$$e_o = -(1 + \frac{R_2}{R_1})\, e_1 = -2e_1$$

This bootstrap technique offers very high input impedance. However, it should be noted that the input must be floating.

Reducing resistor R_2 to zero, and removing resistor R_1 will result in the output duplicating the input. The network serves as a buffer for a "floating" input signal.

Current-to-Voltage Converter

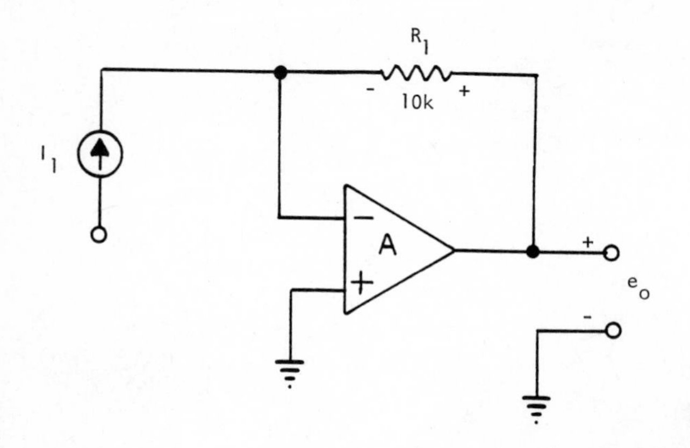

$$e_o = I_1 R_1$$

The current source (I_1) produces a voltage drop across resistor R_1. This voltage value is seen at the output.

For current sources less than 1 ma, FET Op Amps are recommended. These devices offer the advantage of very low input bias currents which help minimize the DC offset error.

Differential Current-to-Voltage Converter

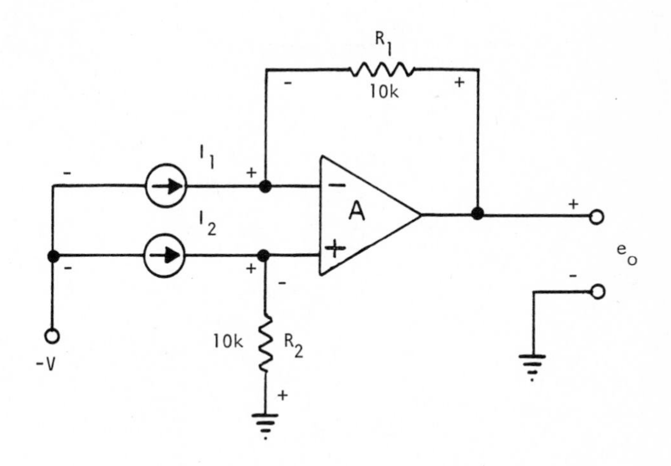

$$e_o = R_1 I_1 - R_2 I_2$$

The I_1 current source works in the same manner as the previous network. The I_2 current source will cause a voltage drop across R_2.

Provided the I_1 current source has a very high output impedance relative to R_1, the voltage across R_2 will be reproduced (unity gain non-inverting circuit) at the output.

The output will be the algebraic sum of the two inputs.

Current-to-Voltage Converter

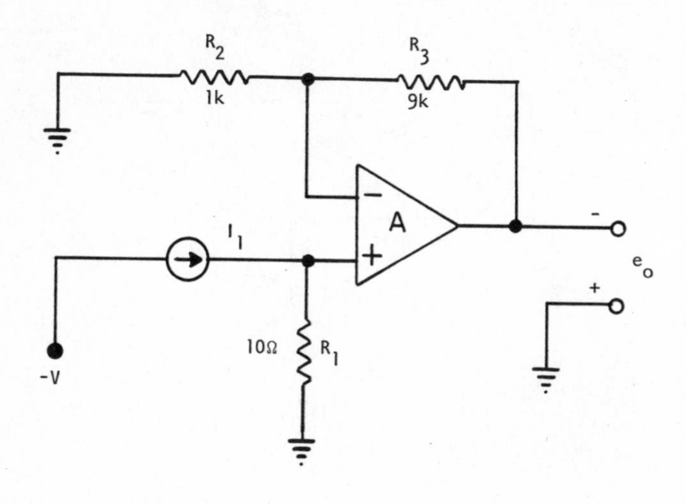

$$e_o = I_1 R_1 \left(1 + \frac{R_3}{R_2}\right)$$

$$= 100\ I_1$$

This approach utilizes a non-inverting gain circuit to amplify the voltage drop across resistor R_1.

Very high current-to-voltage gains can be achieved with low resistor values using this technique.

Inverting Voltage Reference

Care should be taken to insure that R_2 does not load down the circuit. The current through R_2 should be no more than one tenth of the Zener circuit D_1.

$$e_o = V_{D1} \frac{R_3}{R_2}$$

$$= (5V) \left(\frac{20k}{10k}\right)$$

$$= 10V$$

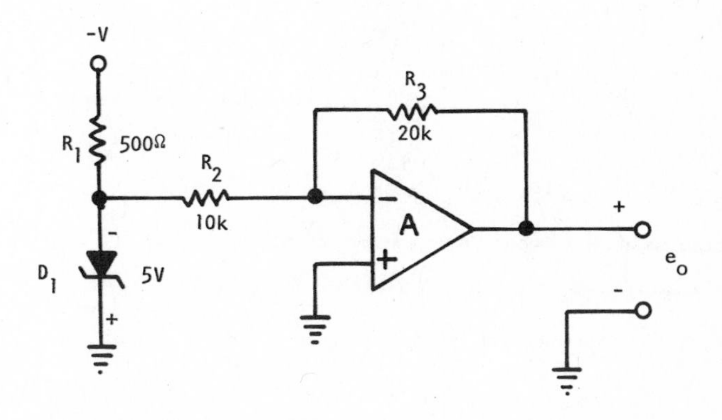

Non-inverting Voltage Reference

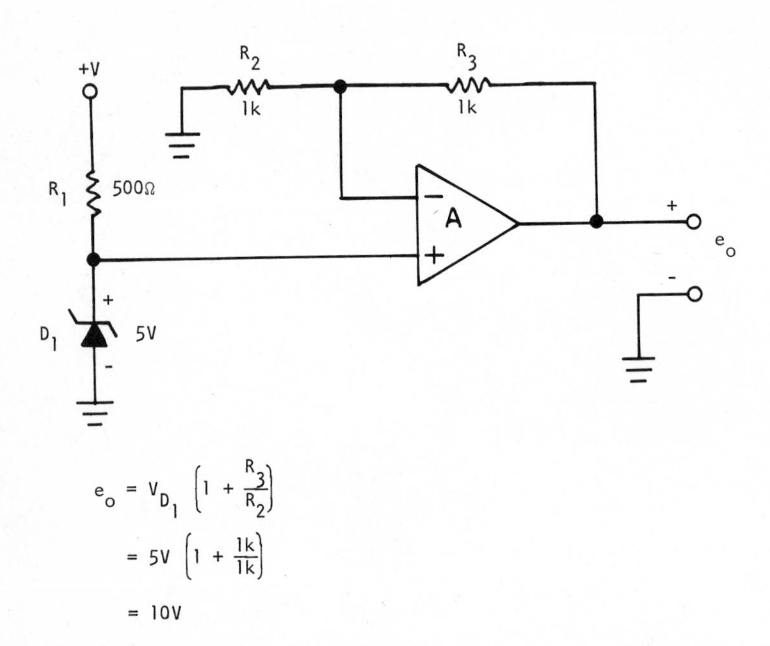

$$e_o = V_{D_1}\left(1 + \frac{R_3}{R_2}\right)$$

$$= 5V\left(1 + \frac{1k}{1k}\right)$$

$$= 10V$$

This circuit has the advantage of not loading the Zener network. The limitation is the input common mode voltage of the Op Amp. Note that the non-inverting (+) input of the Op Amp must be capable of accepting the voltage across the zener diode D_1.

Standard Cell Voltage Reference

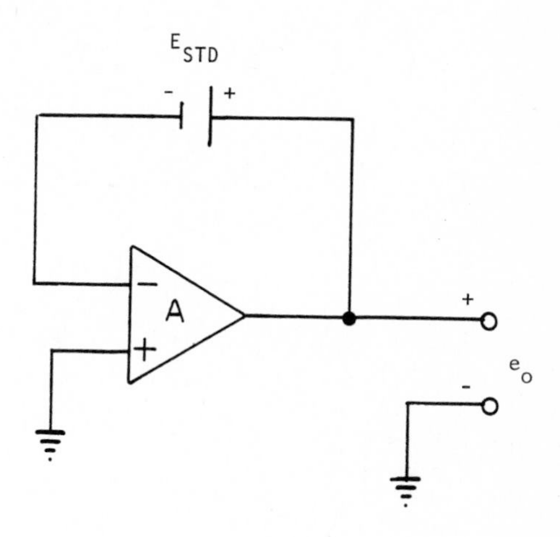

$e_o = E_{STD}$

This network gives excellent isolation to the sensitive standard cell. The load current through the cell is virtually eliminated. ($I_{STD} = I_{BIAS}$ of Op Amp.)

This circuit is ideal for use with Chopper Stabilized Op Amps because the Op Amp is used in the inverting configuration.

Standard Cell Voltage Reference (continued)

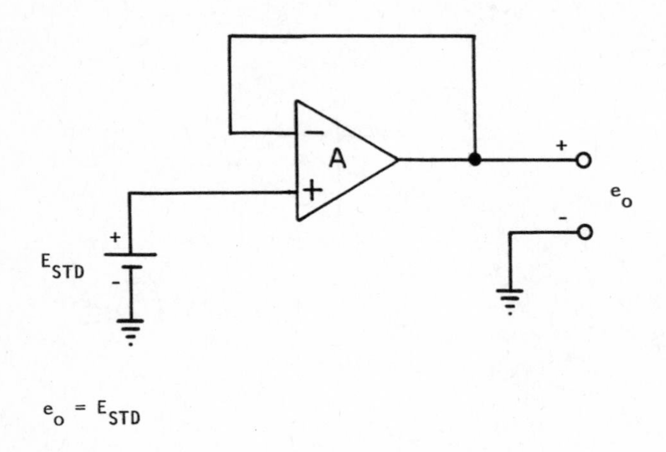

$e_o = E_{STD}$

Complete output isolation is achieved by this technique. The circuit is ideal for FET input Op Amps. The only major source of error is input voltage offset and temperature drift of the Op Amp.

Standard Cell Voltage Reference (continued)

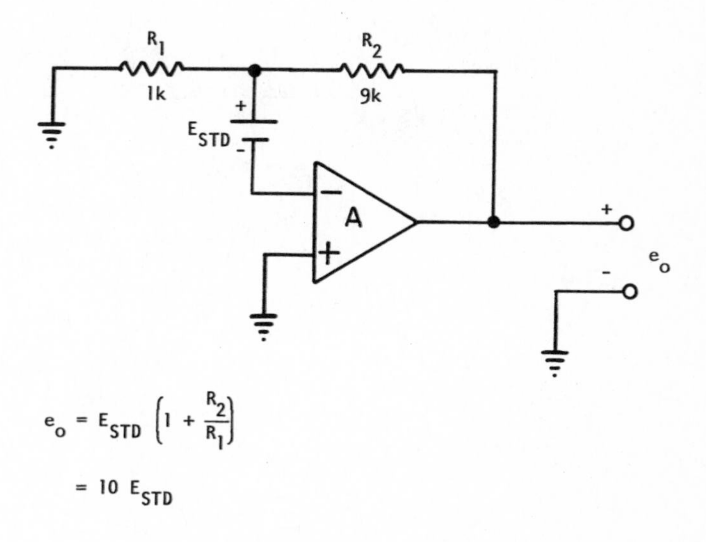

$$e_o = E_{STD}\left(1 + \frac{R_2}{R_1}\right)$$

$$= 10\ E_{STD}$$

This is a method for achieving output gain while providing excellent isolation.

The technique is especially useful with Chopper Stabilized Op Amps since they are usually limited to the inverting mode only.

Buffered Variable Voltage Source

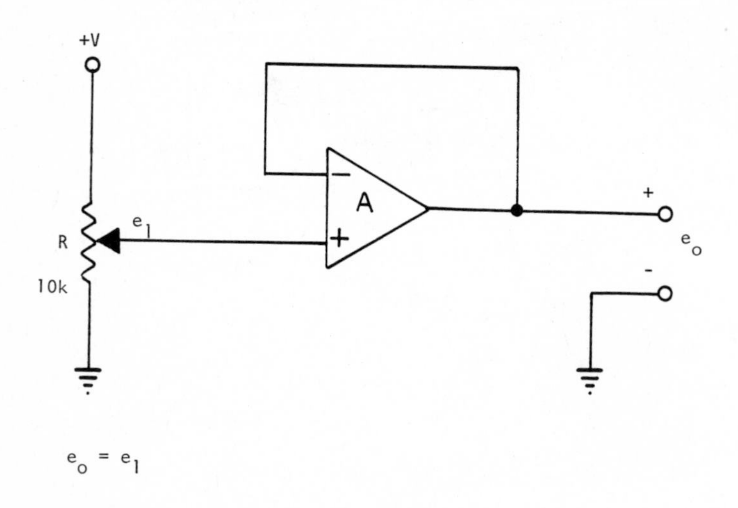

$$e_o = e_1$$

This is a method for making a precision voltage divider. The output load does not affect the input bleeder circuit. The resistor value R should be small enough so that the Op Amp input bias current does not produce a significant voltage drop error. The low-bias current FET Op Amp is an excellent candidate for this type of network.

Standard Cell Bi-polar Voltage Reference

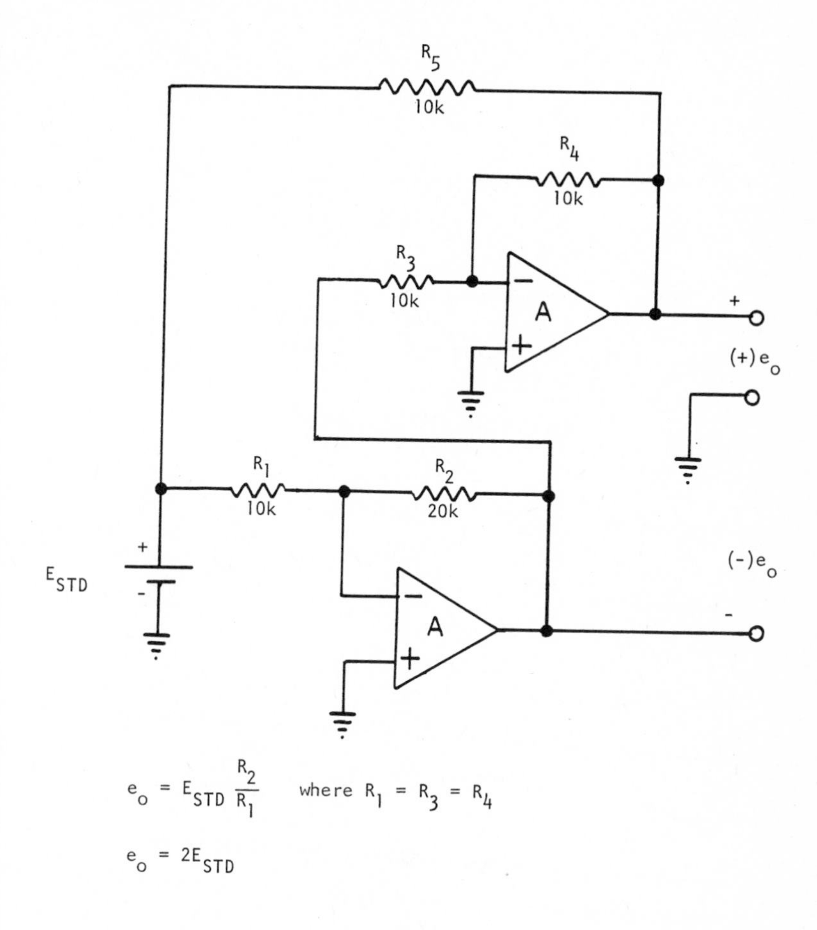

$$e_o = E_{STD} \frac{R_2}{R_1} \quad \text{where } R_1 = R_3 = R_4$$

$$e_o = 2E_{STD}$$

The value of R_5 should be trimmed to exactly cancel the current in R_1, leaving the cell unloaded.

Inverting Variable Gain Control

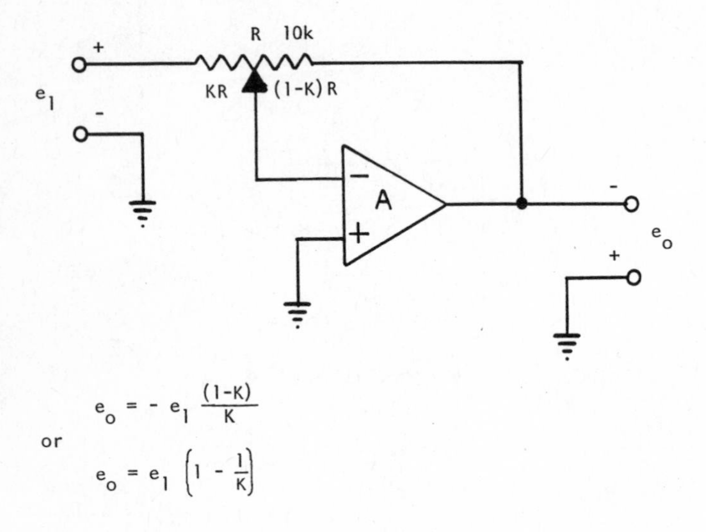

$$e_o = -e_1 \frac{(1-K)}{K}$$

or

$$e_o = e_1 \left(1 - \frac{1}{K}\right)$$

Note that the input is heavily loaded for high gain settings. The disadvantage of this network is the low input impedance (which also varies with the gain control). The advantage of this approach is its simplicity.

Non-inverting Variable Gain Control

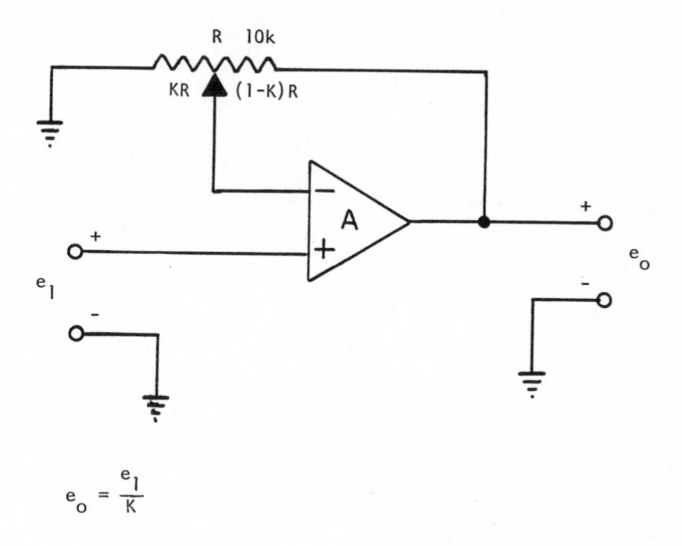

$$e_o = \frac{e_1}{K}$$

This circuit has the advantage of high input impedance isolation as well as a simple gain control.

This disadvantage of the circuit is that the potentiometer adjustment is not linear with gain.

DC VOLTAGE OUTPUT

DC Voltmeter Using Gain Control

$$e_o = \frac{e_1}{K}$$

$$R_T = R_2 + R_3 + R_4 + R_5 + R_6$$

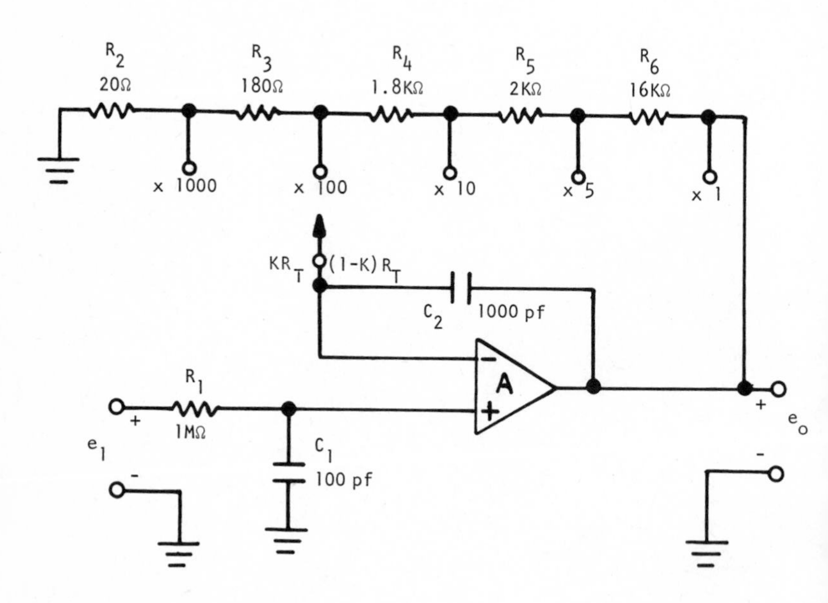

The components R_1, C_1 are used for input protection and isolation of transients. Capacitor C_2 stabilizes the network from strong oscillations. For best results the Op Amp should be a FET input type (for high input impedance).

Variable Gain Control

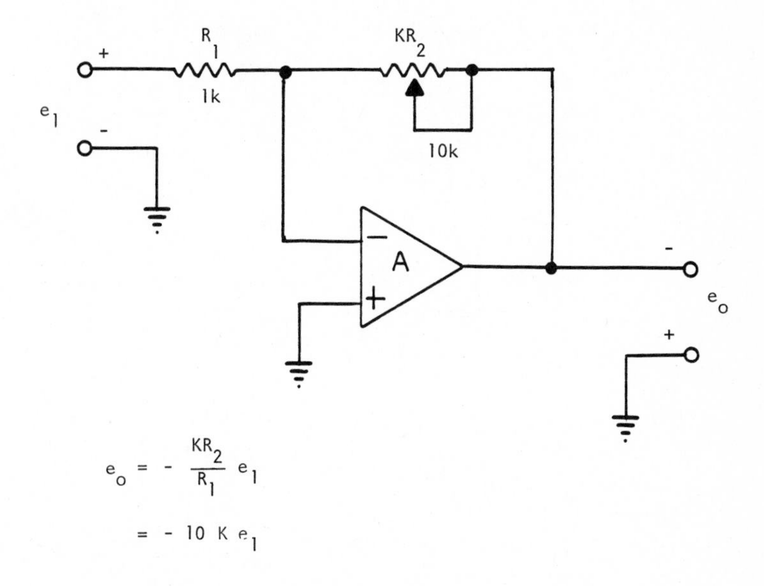

$$e_o = -\frac{KR_2}{R_1} e_1$$

$$= -10\ K\ e_1$$

This network gives a linear gain control, but input impedance is low (R_1).

The low input impedance can be overcome by adding a unity gain non-inverting Op Amp at the input to the network.

The gain range is limited by the value of R_2.

Variable Gain Control (continued)

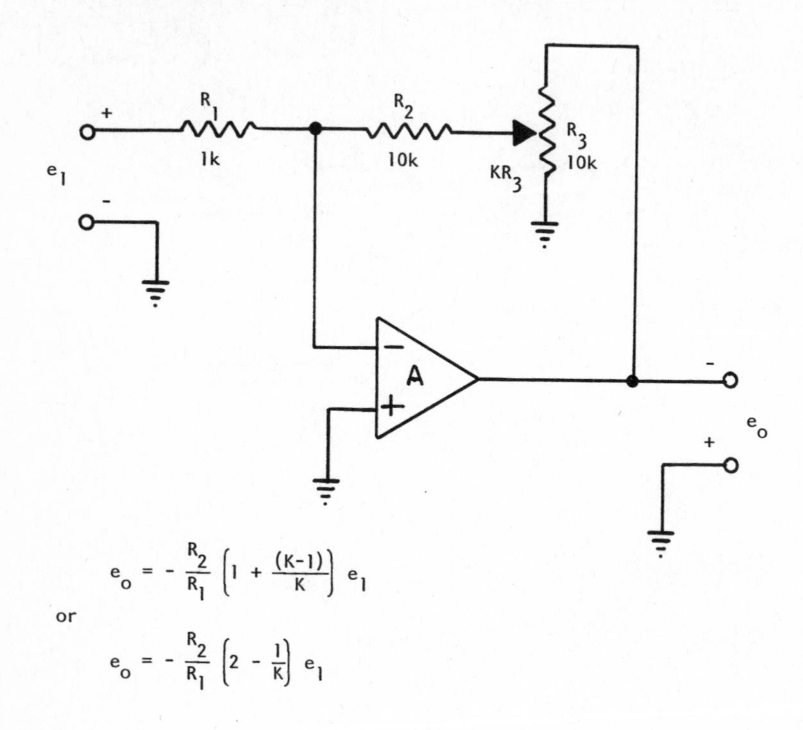

$$e_o = -\frac{R_2}{R_1}\left(1 + \frac{(K-1)}{K}\right) e_1$$

or

$$e_o = -\frac{R_2}{R_1}\left(2 - \frac{1}{K}\right) e_1$$

This approach offers a wide gain range without the need of large feedback resistor values.

The basic disadvantage is that the equation is relatively complex and non-linear.

Bi-polar Variable Gain Control

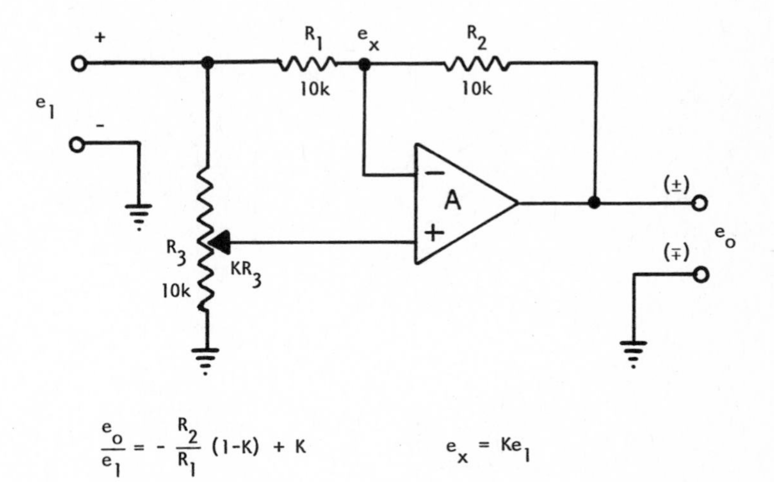

$$\frac{e_o}{e_1} = -\frac{R_2}{R_1}(1-K) + K \qquad\qquad e_x = Ke_1$$

$$\frac{e_o}{e_1} = +1, \qquad \text{for } K = 1$$

$$\frac{e_o}{e_1} = -1, \qquad \text{for } K = 0$$

This unique network allows the gain to be changed from inverting to non-inverting by adjusting the position of the potentiometer.

The relatively low input impedance can be overcome by adding a unit gain non-inverting Op Amp at the input.

Variable Gain Control

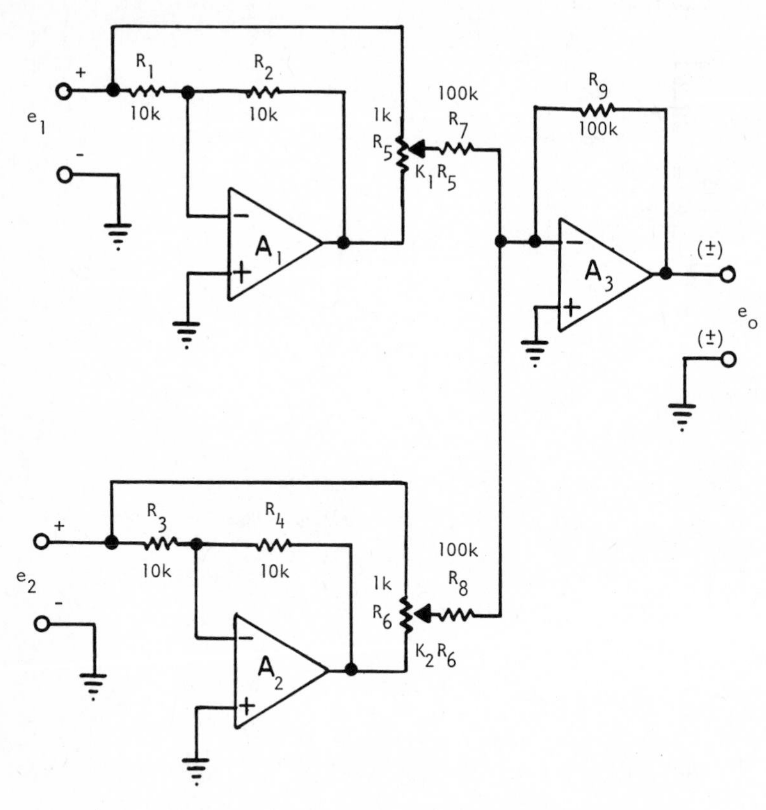

$$e_o = e_1(1-2K_1) + e_2(1-2K_2)$$

If the pots R_5 and R_6 are "ganged" together, then $K_1=K_2=K$, and

$$e_o = (e_1+e_2)(1-2K)$$

Note that

$$e_o = e_1+e_2 , \qquad \text{for } K=0$$

$$eo = -(e_1+e_2) , \qquad \text{for } K=1$$

Variable Gain Control

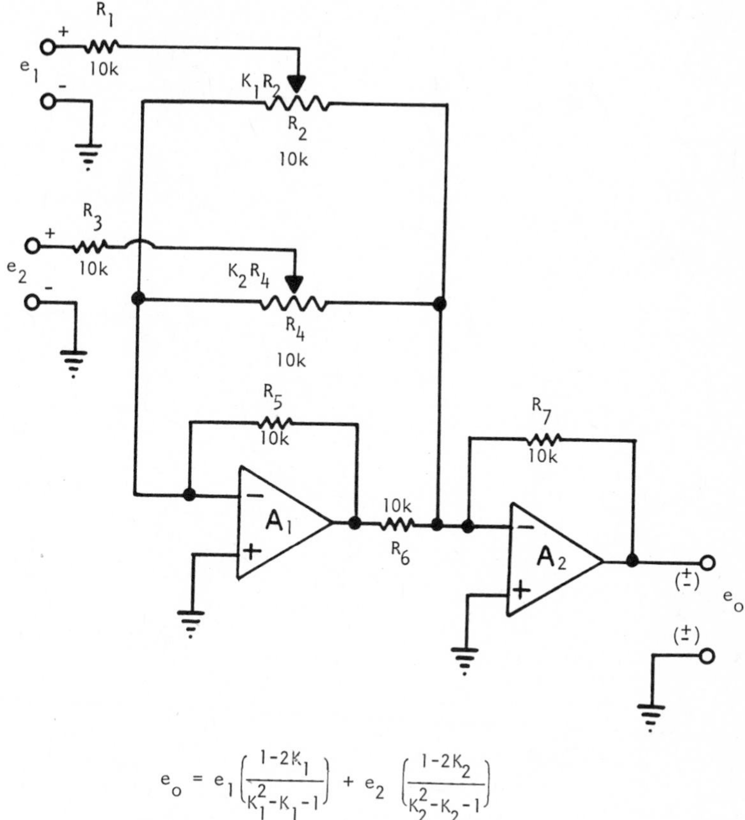

$$e_o = e_1\left(\frac{1-2K_1}{K_1^2-K_1-1}\right) + e_2\left(\frac{1-2K_2}{K_2^2-K_2-1}\right)$$

If pots R_2 and R_4 are "ganged" together, then $K_2 = K_4 = K$, and

$$e_o = (e_1+e_2)\ \frac{1-2K}{K^2-K-1}$$

Note that

$$e_o = e_1 + e_2\ , \qquad \text{for } K=1$$

$$e_o = -(e_1+e_2)\ , \qquad \text{for } K=0$$

Input Capacitance Neutralizing

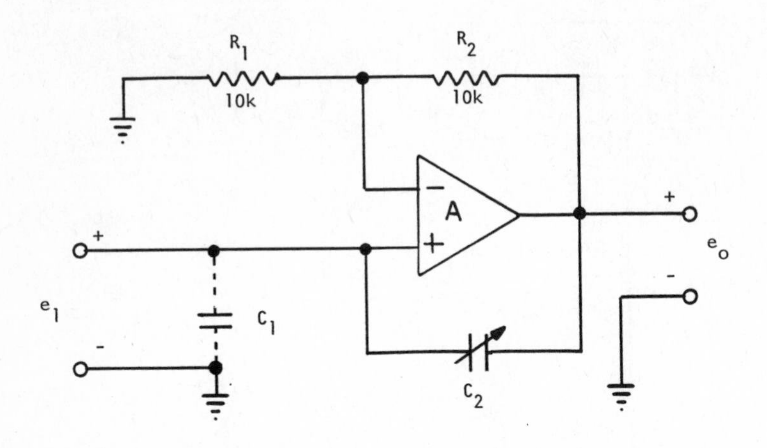

for neutralization $I_{C1} = I_{C2}$, where $I_{C1} = \frac{e_1}{X_{C1}}$, and $I_{C2} = \frac{e_o - e_1}{X_{C2}}$

$$\frac{e_1}{X_{C1}} = \frac{e_o - e_1}{X_{C2}} \text{ , and } e_o = \left(1 + \frac{R_2}{R_1}\right) e_1$$

therefore,

$$C_2 = \frac{R_1}{R_2} C_1$$

$$= C_1 \text{ , for } R_1 = R_2$$

Capacitor C_2 can be adjusted to neutralize capacitor C_1. This is an Op Amp technique utilizing capacitance to exhibit an inductive effect.

Output Current Booster (Unipolar Buffer)

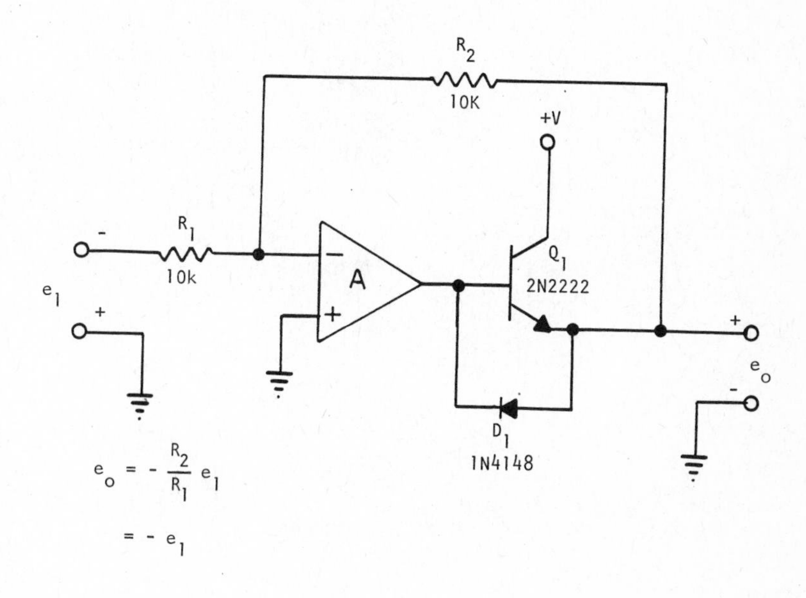

$$e_o = -\frac{R_2}{R_1} e_1$$

$$= -e_1$$

Additional output current capability (positive only) is carried by Q_1. The diode D_1 protects the transistor when the output reverses polarity. Reversing the diode and using a PNP transistor will give negative output current capability.

Output Current Booster (Bi-polar Buffer)

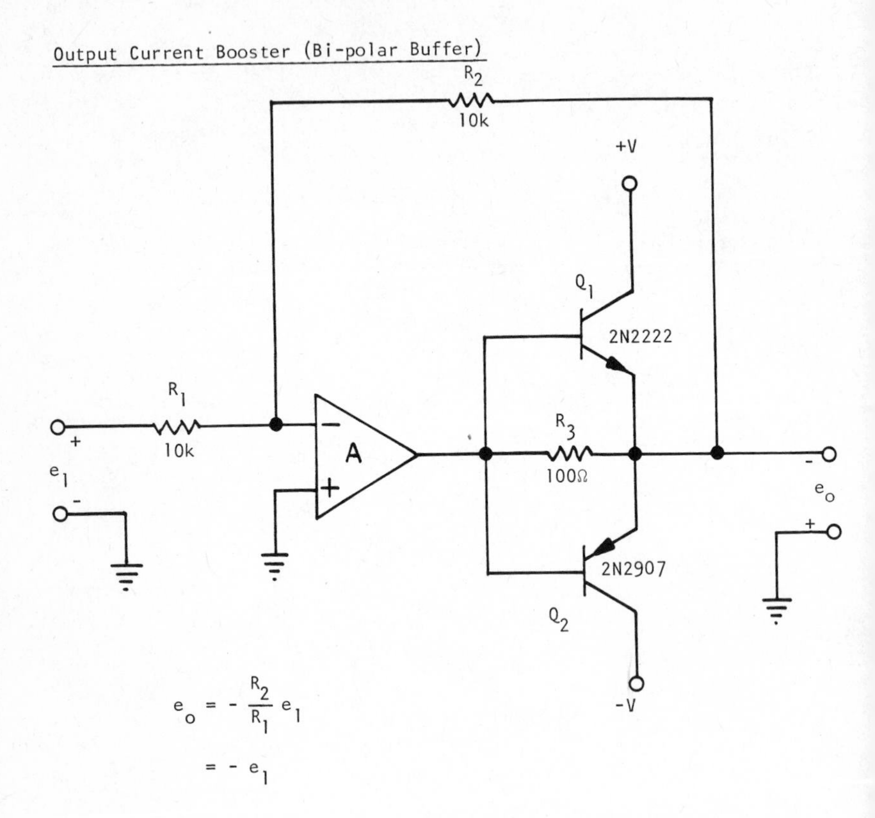

$$e_o = -\frac{R_2}{R_1} e_1$$

$$= - e_1$$

This circuit is simple and works for both output polarities. At high output gains (1000) the output may show slight distortion as it passes through zero. This is because both transistors Q_1, Q_2 are biased off at this point. This phenomenon is not usually noticeable at lower gains because the feedback (high loop gain) heavily compensates for any errors within the loop. A more sophisticated approach is shown in Appendix M. These circuits may be purchased in modular packages from most module Op Amp manufacturers.

DC VOLTAGE OUTPUT

Unity Gain Buffer

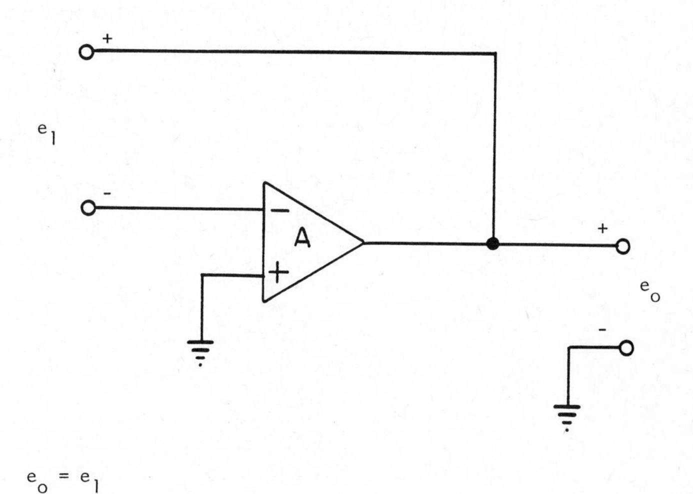

$e_o = e_1$

Input must be floating. The input voltage is simply reproduced at the output. The use of a FET Op Amp minimizes any current flow to the voltage source e_1.

This is a successful technique used to store voltage across a capacitor for an extended period of time.

Push-Pull Output Amplifier

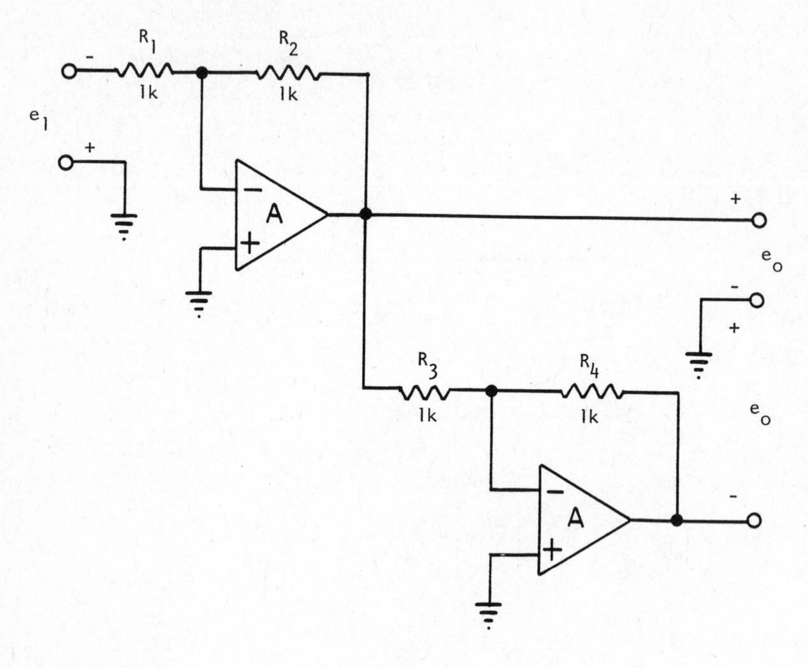

$$e_o = \pm e_1 \ , \ \text{for} \ \frac{R_2}{R_1} = \frac{R_4}{R_3}$$

The input voltage e_1 is reproduced symmetrically above and below common ground at the output. The output signals e_o are 180° out of phase with each other.

Null Detection Amplifier (Floating Supply)

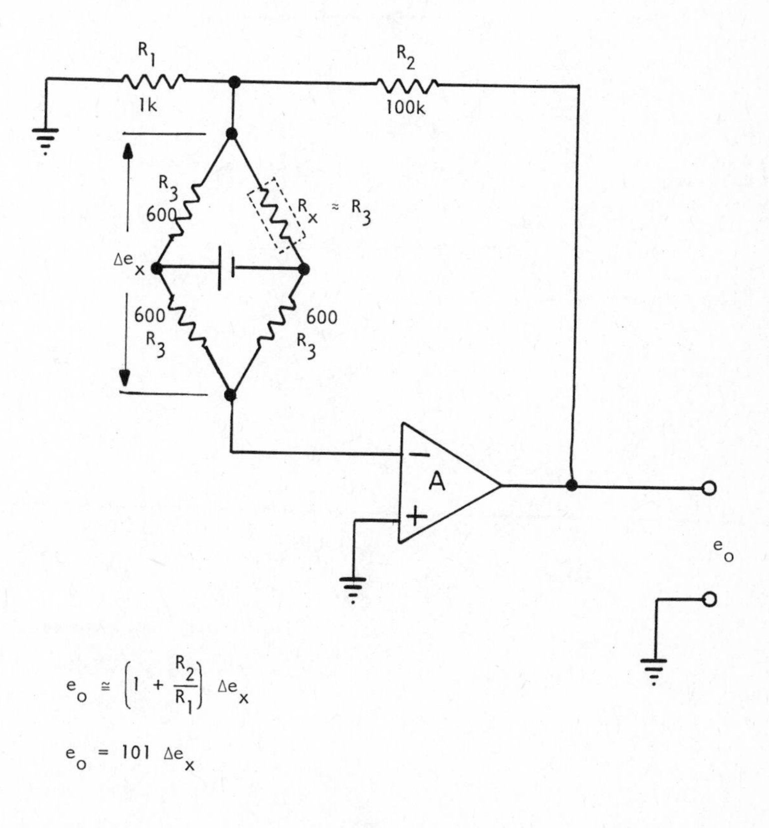

$$e_o \cong \left(1 + \frac{R_2}{R_1}\right) \Delta e_x$$

$$e_o = 101\ \Delta e_x$$

In this circuit when $R_x = R_3$ the output will null.

Null Detection Amplifier (Grounded Supply)

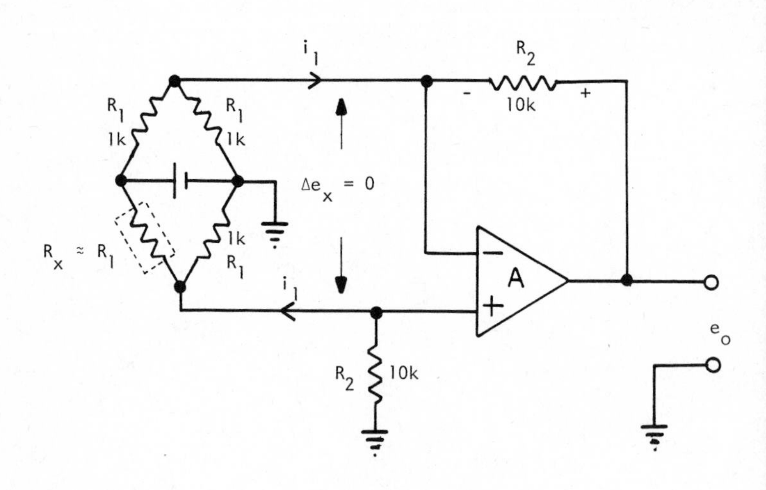

$e_o = 2 i_1 R_2$

Since $\Delta e_x = 0$,

$$i_1 = \frac{V}{2R_1} - \frac{V}{2R_1 \pm \Delta R_x}$$

$$\Delta R_x = R_1 - R_x$$

$$\text{Gain} \approx \frac{2R_2}{R_1}$$

$$= 20$$

In this network when $R_x = R_1$ the output will null.

Null Detection Amplifier (Bridge Loop)

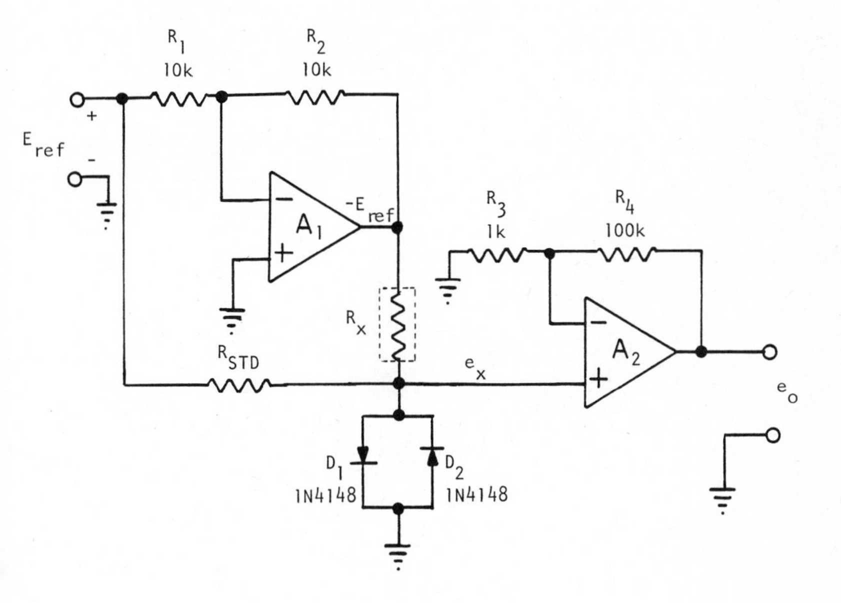

When R_x is equal to R_{STD}, the voltage e_x will null to zero. By connecting e_x to a high input impedance Op Amp network the "null region" can be magnified by the gain of the circuit.

$$e_o = e_x \left(1 + \frac{R_4}{R_3}\right)$$

$$e_o \approx 100\ e_x$$

Diodes D_1 and D_2 simply clamp the input of A_2 for protection. The output of A_2 may also be clamped if desired (see subsection on Clips, Clamps).

Null Detection Amplifier (Balanced Supplies)

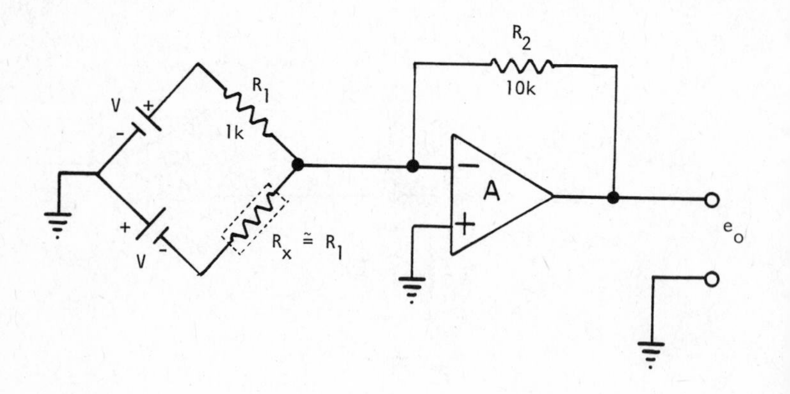

$$e_o = (i_x - i_1) R_2$$

$$\text{Gain} = \frac{R_2}{R_1 || R_x} \approx \frac{2R_2}{R_1}$$

$$= 20$$

The accuracy of this network depends heavily on the voltage supplies (V) being precisely matched.

Null Detection (Open Bridge)

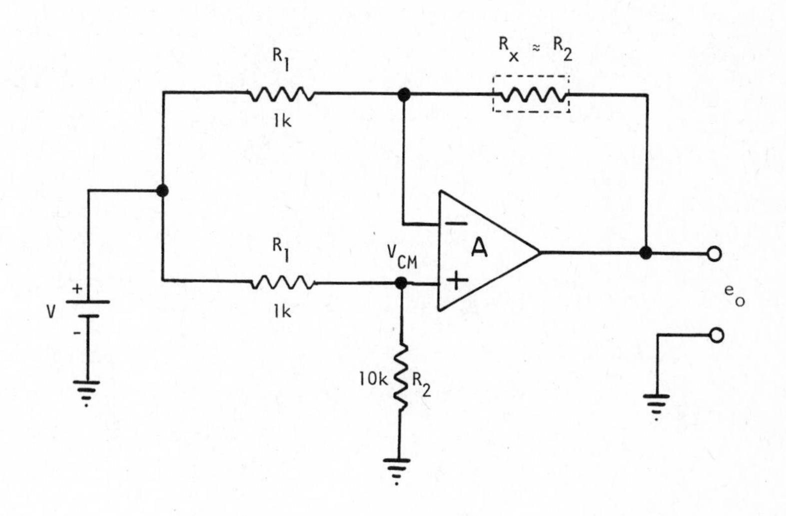

$$e_o = V \frac{R_2}{R_1+R_2} - i_1 R_x$$

$$\text{Gain} = \frac{R_2}{R_1}$$

$$= 10$$

The accuracy of this network depends on a precise matching of R_1 resistors. The Op Amp should also have good input common mode rejection because input voltage (V) is also seen as a common mode voltage V_{CM} (through R_1, R_2).

Null Detection Grounded Load (Open Bridge)

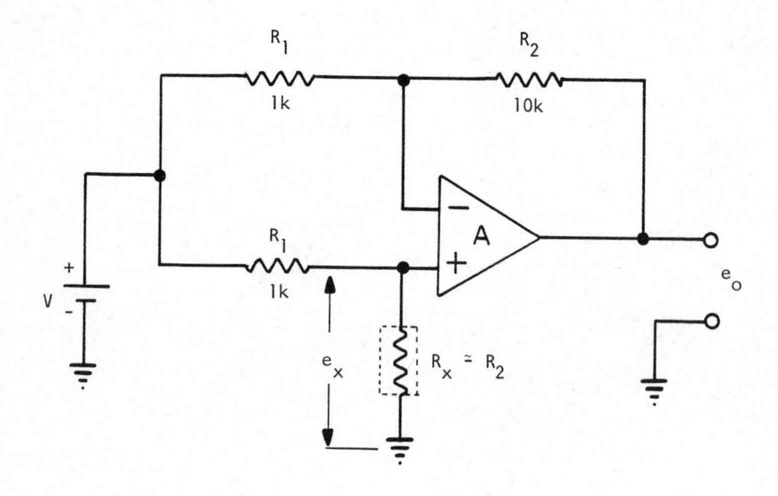

$$e_o = e_x \left(1 + \frac{R_2}{R_1}\right) - \frac{VR_2}{R_1}$$

$$\text{Gain} \approx \frac{R_2}{R_1}$$

The same precautions must be taken with this network as with the previous network. The design concept is basically the same.

This approach offers the advantage of grounding the resistor under test (R_x).

Inverting Adder

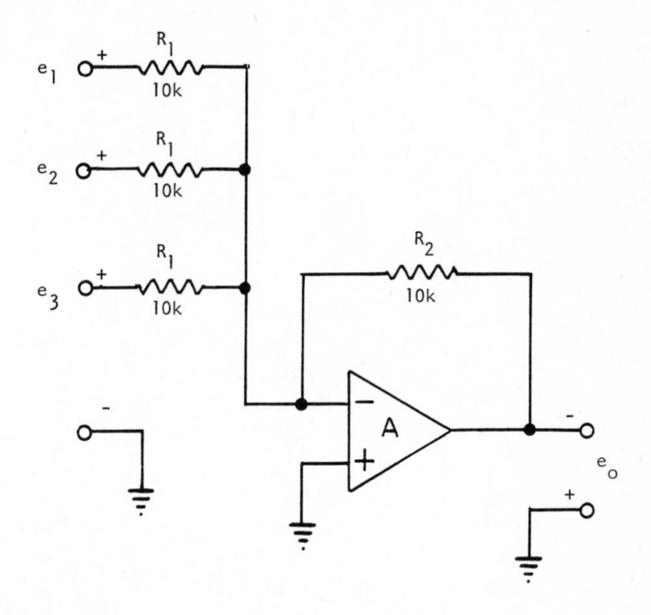

$$e_o = -\frac{R_2}{R_1}(e_1 + e_2 + e_3)$$

$$= -(e_1 + e_2 + e_3)$$

The accuracy depends on the matching of the input resistors R_1. Resistor R_2 determines the gain of the network.

Inverting Scaled Adder

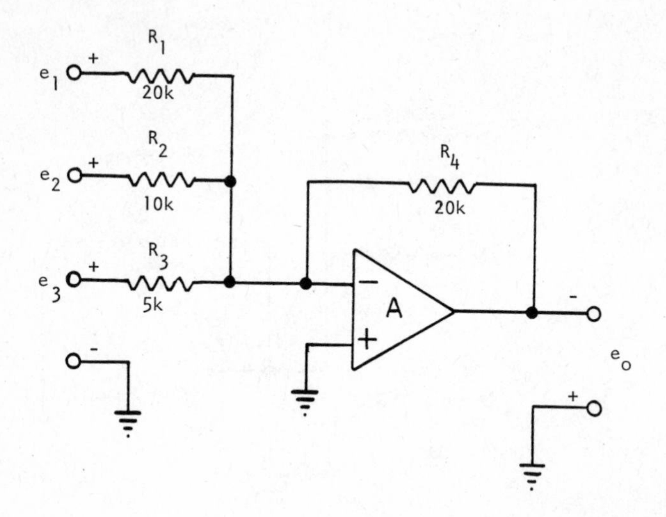

$$-e_o = \frac{R_4}{R_1} e_1 + \frac{R_4}{R_2} e_2 + \frac{R_4}{R_3} e_3$$

$$= e_1 + 2e_1 + 4e_3$$

This technique is used for algebraic addition, averaging, or scaled addition. It also can be used for digital-to-analog conversion when each input voltage represents a digital logic level corresponding to a weighted bit.

For 3-bit binary weighting

$-e_o + e_1 + 2e_1 + 4E_1$, where $e_1 + e_2 + e_3$ = on/off logic levels

Inverting Average

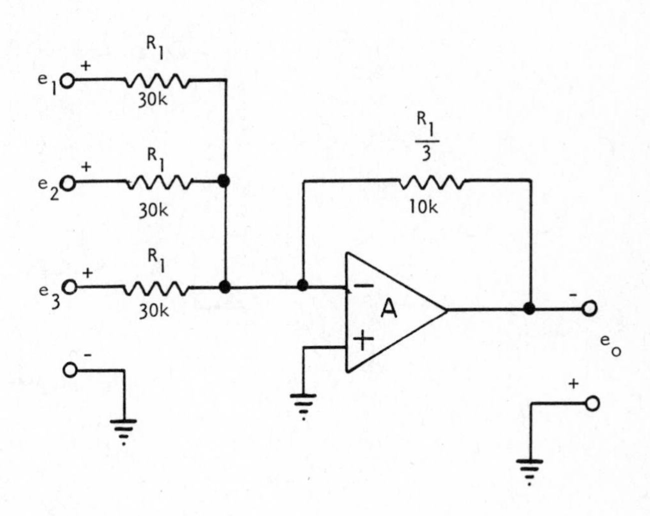

$$-e_o = \frac{e_1 + e_2 + e_3}{3}$$

The accuracy of this network depends on a precise matching of input resistors. The output feedback resistor must also be calibrated to the correct value, $\frac{R_1}{3}$.

Inverting Weighted Averager

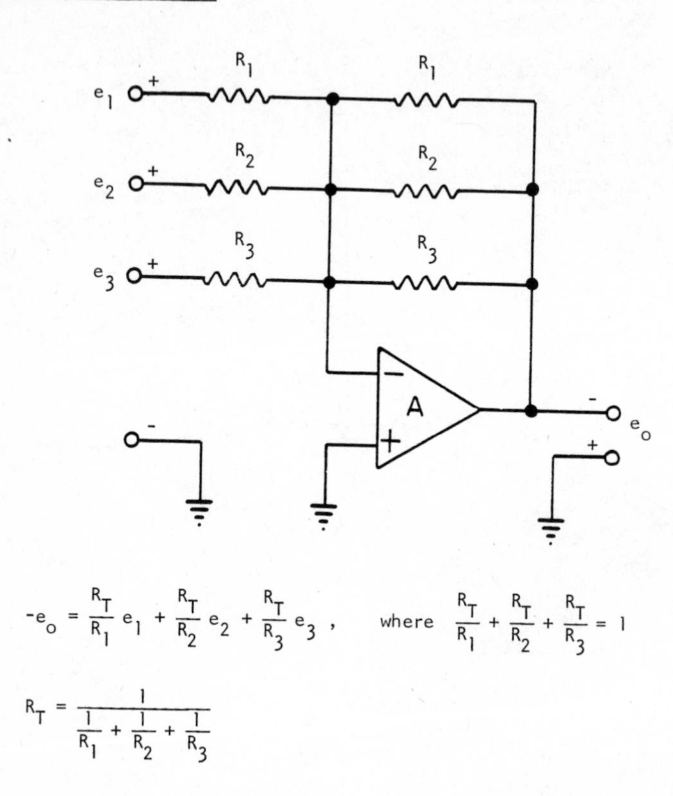

$$-e_o = \frac{R_T}{R_1} e_1 + \frac{R_T}{R_2} e_2 + \frac{R_T}{R_3} e_3 , \qquad \text{where} \quad \frac{R_T}{R_1} + \frac{R_T}{R_2} + \frac{R_T}{R_3} = 1$$

$$R_T = \frac{1}{\frac{1}{R_1} + \frac{1}{R_2} + \frac{1}{R_3}}$$

This method eliminates calibration of the feedback resistor, provided each resistor pair is accurately matched.

Non-Inverting Adder

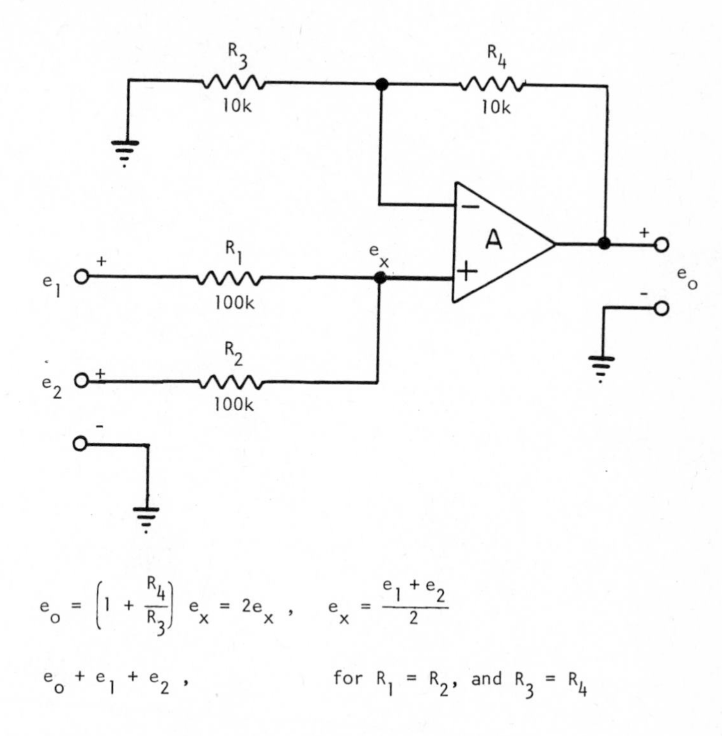

$$e_o = \left(1 + \frac{R_4}{R_3}\right) e_x = 2e_x , \quad e_x = \frac{e_1 + e_2}{2}$$

$$e_o + e_1 + e_2 , \qquad \text{for } R_1 = R_2, \text{ and } R_3 = R_4$$

To establish the correct output relationship, the gain must equal 2. When more inputs are needed, the e_x expression will change according to the number of inputs.

Non-inverting Adder (continued)

For example, consider three inputs. Let all input resistors be identical.

$$i_1 + i_2 + i_3 = 0$$

$$\frac{e_1 - e_x}{R_1} + \frac{e_2 - e_x}{R_2} + \frac{e_3 - e_x}{R_3} = 0$$

$$e_x = \frac{e_1 + e_2 + e_3}{3}, \qquad \text{for } R_1 = R_2 = R_3$$

The maximum amplitude of e_x must be within the common mode voltage specification limits of the Op Amp for proper linear operation.

The equation for output gain remains the same except the ratios of R_3 and R_4 must be changed to accommodate the new weighted inputs. The new value of gain must now be 3.

$$e_o = \left(1 + \frac{R_4}{R_3}\right)\left(\frac{e_1 + e_2 + e_3}{3}\right)$$

if

$$1 + \frac{R_4}{R_3} = 3, \qquad \text{then } \frac{R_4}{R_3} = 2$$

and

$$e_o = e_1 + e_2 + e_3$$

Non-inverting Adder (continued)

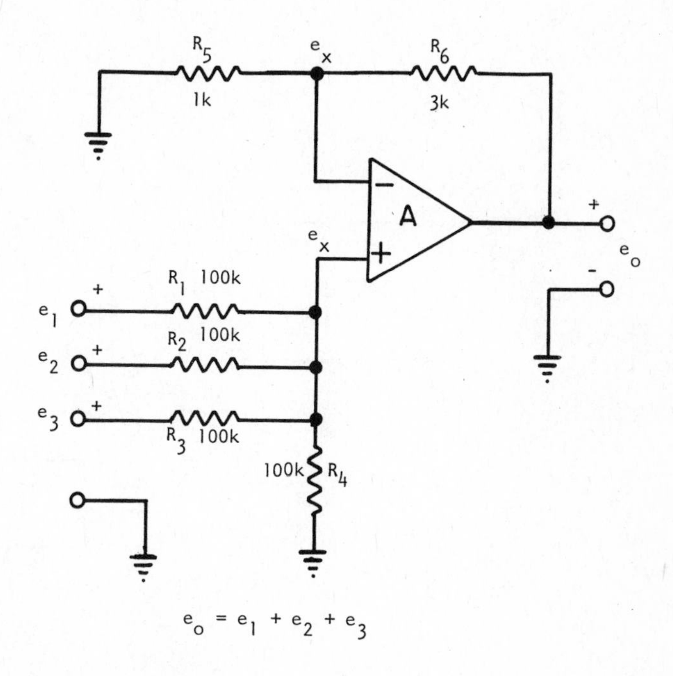

$$e_o = e_1 + e_2 + e_3$$

This network is designed in the same manner as the previous circuit with the exception that R_4 has been added.

$$e_o = \left(1 + \frac{R_6}{R_5}\right) e_x , \qquad e_x = \frac{e_1 + e_2 + e_3}{4}$$

Substituting for e_x yields the output equation provided that $1 + \frac{R_6}{R_5} = 4$, or $\frac{R_6}{R_5} = 3$. Notice that the ratio of R_6 to R_5 should always be equal to the number of inputs (all input resistors are identical). Voltage e_x must be within common mode voltage specification.

Subtractor

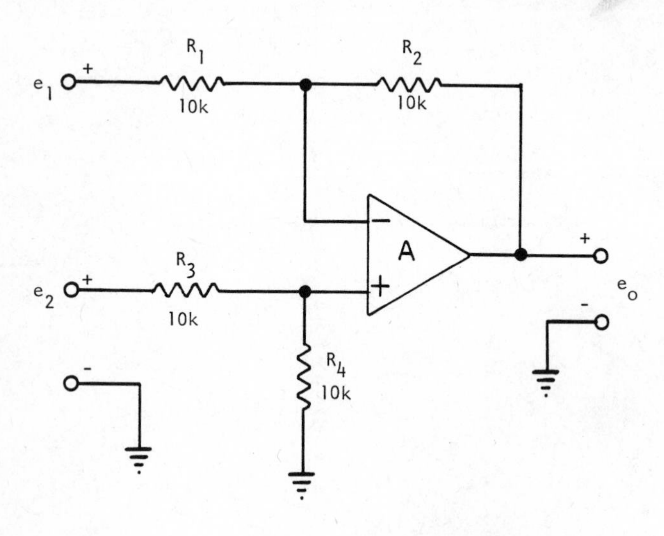

$$e_o = \frac{R_2}{R_1}(e_2 - e_1) , \qquad \text{where } \frac{R_4}{R_3} = \frac{R_2}{R_1}$$

$$e_o = e_2 - e_1 , \qquad \text{where } R_1 = R_2 = R_3 = R_4$$

The general expression is

$$e_o = -\frac{R_2}{R_1} e_1 + \frac{\left(1 + \frac{R_2}{R_1}\right)}{\left(1 + \frac{R_3}{R_4}\right)} e_2$$

Adder-Subtractor

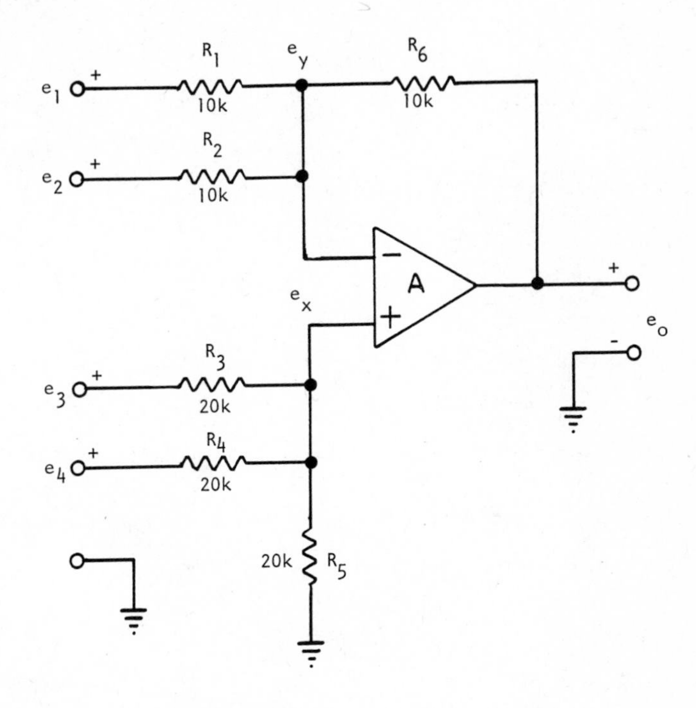

$e_o = e_3 + e_4 - e_1 - e_2$, where $R_1 = R_2 = R_6$, and $R_3 = R_4 = R_5$

This network is the logical extension of the previous adder and subtractor circuits.

Adder-Subtractor (continued)

$$e_y = \frac{e_1 + e_2 + e_o}{3}, \qquad \text{for } R_1 = R_2 = R_6$$

$$e_x = \frac{e_3 + e_4}{3}, \qquad \text{for } R_3 = R_4 = R_5$$

Equating e_x and e_y will yield the output expression shown with the network. Notice that the R_1, R_2, R_6 combination is independent from the R_3, R_4, R_5 combination; only the ratios must be equal in order to obtain the correct arithmetic output function.

Again, care must be taken to insure that e_x does not exceed the common mode voltage specification of the Op Amp.

DC VOLTAGE OUTPUT

Balanced Differential Amplifier

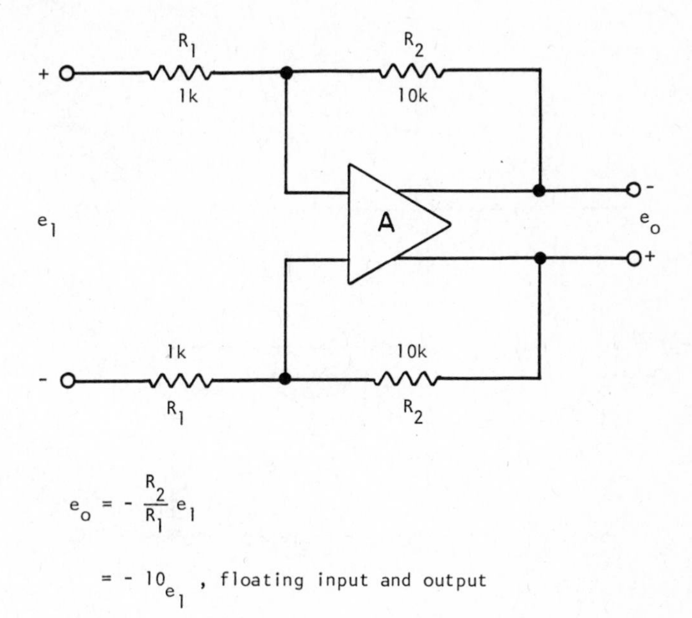

$$e_o = -\frac{R_2}{R_1} e_1$$

$$= -10 e_1 \text{ , floating input and output}$$

This type Op Amp is designed specifically for a differential floating output. It is no longer very common in modern instrumentation design.

The floating input e_1 must be referenced to ground (power supply ground) within the limits of the common mode voltage specification of the Op Amp.

Unbalanced Differential Amplifier

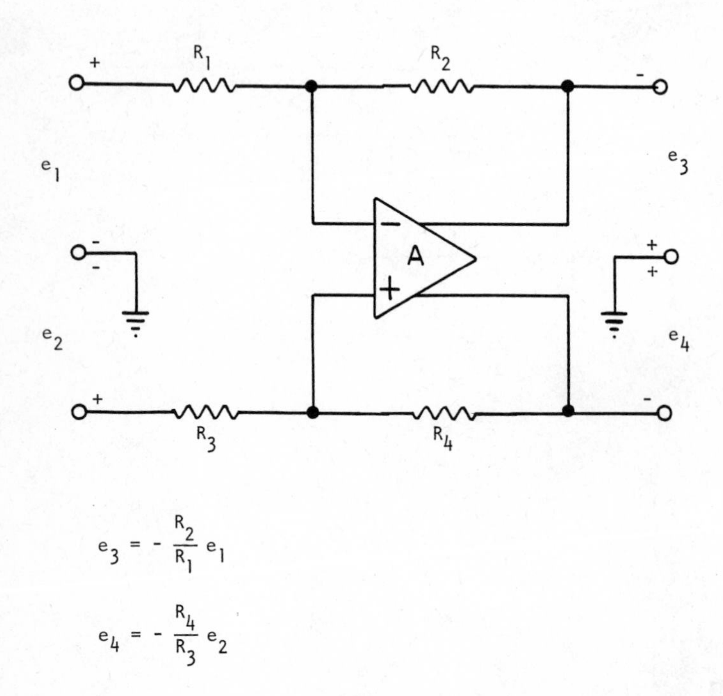

$$e_3 = -\frac{R_2}{R_1} e_1$$

$$e_4 = -\frac{R_4}{R_3} e_2$$

This circuit is an unsymmetrical version of the previous network.

Differential Instrumentation Amplifier

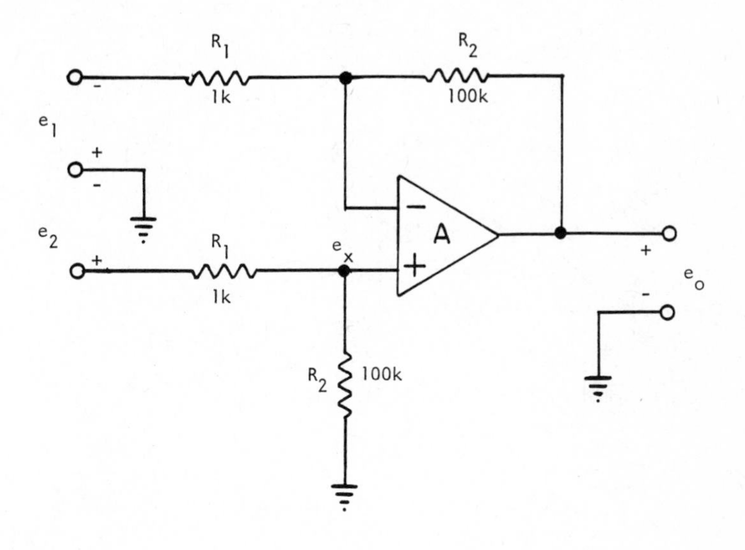

$$e_o = \frac{R_2}{R_1}(e_2 - e_1)$$

$$= 100\ (e_2 - e_1)$$

The advantage of this network is its simplicity.

The disadvantage of this approach is the low input impedance of R_1. Also the maximum amplitude of the input signal is limited by the maximum common mode voltage e_x of the Op Amp.

The resistors should be precisely matched, and the Op Amp should have high input common mode rejection for best accuracy results.

Differential Instrumentation Amplifier

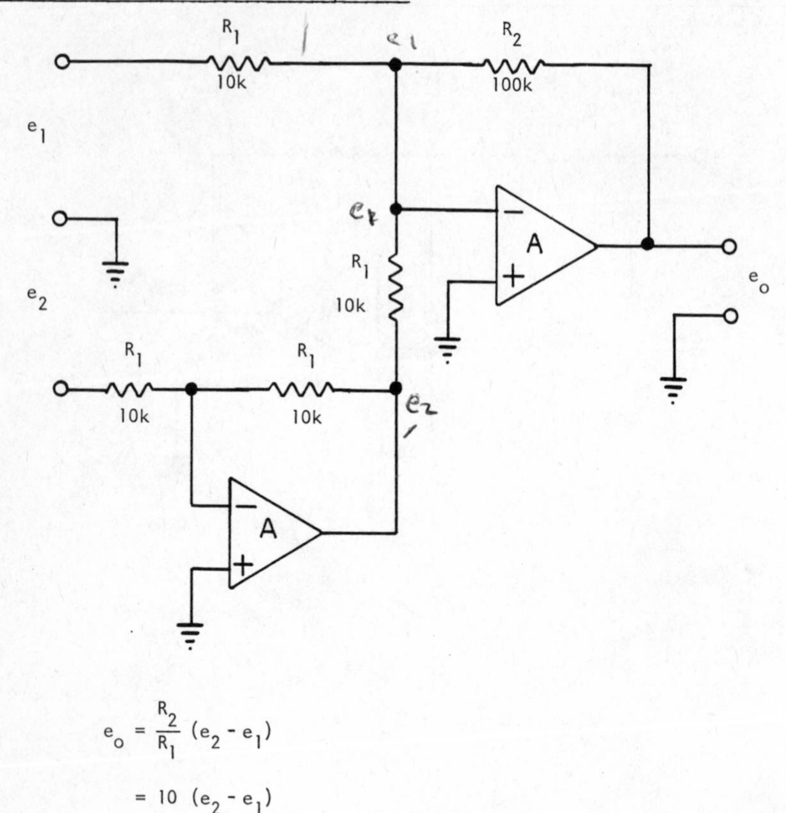

$$e_o = \frac{R_2}{R_1}(e_2 - e_1)$$

$$= 10\ (e_2 - e_1)$$

The advantage of this approach is that large voltage signals may be applied to the inputs. The inputs are not affected by the common mode voltage limits of the Op Amps. The disadvantage is the input impedance is equal to the value of R_1 (relatively low).

Differential Instrumentation Amplifier (Variable Gain)

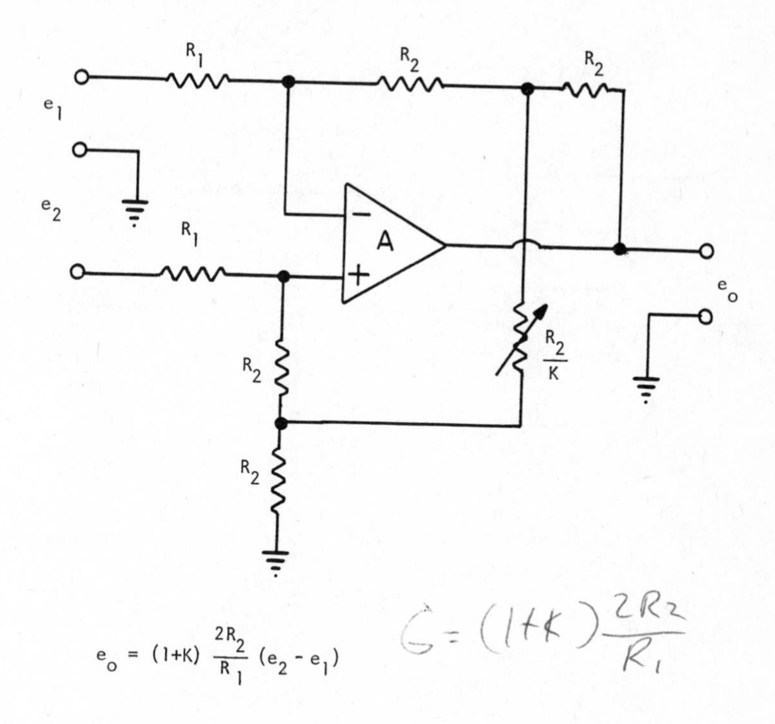

$$e_o = (1+K)\,\frac{2R_2}{R_1}\,(e_2 - e_1)$$

This approach offers a simple method for changing gain by using a single control pot.

Improved Differential Instrumentation Amplifier

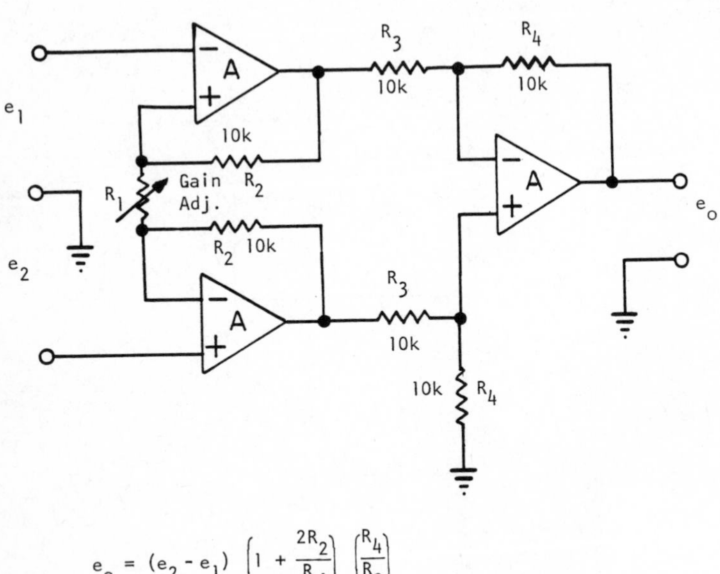

$$e_o = (e_2 - e_1)\left(1 + \frac{2R_2}{R_1}\right)\left(\frac{R_4}{R_3}\right)$$

$$= (e_2 - e_1)\left(1 + \frac{20k}{R_1}\right)$$

This approach offers the advantage of a high input impedance as well as a single variable gain control. It is an excellent technique.

DC CURRENT OUTPUT

Introduction

The networks shown in this category are generally referred to as DC Current Amplifiers. The inputs to the networks may be either voltage or current sources. The outputs of the circuits in this subsection, however, are those which deliver an output current only.

Networks illustrated include:

- Voltage-to-Current Converters
- Difference Voltage-to-Current Converters
- Transistor Gain Measurement Circuit

Voltage-to-Current Converter

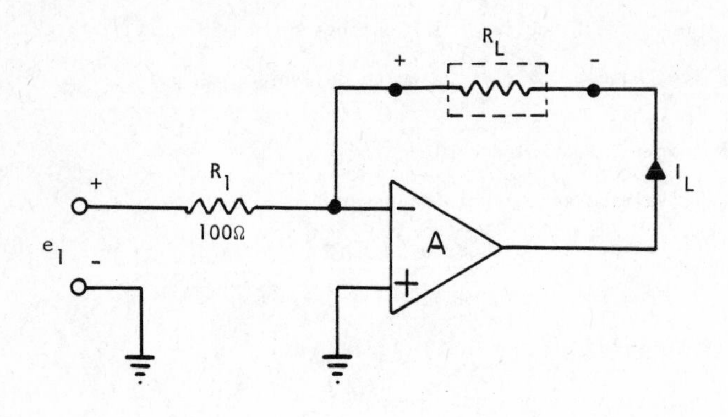

$$I_L = \frac{e_1}{R_1}$$

$$= \frac{e_1}{100}$$

With the input voltage e_1 remaining constant, the current through the load resistor R_L will stay the same regardless of its resistance value.

The advantage of this network is its simplicity. The disadvantage is that the input voltage source must be capable of supplying the value of the full load current.

Voltage-to-Current Converter

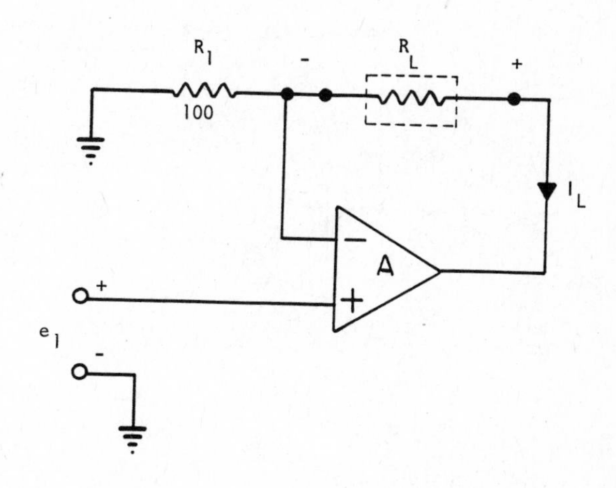

$$I_L = \frac{e_1}{R_1}$$

$$= \frac{e_1}{100}$$

This approach offers the advantage of high input impedance. The voltage e_1 is limited to the maximum input common mode voltage of the Op Amp.

Voltage-to-Current Converter

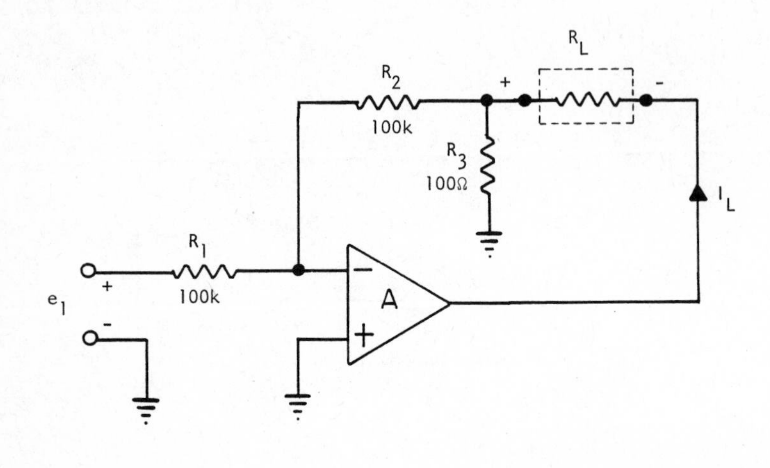

$$I_L = \frac{e_1}{R_1}\left(1 + \frac{R_2}{R_3}\right)$$

$$= \frac{e_1}{100}$$

The network offers high input impedance (R_1) while still achieving high gain.

This approach is desirable where Chopper Stabilized (inverting mode only) Op Amps are used for very low drift applications.

Voltage-to-Current Converter

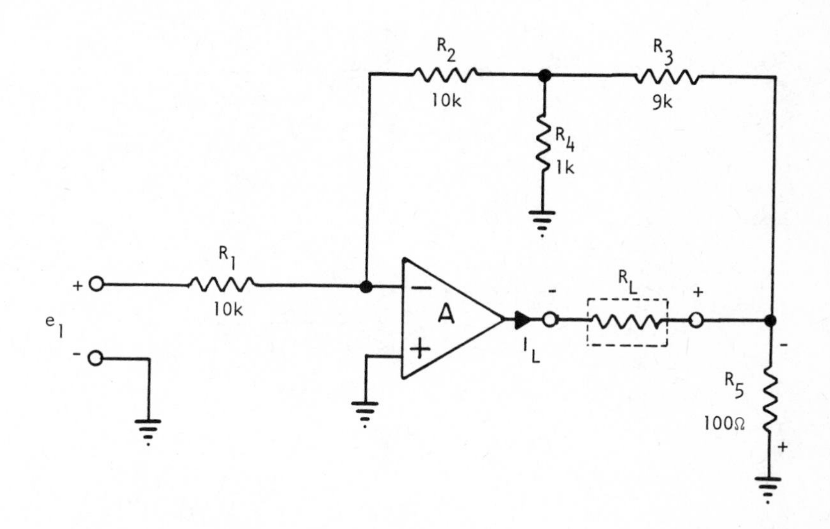

$$I_L = \frac{e_1 R_2}{R_1 R_5}\left(1 + \frac{R_3}{R_4}\right)$$

$$= \frac{e_1}{10}$$

This approach offers the advantage of using smaller value feedback resistors (T-network) to achieve high gain. See Appendix F for a more detailed analysis of the feedback T network.

Voltage-to-Current Converter (Floating Input)

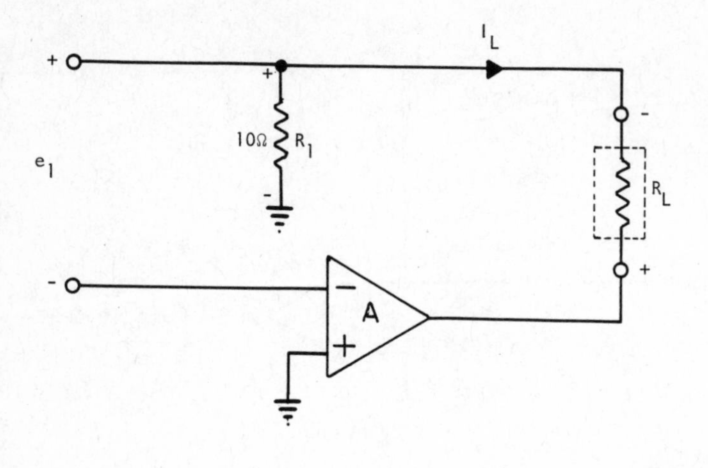

$$I_L = \frac{e_1}{R_1}$$

$$= \frac{e_1}{10}$$

The advantage of this circuit is the simplicity of the approach and the high input impedance.

The disadvantage is that both input and output must be floating.

Voltage-to-Current Amplifier

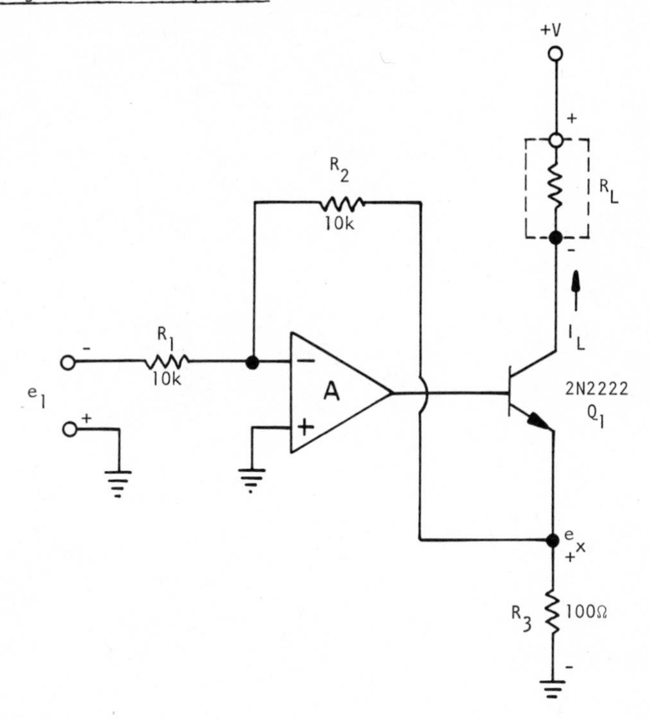

$$I_L = \frac{e_1 R_2}{R_1 R_3} \quad \text{where} \quad e_x = e_1 \frac{R_2}{R_1} \quad \text{and} \quad I_L = \frac{e_x}{R_3}$$

$$= \frac{e_1}{100}\,,$$ for negative input voltage only.

This technique uses the external transistor Q_1 to provide the power to drive the load R_L. A low output current Op Amp can be used to drive the transistor which, in turn, drives the hi-current load. The polarity can be reversed by using a PNP output configuration (-V).

Voltage-to-Current Converter (Grounded Load)

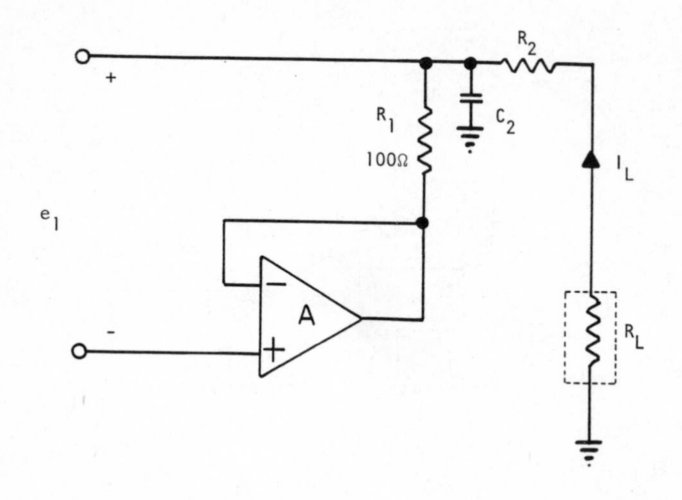

$$I_L = \frac{e_1}{R_1}$$

$$= \frac{e_1}{100}$$

The input must be floating. Components R_2 and C_2 are used for dynamic stability when using reactive loads ($R_L = Z_L$).

The advantage of this technique is that the load R_1 is grounded rather than floating as in the previous networks.

Voltage-to-Current Converter (Grounded Load)

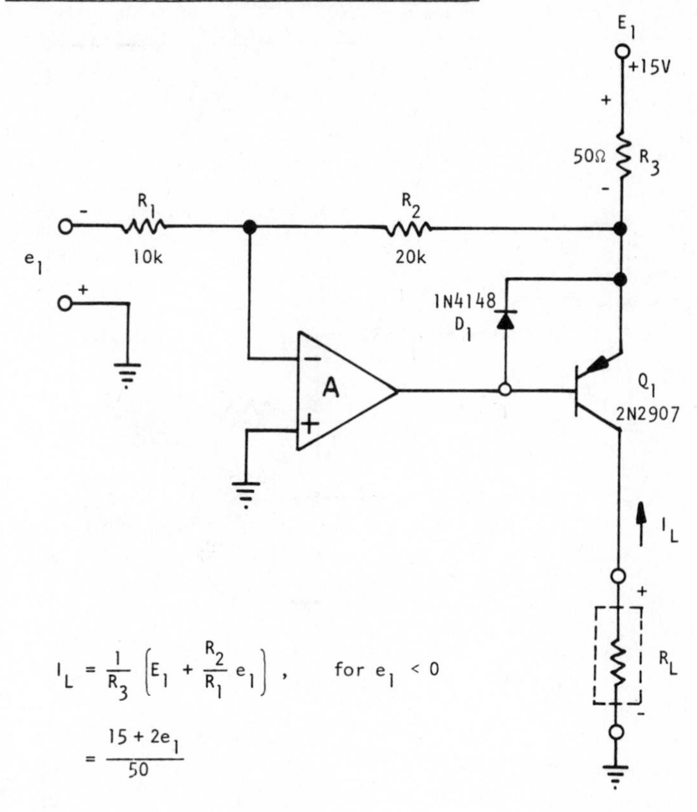

$$I_L = \frac{1}{R_3}\left(E_1 + \frac{R_2}{R_1} e_1\right), \qquad \text{for } e_1 < 0$$

$$= \frac{15 + 2e_1}{50}$$

for $e_1 = -5V$, $\quad I_L = 100$ ma

The input must always be negative for this circuit to function properly. Diode D_1 is used to protect the transistor against reverse overvoltage.

This approach offers the advantage of single ended input and output. The major disadvantage is that voltage E_1 directly affects the accuracy of the load current.

Precision Voltage-to-Current Converter (Grounded Load)

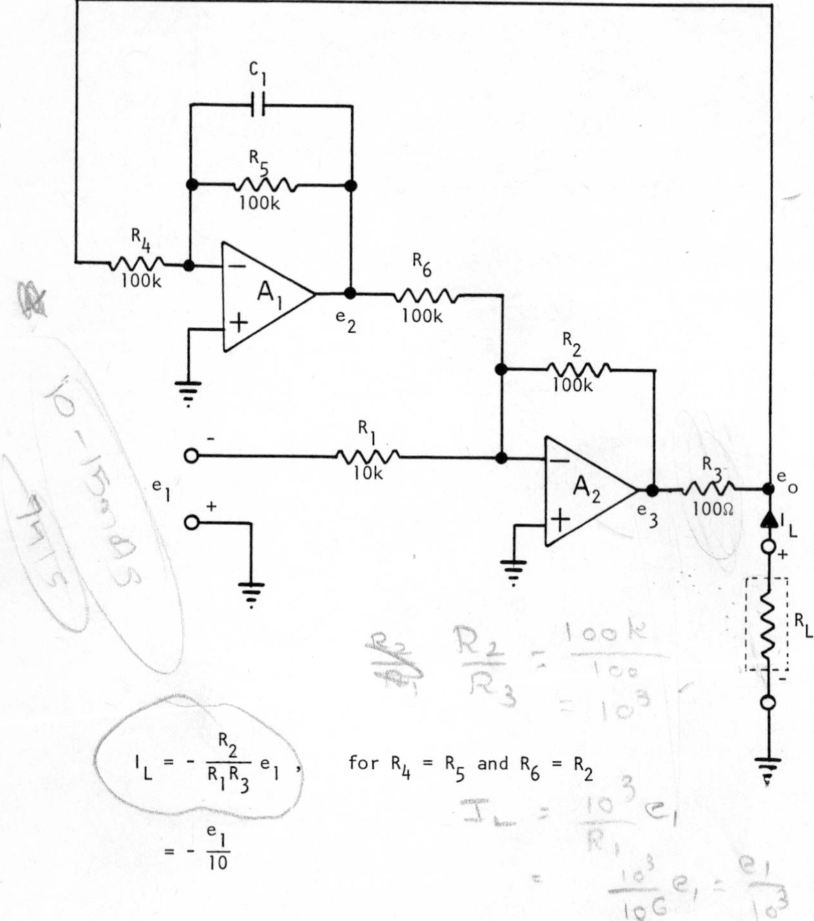

$$I_L = -\frac{R_2}{R_1 R_3} e_1 \, , \quad \text{for } R_4 = R_5 \text{ and } R_6 = R_2$$

$$= -\frac{e_1}{10}$$

This approach offers precision along with the advantage of single ended input and output. Capacitor C_1 prevents oscillations. Resistors R_4 and R_5 should be precisely matched and track closely over temperature for highest accuracy.

Precision Voltage-to-Current Converter (Grounded Load) (continued)

The output current can be most simply calculated by shorting the load R_L to ground. The load current is now easily determined by the output voltage e_3 (A_2 output) and resistor R_3.

$$I_L = i_3 = \frac{e_3}{R_3} \quad \text{and} \quad e_3 = -\frac{R_2}{R_1} e_1 , \quad \text{since } e_2 = 0$$

$$I_L = -\frac{R_2}{R_1} e_1$$

The remainder of the network (A_1, R_4, R_5, R_6) serves as a positive feedback circuit used to sense e_o. This feedback circuit must be unity gain in order to change e_3 exactly the same amount that e_o changes. This keeps the voltage across R_3 constant, thus keeping i_3 constant. If i_4 is negligible with respect to i_3, then I_L is virtually a constant current for all values of e_o (within Op Amp saturation limits of e_3).

Capacitor C_1 is used to improve dynamic stability.

A non-inverting input can be provided by placing an additional input circuit into the A_1 Op Amp summing junction if desired. This will produce an output proportional to the ratio of R_5 to the new input summing resistor. This arrangement would work in exactly the same manner as the A_2 amplifier, except the output will not invert the input signal.

$$I_L = \frac{R_5}{R_x R_3} e_x$$

Difference Voltage-to-Current Converter (Grounded Load) (continued)

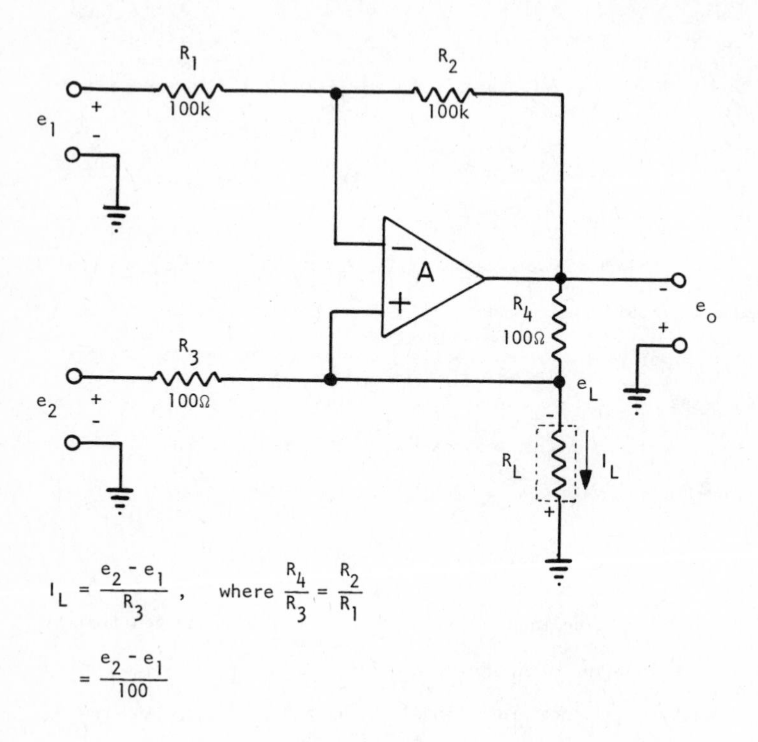

$$I_L = \frac{e_2 - e_1}{R_3}, \quad \text{where } \frac{R_4}{R_3} = \frac{R_2}{R_1}$$

$$= \frac{e_2 - e_1}{100}$$

If only one input is needed, ground e_2. Notice that input e_2 will pull heavy current compared to input e_1. This input is free to use high value resistors without affecting output current.

Difference Voltage-to-Current Converter (Grounded Load) (continued)

The maximum resistance of the load R_L is limited by the output saturation voltage and the input common mode voltage of the Op Amp.

See Appendix K for the derivation of these equations.

Difference Voltage-to-Current Converter (Grounded Load)

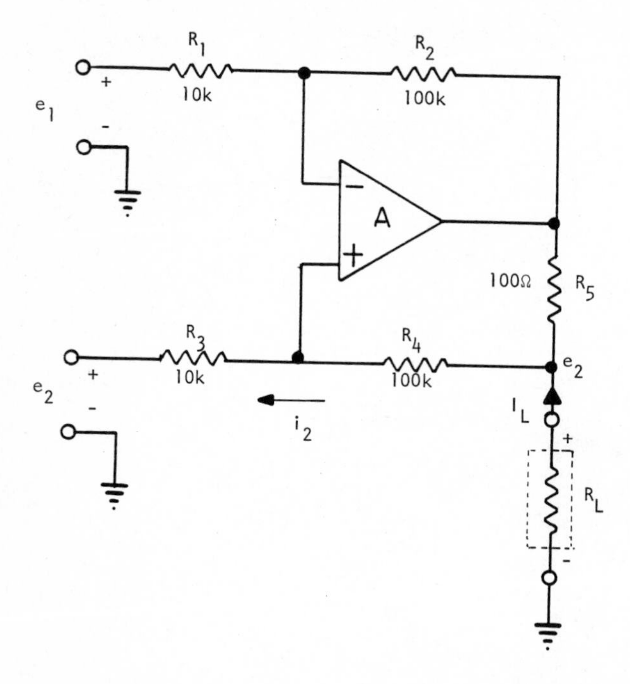

$$I_L = \frac{R_2}{R_1 R_5}(e_2 - e_1) \quad , \quad \text{for} \quad \frac{R_2}{R_1} = \frac{R_4}{R_3}$$

$$= \frac{e_2 - e_1}{10}$$

Difference Voltage-to-Current Converter (continued)

This network is useful where it is required that input e_2, as well as e_1, have a high input impedance. The maximum value of the current i_2 should be kept at least 100 times smaller than the load current I_L ($i_{s(max)} << I_L$).

$$i_{2(max)} = \left| \frac{e_{2(max)} - e_{L(max)}}{R_3 + R_4} \right|$$

Refer to Appendix L for the theoretical derivations of this network.

Transistor Gain (β) Measurement Circuit

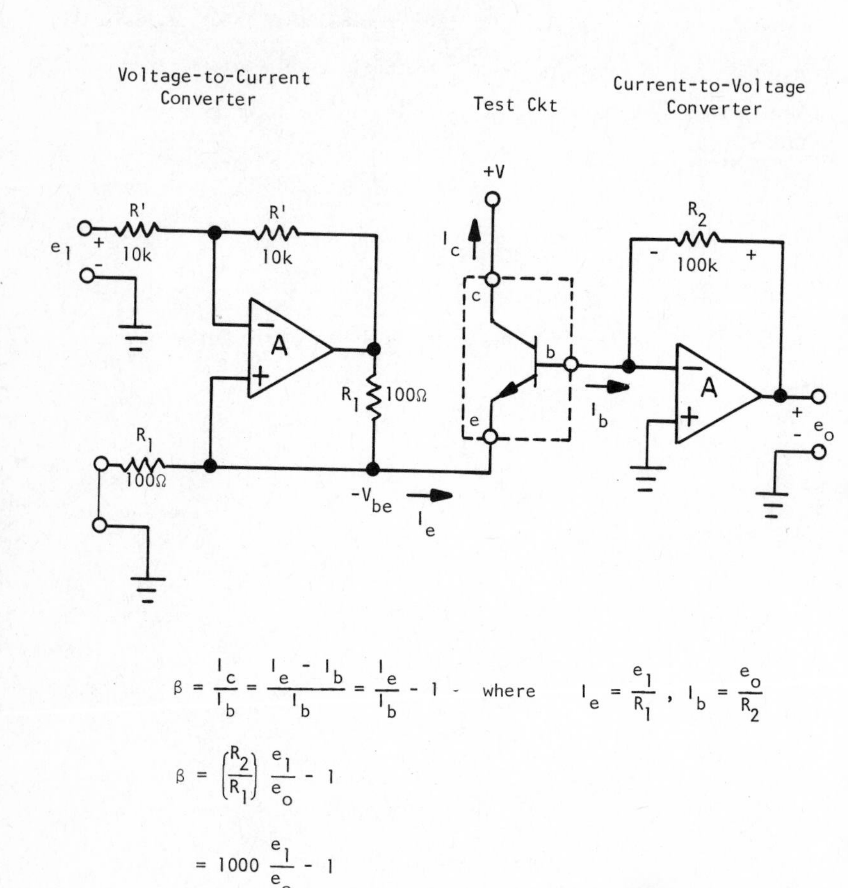

$$\beta = \frac{I_c}{I_b} = \frac{I_e - I_b}{I_b} = \frac{I_e}{I_b} - 1 \quad \text{where} \quad I_e = \frac{e_1}{R_1}, \; I_b = \frac{e_o}{R_2}$$

$$\beta = \left(\frac{R_2}{R_1}\right) \frac{e_1}{e_o} - 1$$

$$= 1000 \frac{e_1}{e_o} - 1$$

$$\beta \approx 1000 \frac{e_1}{e_o}$$

AC AMPLIFIERS

Introduction

AC operational amplifier circuits will amplify low frequency signals that are below the range of more conventional amplifiers (0.1 to 20 Hz) while DC blocking still is present. High-gain closed loop operation is also practical. Only a single ended supply is needed in some circuit applications. DC and AC operational amplifier circuits function similarly at higher frequencies from a network analysis standpoint.

Simple Inverting

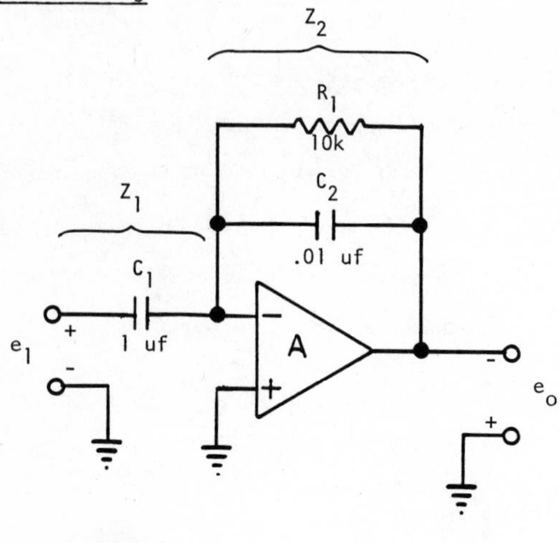

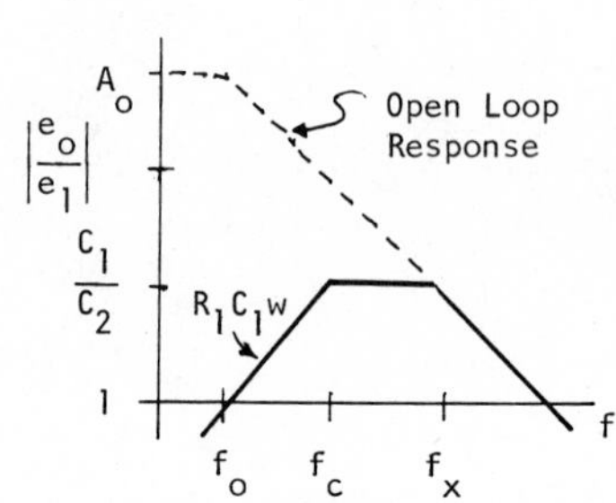

$$e_o = -\frac{Z_2}{Z_1}e_1$$

$$\frac{e_o(S)}{e_1(S)} = -\frac{R_1C_1S}{1+R_1C_2S} \qquad w = 2\pi f\ , \quad S = jw$$

$$\left|\frac{e_o}{e_1}\right| = R_1C_1w\ , \qquad \text{for low frequency } f < f_c$$

$$\left|\frac{e_o}{e_1}\right| = \frac{C_1}{C_2}\ , \qquad \text{for mid frequency } f_c < f < f_x$$

Simple Inverting (continued)

The unit gain crossover point f_o occurs when $R_1 = X_{C_1}$

$$f_o = \frac{1}{2\pi R_1 C_1}$$

$$= 16 \text{ Hz}$$

The breakpoint f_c occurs when $R_1 = X_{C_2}$

$$f_c = \frac{1}{2\pi R_1 C_2}$$

$$= 1600 \text{ Hz}$$

AC AMPLIFIERS

Inverting with Frequency Stop

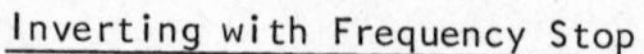

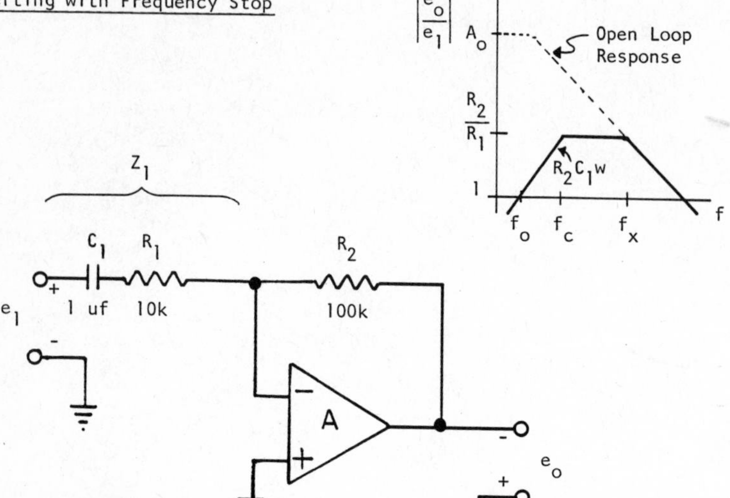

$$e_o = -\frac{R_2}{R_1} e_1 \qquad\qquad w = 2\pi f$$

$$\frac{e_o(S)}{e_1(S)} = -\frac{R_2 C_1 S}{1 + R_1 C_1 S}$$

$$\left|\frac{e_o}{e_1}\right| = R_2 C_1 w \,, \qquad \text{for low frequencies} \quad f < f_c$$

$$\left|\frac{e_o}{e_1}\right| = \frac{R_2}{R_1} \,, \qquad \text{for mid frequencies} \quad f_c < f < f_x$$

$$= 10$$

Inverting with Frequency Stops (continued)

Low frequency roll-off f_c occurs when $R_1 = X_{C_1}$

$$f_c = \frac{1}{2\pi R_1 C_1}$$

$$= 16 \text{ Hz}$$

The low frequency unity gain crossover f_o occurs when

$$\left|\frac{e_o}{e_1}\right| = R_2 C_1 w_o = 1$$

$$f_o = \frac{1}{2\pi R_2 C_1}$$

$$= 1.6 \text{ Hz}$$

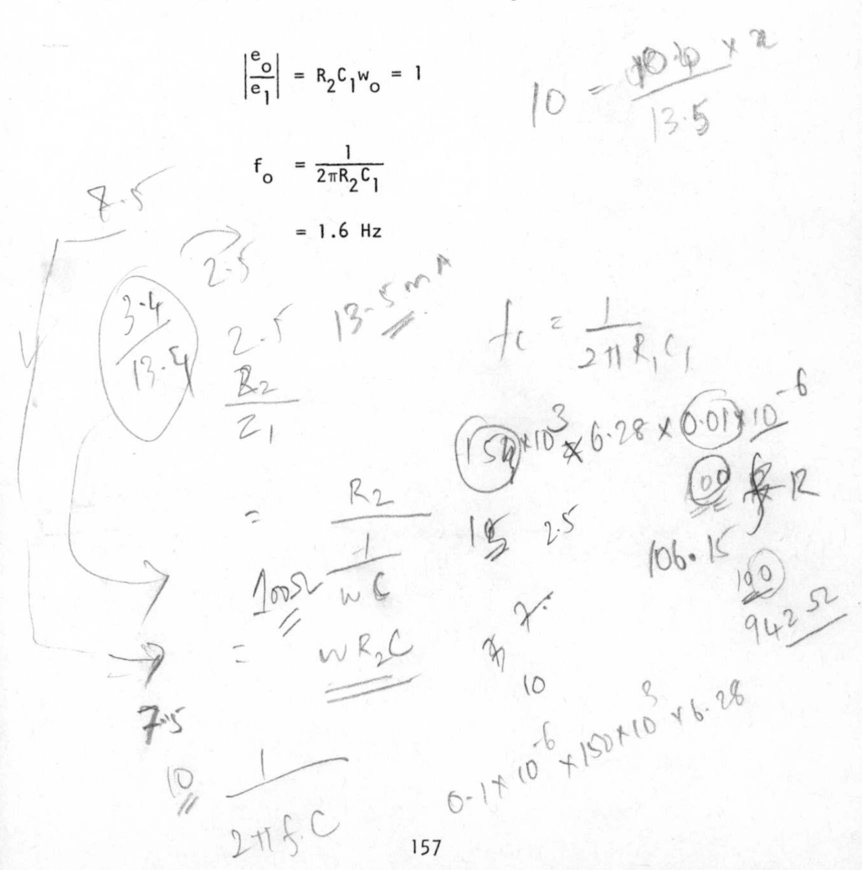

Inverting with Frequency Stop (Single Ended Supply)

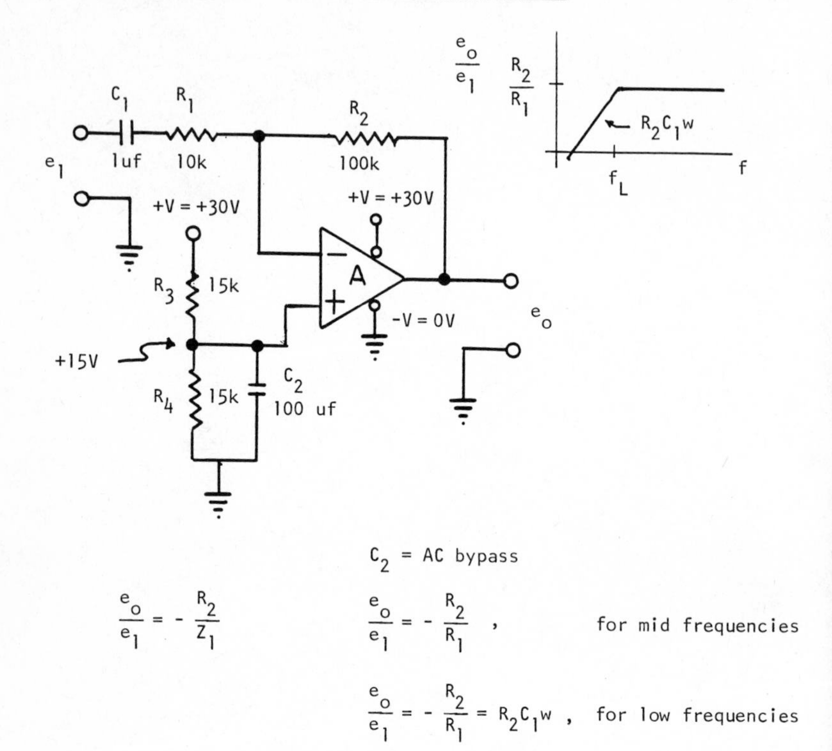

C_2 = AC bypass

$$\frac{e_o}{e_1} = -\frac{R_2}{Z_1} \qquad \frac{e_o}{e_1} = -\frac{R_2}{R_1}, \quad \text{for mid frequencies}$$

$$\frac{e_o}{e_1} = -\frac{R_2}{R_1} = R_2C_1w, \quad \text{for low frequencies}$$

This circuit is equivalent to the simple AC amplifier except that the supply is "floated" above ground.

C_2 is an AC bypass. The bypass time constant should be at least 10 times R_1C_1.

$$f_L = \frac{1}{2\pi R_1C_1} = 16 \text{ Hz}$$

Non-inverting

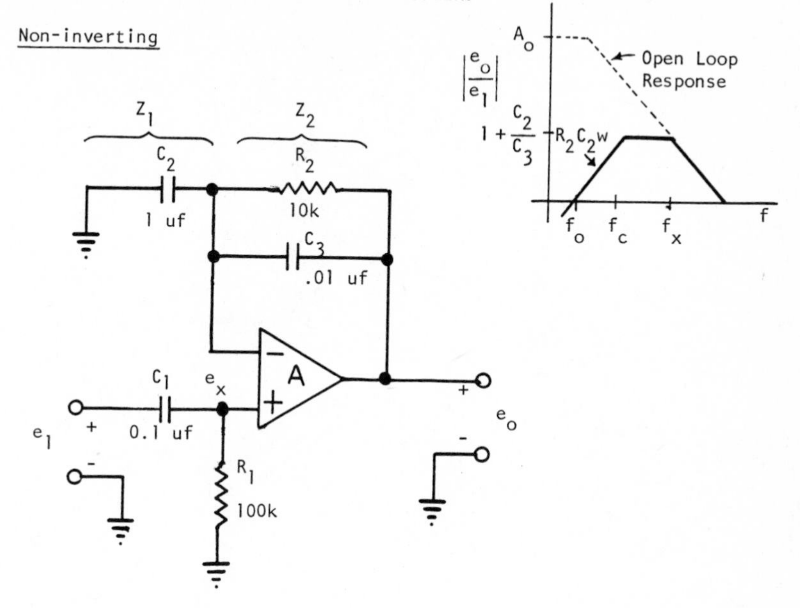

$$e_o = \left(1 + \frac{Z_2}{Z_1}\right) e_x \qquad\qquad w = 2\pi f \ , \ S = jw$$

or $$e_o(S) = \underbrace{\left(\frac{R_1C_1S}{1 + R_1C_1S}\right) e_1(S)}_{e_x} \left(1 + \frac{R_2C_2S}{1 + R_2C_3S}\right) \ , \quad \text{for } f < f_c$$

Letting $R_1C_1 = R_2C_2$ simplifies the design as well as the response characteristics.

$$\left|\frac{e_o}{e_1}\right| = R_1C_1w = R_2C_2w \ , \qquad \text{for low frequenicies } f < f_c$$

Non-inverting (continued)

The low frequency unity gain crossover point f_o occurs when

$$\left|\frac{e_o}{e_1}\right| = R_2C_2w_o = 1$$

$$f_o = \frac{1}{2\pi R_2C_2}$$

$$= 16 \text{ Hz}$$

At mid frequencies the gain simplifies to

$$\left|\frac{e_o}{e_1}\right| = 1 + \frac{C_2}{C_3} \qquad \text{for} \quad f_c < f < f_x$$

Non-inverting (continued)

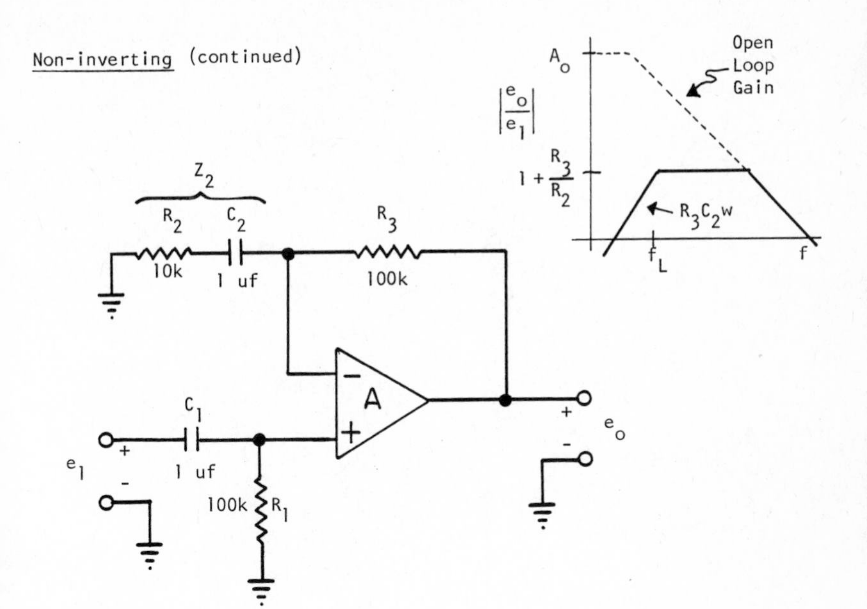

$$e_o = \left(\frac{R_1}{R_1 + X_{C_1}}\right)\left(1 + \frac{R_3}{Z_2}\right) e_1$$

$w = 2\pi f$, $S = jw$

or $$\frac{e_o(S)}{e_1(S)} = \frac{R_1C_1(1 + R_3C_2S)S}{(1 + R_1C_1S)} ,$$ for $f < f_L$

and if $R_1C_1 = R_3C_2$,

$$\frac{e_o(S)}{e_1(S)} = R_3C_2S$$

$$\left|\frac{e_o}{e_1}\right| = R_3C_2w ,$$ for $f < f_L$

Non-inverting (continued)

The low frequency breakpoint f_L occurs when $R_2 = X_{C_2}$

$$f_L = \frac{1}{2\pi R_2 C_2}$$

$$= 16 \text{ Hz}$$

After the breakpoint f_L is passed, the network acts as a simple non-inverting amplifier.

$$e_o = \left(1 + \frac{R_3}{R_2}\right) e_1$$

$$= 11\ e_1$$

Non-inverting Bootstrap Network

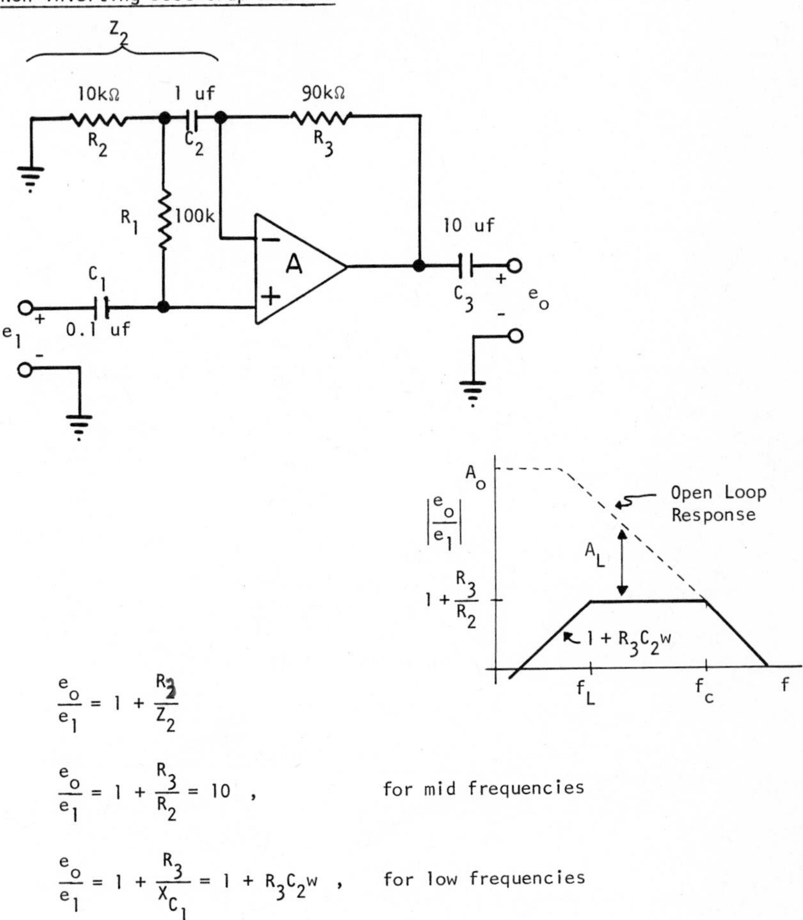

$$\frac{e_o}{e_1} = 1 + \frac{R_2}{Z_2}$$

$$\frac{e_o}{e_1} = 1 + \frac{R_3}{R_2} = 10 \quad , \qquad \text{for mid frequencies}$$

$$\frac{e_o}{e_1} = 1 + \frac{R_3}{X_{C_1}} = 1 + R_3 C_2 w \quad , \qquad \text{for low frequencies}$$

Non-inverting Bootstrap Network (continued)

The low frequency cutoff occurs when $R_2 = X_{C_2}$, or $R_1 = X_{C_1}$, since $R_1C_1 = R_2C_2$.

$$f_L = \frac{1}{2\pi R_2 C_2}$$

$$= 16 \text{ Hz}$$

Extremely high input impedance can be obtained at mid frequencies using this technique.

$$Z_{IN} = R_1(A_L + 1) \text{ , where } A_L \text{ is Loop Gain}$$

For $A_L = 10{,}000$

$$Z_{IN} = (100\text{k}\Omega)(10{,}000) = 1000 \text{ M}\Omega$$

Note that at frequency f_c that loop gain A_L has decreased to zero. The input impedance beyond this frequency will retain the value of R_1.

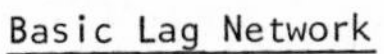

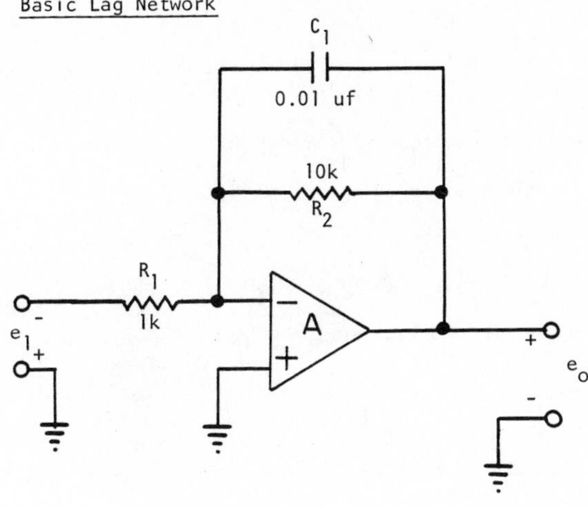

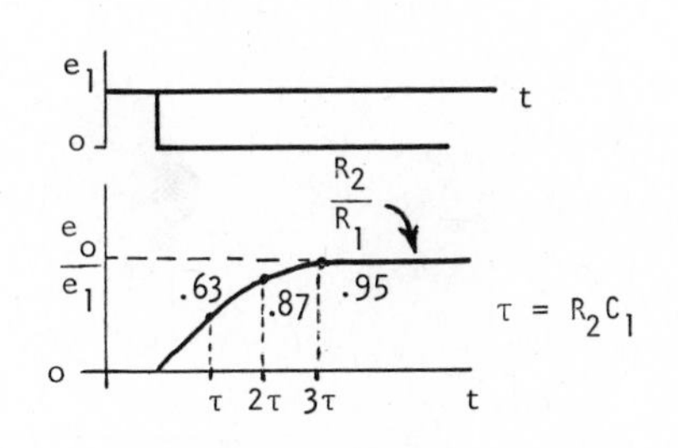

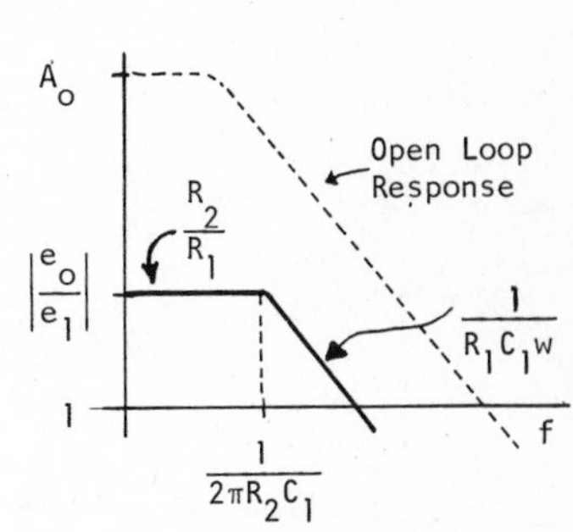

$$e_o(S) = -\frac{R_2}{R_1}\left(\frac{1}{1 + R_2C_1S}\right)e_1(S)$$

$w = 2\pi f$, $S = jw$

for a step function input.

$$e_o(t) = -\frac{R_2}{R_1}\left(1 - \varepsilon^{-t/R_2C_1}\right)e_1(t)$$

Time Constant
$\tau = R_2C_1$

Basic Lag Network (continued)

For a step function input, the output initially behaves like an integrator and then exponentially approaches the inverter response at steady state. The overall effect is to delay or "lag" the appearance of the input. Note that the time constant is a function only of the feedback components, R_2C_1. Therefore, the input resistor R_1 can be used to independently "weight" the output gain.

Controlled Response Band Pass Network

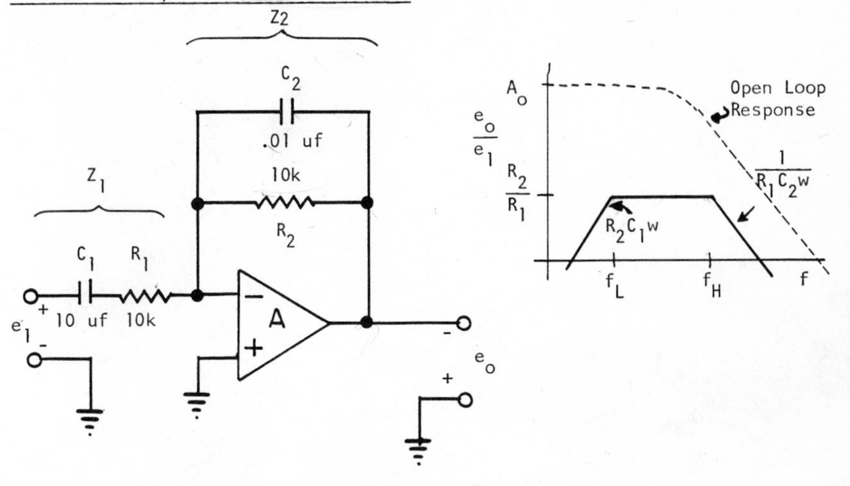

$$\frac{e_o}{e_1} = -\frac{Z_2}{Z_1} \qquad\qquad w = 2\pi f$$

$$\frac{e_o}{e_1} = -\frac{R_2}{X_{C_1}} = -R_2C_1w \quad , \qquad \text{for low frequencies}$$

$$\frac{e_o}{e_1} = -\frac{R_2}{R_1} \quad , \qquad \text{for mid frequencies}$$

$$\frac{e_o}{e_1} = -\frac{X_{C_2}}{R_1} = -\frac{1}{R_1C_2w} \quad , \qquad \text{for high frequencies}$$

The low frequency breakpoint f_L occurs when $R_1 = X_{C_1}$

$$f_L = \frac{1}{2\pi R_1C_1} = 1.6 \text{ Hz}$$

The high frequency cutoff point f_H occurs when $R_2 = X_{C_2}$

$$f_H = \frac{1}{2\pi R_2C_2} = 1600 \text{ Hz}$$

AC Buffer (Floating Input)

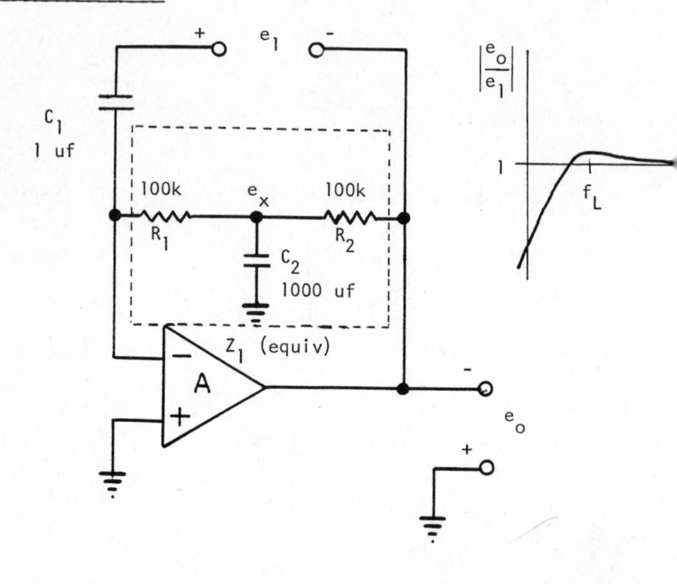

$$e_o = e_1$$

This approach offers an advantage of obtaining high input impedance at low and high frequencies.

$$f_L = \frac{1}{2\pi R_1 C_1} \qquad \text{where } R_1 = X_{C_1}$$

$$= 1.6 \text{ Hz}$$

AC Buffer (continued)

The network response is somewhat more complicated than it might first appear.

$$\frac{e_o(S)}{e_1(S)} = \frac{1 + R_2C_2S}{\frac{1}{R_1C_1S} + 1 + R_2C_2S}$$

To avoid excessive peaking at low frequency cutoff, C_2 should be much larger than C_1. Due to the bootstrap arrangement of this circuit, e_x will always be at a very low potential (virtually grounded). A large capacitor for C_2 is therefore practical.

This circuit offers the advantage of obtaining a very high input impedance at both low and high frequencies.

$$Z_{in} = R_1 \left(\frac{1}{R_1C_1S} + 1 + R_2C_2S \right)$$

For the derivation of this equation see Appendix J.

Multiple Controlled Response (Band Pass)

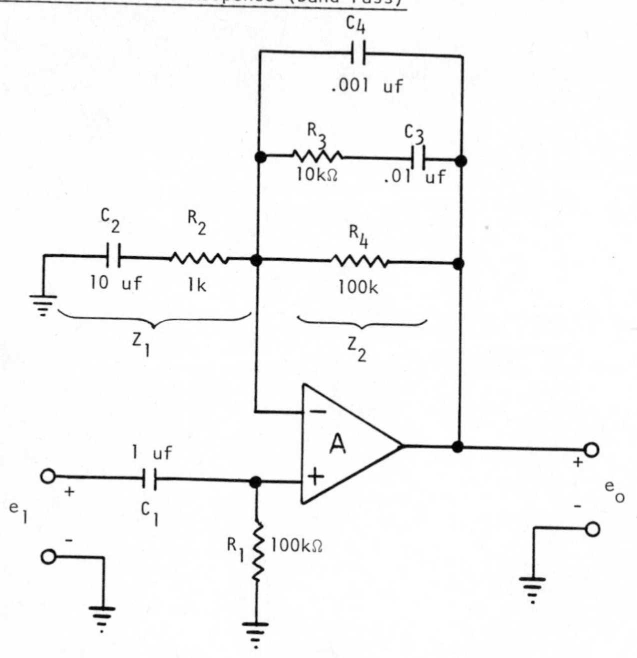

$$e_o = \left(1 + \frac{Z_2}{Z_1}\right) e_1$$

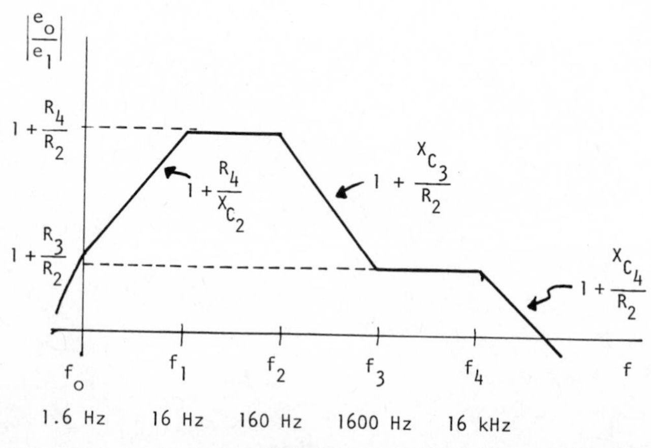

Multiple Controlled Response (continued)

For $$e_o = \left(1 + \frac{Z_z}{Z_1}\right) e_o$$

the frequency breakpoints are

$$f_o = \frac{1}{2\pi R_1 C_1} : f_1 = \frac{1}{2\pi R_2 C_2} : f_2 = \frac{1}{2\pi R_4 C_3} : f_3 = \frac{1}{2\pi R_3 C_3} : f_4 = \frac{1}{2\pi R_3 C_4}$$

This method is an excellent example showing the versatility of the Op Amp when used as an AC amplifier.

The initial frequency breakpoint f_o occurs when R_1 is equal to the reactance of C_1. Similarly, the next breakpoint f_1 occurs when R_2 equals the reactance of C_2. Breakpoint f_2 responds to R_4 and C_3. This same concept is applied throughout the network's frequency band.

By controlling the values of the R and C components, the designer has the freedom to manipulate the locations of the frequency roll-off breakpoints and corresponding output voltage gains of the network.

Multiple Controlled Response (Band Reject)

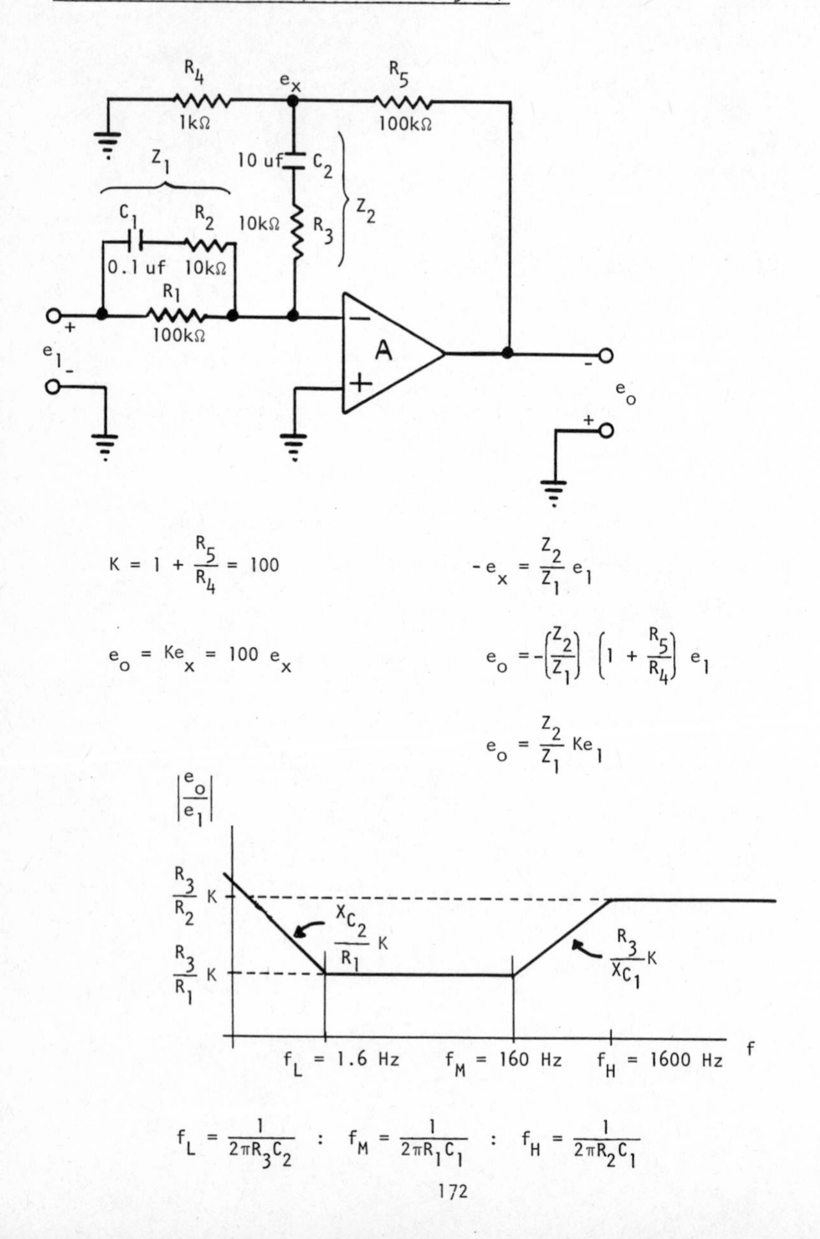

$$K = 1 + \frac{R_5}{R_4} = 100$$

$$-e_x = \frac{Z_2}{Z_1} e_1$$

$$e_o = Ke_x = 100\ e_x$$

$$e_o = -\left(\frac{Z_2}{Z_1}\right)\left(1 + \frac{R_5}{R_4}\right) e_1$$

$$e_o = \frac{Z_2}{Z_1} Ke_1$$

$$f_L = \frac{1}{2\pi R_3 C_2} \quad : \quad f_M = \frac{1}{2\pi R_1 C_1} \quad : \quad f_H = \frac{1}{2\pi R_2 C_1}$$

Multiple Controlled Response (continued)

Resistors R_4 and R_5 can be substituted by a potentiometer with the center arm at point e_x. This will act as an overall gain control K.

$$K = 1 + \frac{R_5}{R_4}$$

Lag T Network (RFI Filter)

This design technique gives excellent Radio Frequency Interference (RFI) rejection while simultaneously allowing linear DC gain (R_2/R_1).

The network can also be used as a basic low-pass filter.

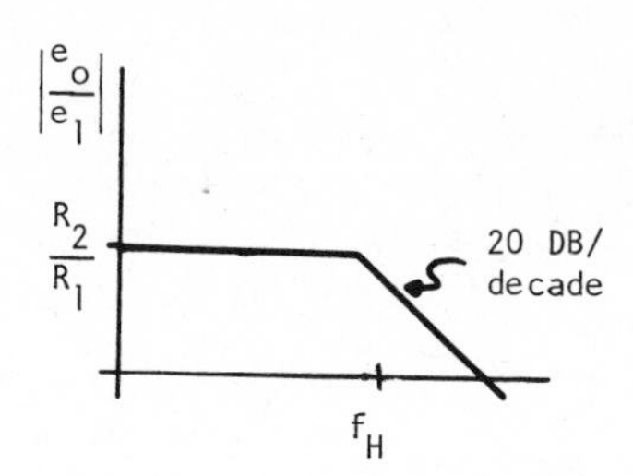

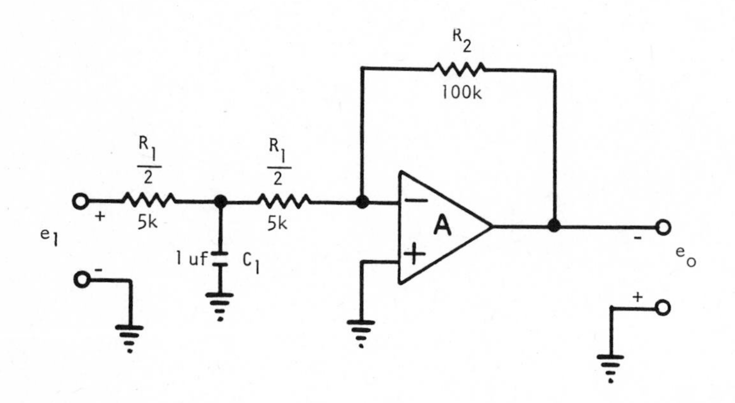

$$e_o(S) = -\frac{R_2}{R_1}\left(\frac{1}{1 + \frac{1}{4}R_1C_1S}\right) e_1(S)$$

The breakpoint occurs when $\frac{R_1}{2} = 2X_{C_1}$.

$$f_H = \frac{2}{\pi R_1 C_1} = 64 \text{ Hz}$$

Lead T Network

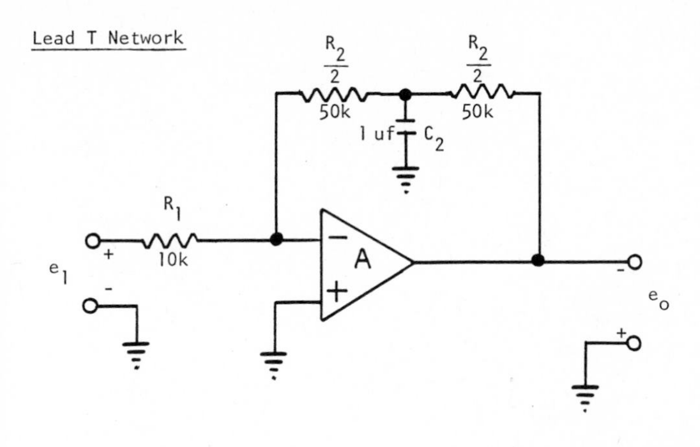

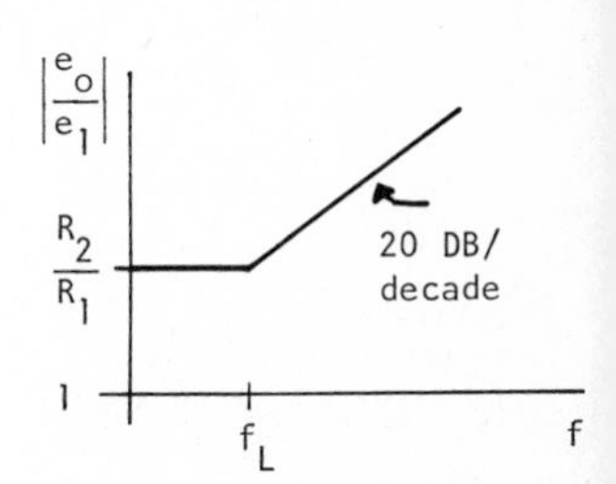

$$e_o(S) = \frac{R_2}{R_1}\left(1 + \tfrac{1}{4}R_2C_2S\right)\, e_1(S)$$

The breakpoint occurs when $\frac{R_2}{2} = 2X_{C_2}$

$$f_L = \frac{2}{\pi R_2 C_2} = 6.4 \text{ Hz}$$

At high frequency the Op Amp open loop response will intersect and over-ride the phase lead network. (Caution: an improperly frequency compensated Op Amp could cause the network to oscillate.)

Lead-Lag T Network

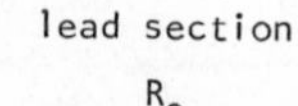

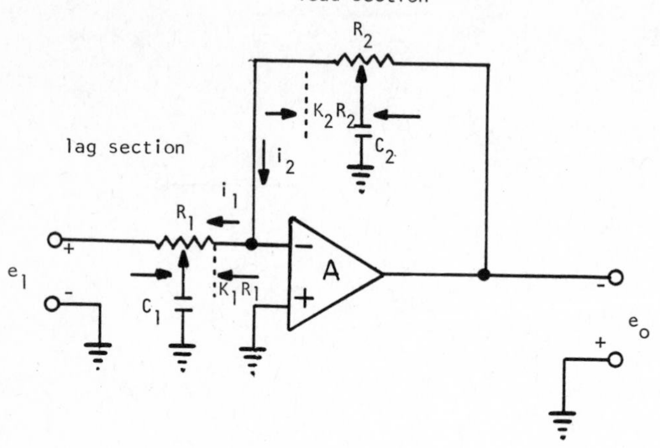

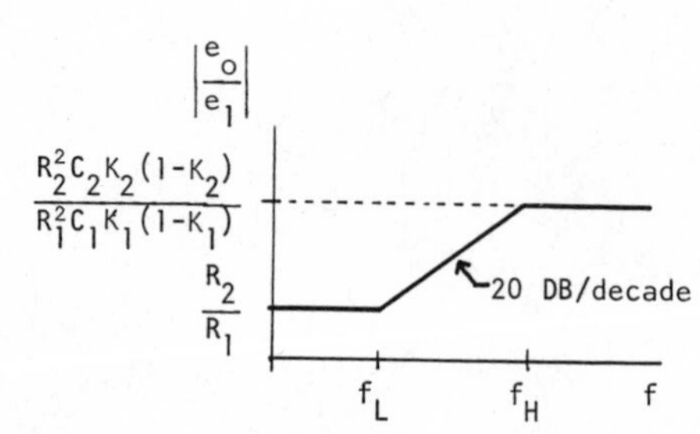

$$e_o(S) = -\frac{R_2}{R_1}\left[\frac{1 + K_2(1-K_2)R_2C_2S}{1 + K_1(1-K_1)R_1C_1S}\right] e_1(S) \quad , \quad \text{for } K_1 < 1,\ K_2 < 1$$

$$f_L = \frac{1}{2\pi K_2(1-K_2)R_2C_2}$$

$$f_H = \frac{1}{2\pi K_1(1-K_1)R_1C_1}$$

Lead-Lag T Network (continued)

The currents into the summing junctions i_1 and i_2 each have identical type equations.

$$i_1(S) = \frac{e_1(S)}{R_1}\left(\frac{1}{1 + K_1(1-K_1)R_1C_1S}\right)$$

$$i_2(S) = \frac{e_o(S)}{R_2}\left(\frac{1}{1 + K_2(1-K_2)R_2C_2S}\right)$$

$$i_1 = -i_2$$

This symmetry comes from the fact that the summing junction is ground and both T networks are constructed the same. The output gain is determined simply by equating these two currents and solving for e_o.

Although this network appears somewhat complicated at first, closer conceptual examination shows that it breaks down into relatively simple parts that can be manipulated independently. This facilitates surprising ease in understanding.

A word of caution: Careful by-passing and minimum capacitance loading is usually necessary to alleviate possible stability problems when designing these type networks.

Lead-Lag T Network (continued)

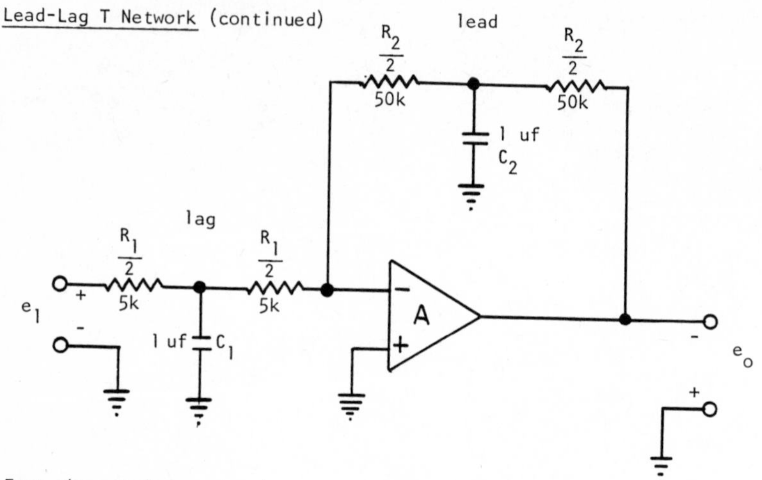

From the previous network

for $K_1 = K_2 = \frac{1}{2}$,

the T network circuits

become standardized.

$$e_o(S) = -\frac{R_2}{R_1}\left(\frac{1 + \frac{1}{4}R_2C_2S}{1 + \frac{1}{4}R_1C_1S}\right) e_1(S)$$

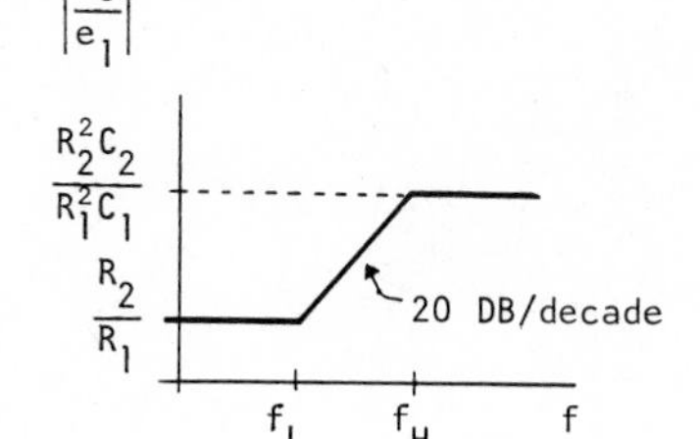

$f_L = \frac{2}{\pi R_2C_2} = 6.4$ Hz , lead section breaks first , $\frac{R_2}{2} = 2X_{C_2}$

$f_H = \frac{2}{\pi R_1C_1} = 64$ Hz , lag section breakpoint, $\frac{R_1}{2} = 2X_{C_1}$

At low frequencies the output is $e_o = -\frac{R_2}{R_1} e_1 = -10\, e_1$ for S = jw

At high frequencies the output is, $e_o = -\frac{R_2^2C_2}{R_1^2C_1} = -100\, e_1$, for $S = jw = \infty$

Lead-Lag T Network (continued)

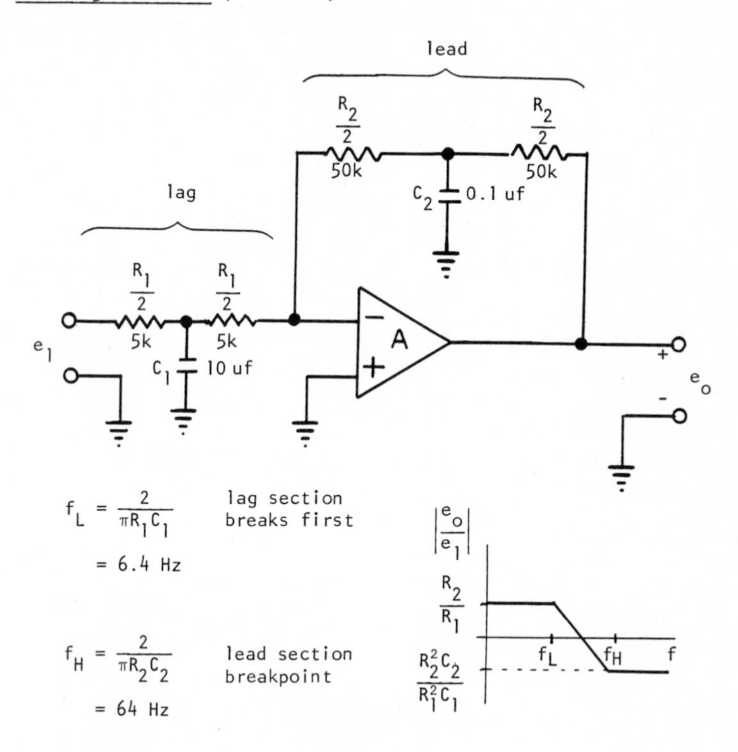

At low frequencies the output is

$$e_o = -\frac{R_2}{R_1} e_1 = -10\ e_1 \ , \text{ for } S = jw = 0$$

At high frequencies the output is

$$e_o = -\frac{R_2^2 C_2}{R_1^2 C_1} e_1 = -0.1\ e_1, \text{ for } S = jw = \infty$$

Notice that the lead and lag breakpoints (f_L, f_H) have been interchanged by assigning the appropriate values to C_1 and C_2 to these two identical networks.

Lead-Lag T Network (continued)

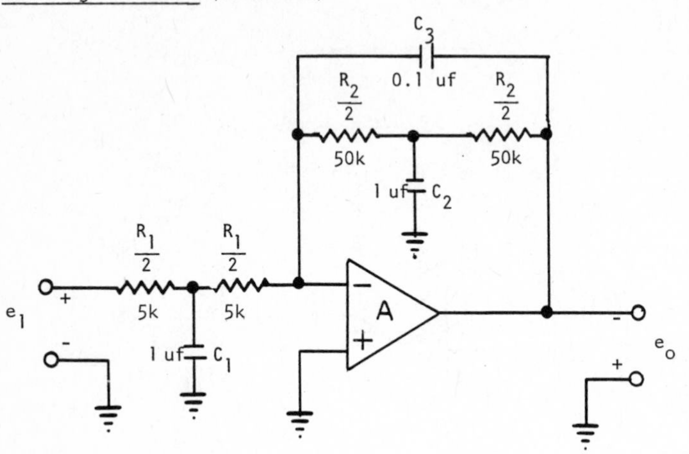

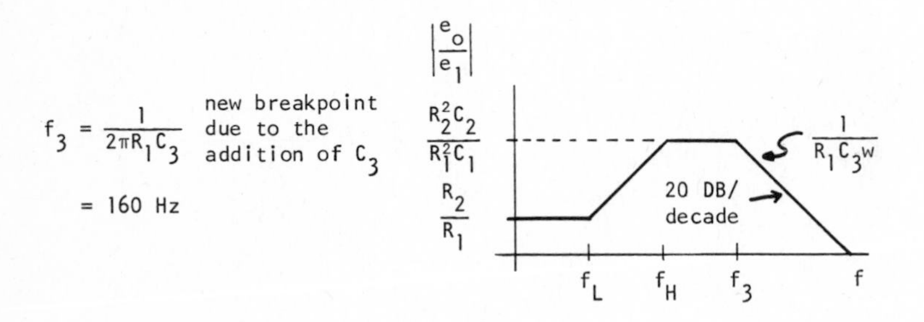

The addition of a feedback capacitor C_3 placed directly across R_2 will improve the stability of the previous network. However, it will not change the existing T network response. This capacitor will add a frequency breakpoint f_3 to the overall network. The new roll-off point will occur where $R_1 = X_{C_3}$.

For the response shown, the value of C_3 must be small enough to cause the roll-off breakpoint to occur beyond f_L and f_H.

Lead-Lag Combination T Network

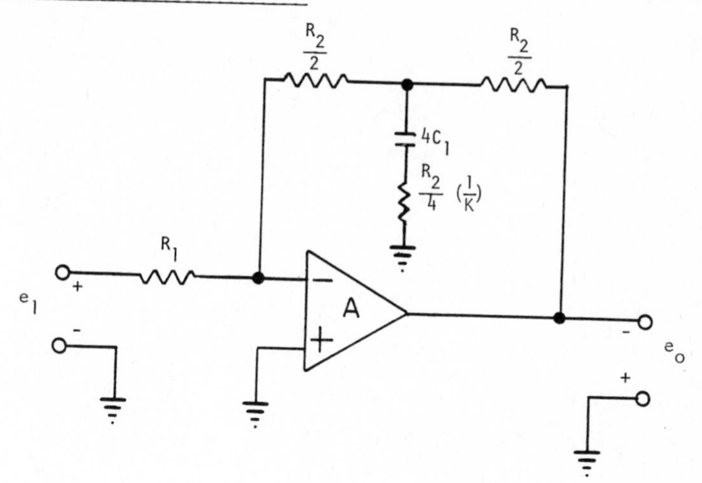

The addition of the small resistor $\frac{R_2}{4}\left(\frac{1}{K}\right)$ creates a lag breakpoint f_2 that improves frequency stability. This approach offers the advantage of a simple resistive input impedance (R_1).

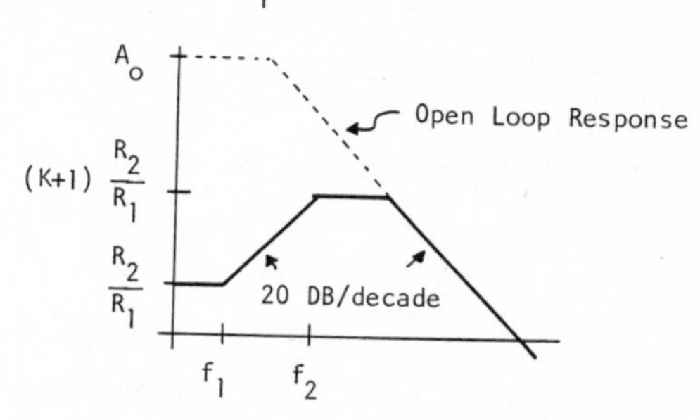

$$e_o(S) = -\frac{R_2\,(1 + R_2C_1S)}{R_1\left[1 + \left(\frac{1}{K+1}\right) R_2C_1S\right]}\,e_1(S)$$

$$f_1 = \frac{1}{2\pi R_2C_1} \qquad\qquad f_2 = \frac{k+1}{2\pi R_2C_1}$$

RECTIFIERS

Introduction

Op Amp rectification techniques offer an excellent approach to designing precision rectifier networks. Using the methods illustrated in this subsection, it becomes relatively simple to design very high accuracy Absolute Value Circuits, RMS Converters, Peak Converters, and various other AC-to-DC converters.

The key to the high accuracy in the majority of these circuits is that the rectifying elements (diodes) can be located inside the feedback loop of the Op Amp network. Therefore, the errors associated with these components (voltage drop, drift, etc.) are drastically reduced because of the high loop gain of the network.

However, it should be noted that the loop gain decreases as frequency increases. This phenomenon can cause a reduction in accuracy particularly at the zero crossover point locations.

Care should be taken in choosing the correct Op Amp that will pass the high frequency components of the rectified waveform. This is especially true when designing full wave rectification.

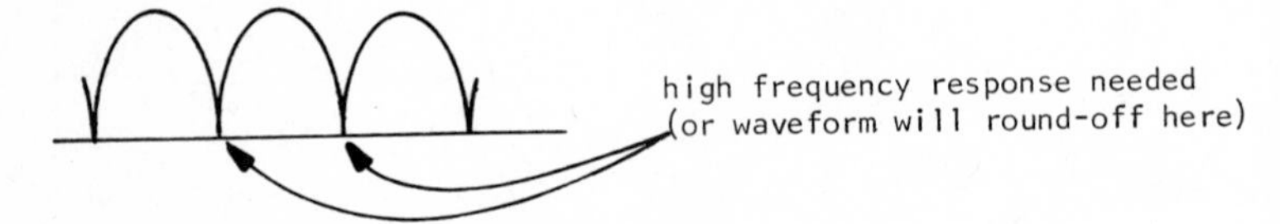

The greater the bandwidth of the Op Amp, the less "rounding-off" will occur at the base (zero axis crossover points) of the waveform.

Full Wave Rectifier

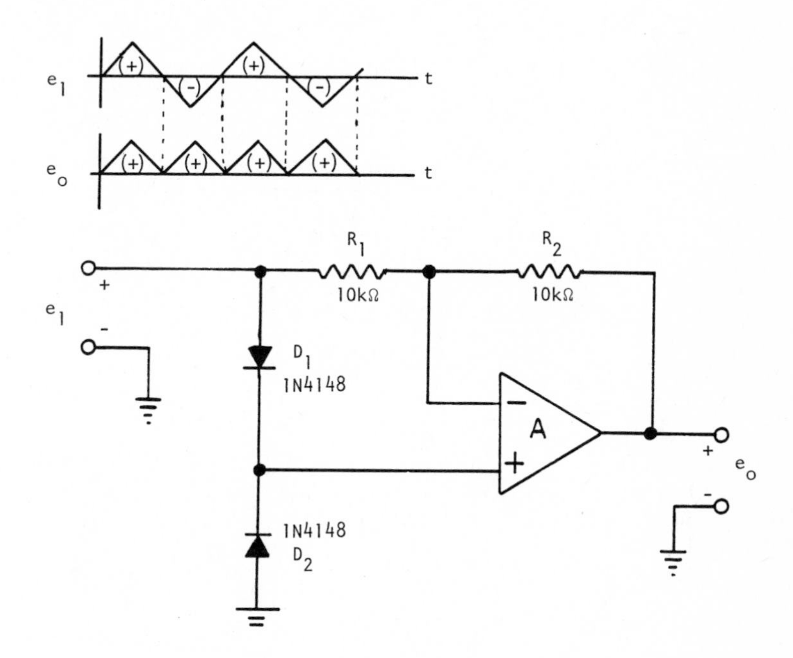

$e_o = +|e_1|$

During $(+)e_1$ the network operates in the non-inverting mode.

During $(-)e_1$ the network changes to the inverting mode.

Reversing the diodes will reverse the output polarity. A disadvantage of this approach is the forward diode voltage drops will cause output errors.

Improved Full Wave Rectifier

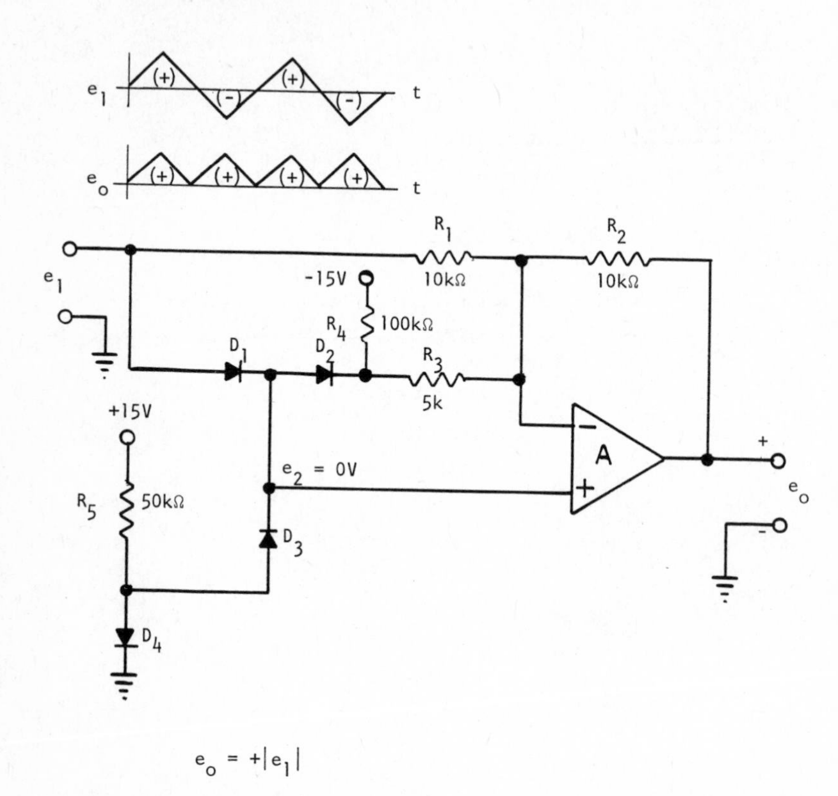

$e_o = +|e_1|$

This circuit is an improved version of the preceding configuration. Diodes D_2 and D_4 are used to cancel the forward voltage drop errors of D_1 and D_3.

Reversing all diode polarities (and associated diode supply voltages) will reverse the output polarity.

Half Wave Precision Rectifier

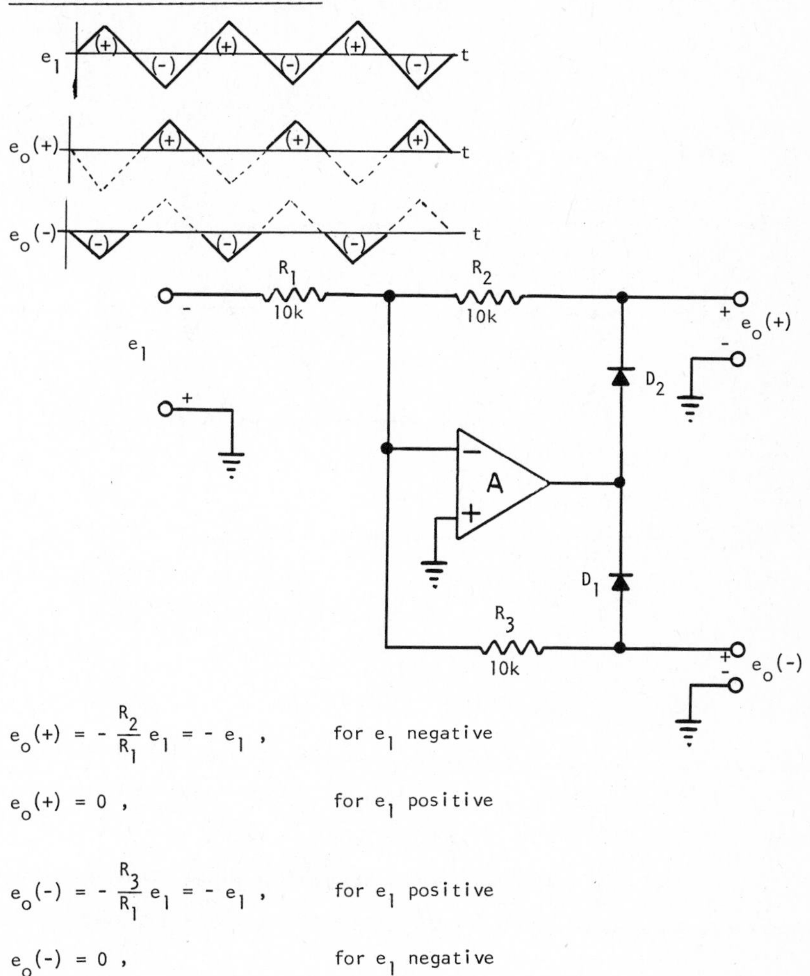

$e_o(+) = -\frac{R_2}{R_1} e_1 = -e_1$, for e_1 negative

$e_o(+) = 0$, for e_1 positive

$e_o(-) = -\frac{R_3}{R_1} e_1 = -e_1$, for e_1 positive

$e_o(-) = 0$, for e_1 negative

Since the diodes are located inside the feedback loop, the forward diode drop error is virtually eliminated by the high loop gain.

Precision Full Wave Rectifier (Absolute Value Circuit)

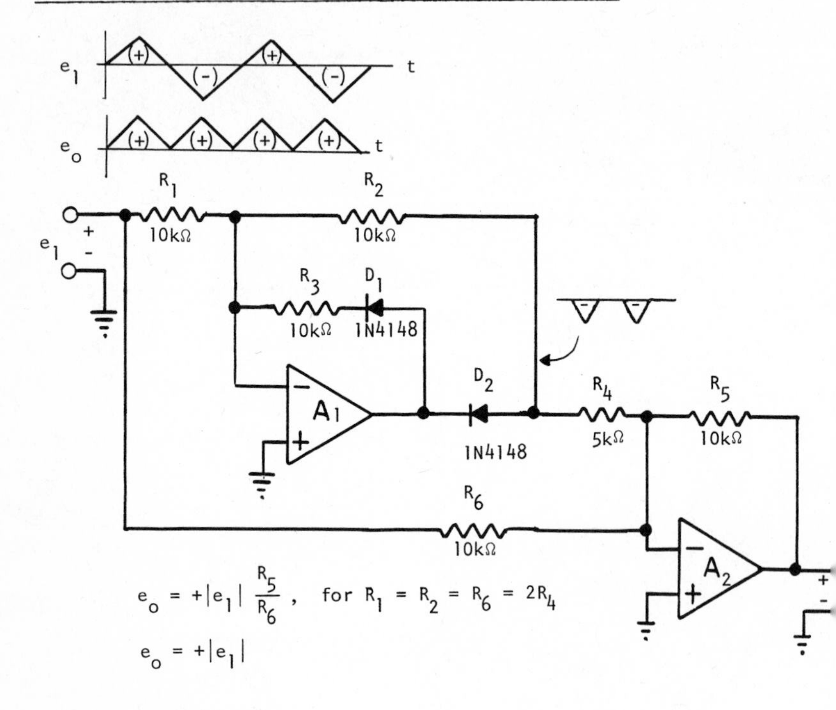

$$e_o = +|e_1| \frac{R_5}{R_6}, \quad \text{for } R_1 = R_2 = R_6 = 2R_4$$

$$e_o = +|e_1|$$

This is an extension of the Precision Half Wave Rectifier. The output Op Amp network (A_2) uses a gain of 2 (R_5/R_4) while summing the input at unity gain (R_5/R_6) to achieve full wave rectification. The value of R_3 is equal to R_2 in order to balance thermal dissipation within the Op Amp itself. Generally, this is unnecessary, and R_3 is completely eliminated (shorted).

The output is virtually ideal since the diodes are inside the feedback loop. Reversing all diodes will reverse the output polarity.

AC-to-DC Converter (Absolute Value Circuit)

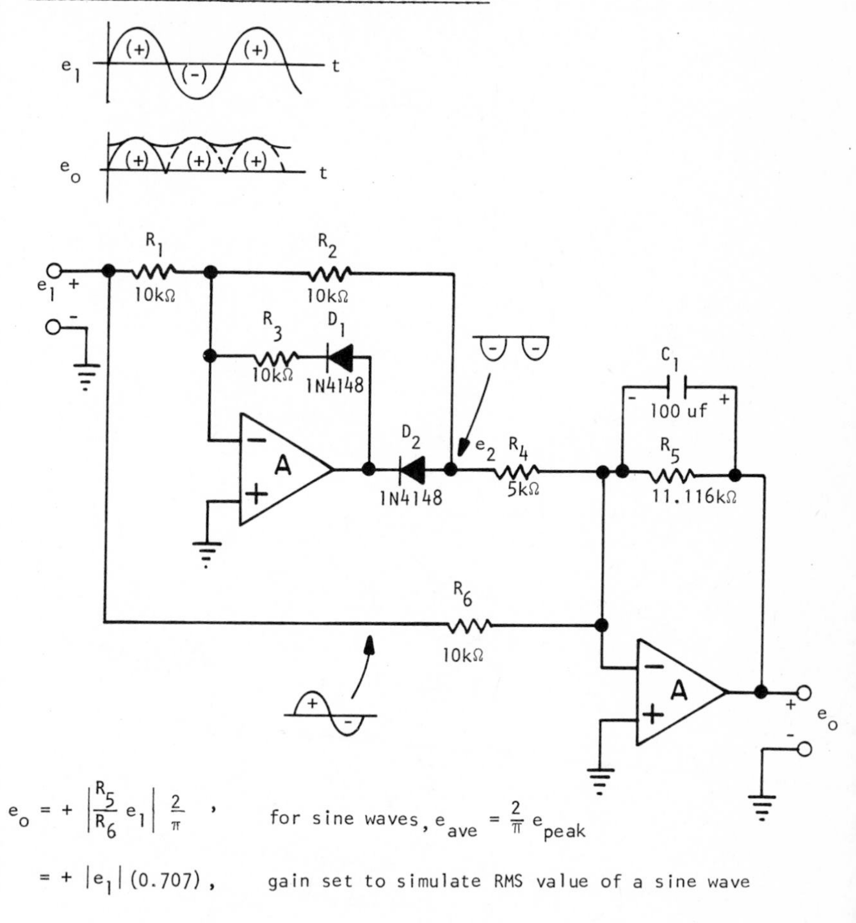

$$e_o = + \left| \frac{R_5}{R_6} e_1 \right| \frac{2}{\pi} \,, \qquad \text{for sine waves, } e_{ave} = \frac{2}{\pi} e_{peak}$$

$$= + |e_1| (0.707) \,, \qquad \text{gain set to simulate RMS value of a sine wave}$$

The value of R_5 can be used to "scale" the output for RMS, PEAK, or other readings for a specific input waveshape. The time constant R_5C_1 should be as long as possible for minimizing the output ripple without jeopardizing the overall required system response time.

AC-to-DC Converter (Full Wave)

This circuit produces a full wave output to a floating load. A DC ammeter may be placed in series with resistor R_2.

$$e_o = \frac{R_2}{R_1} |e_1|$$

$$I_o = \left|\frac{e_1}{R_1}\right|$$

$$= \frac{e_1}{10} \text{ ma}$$

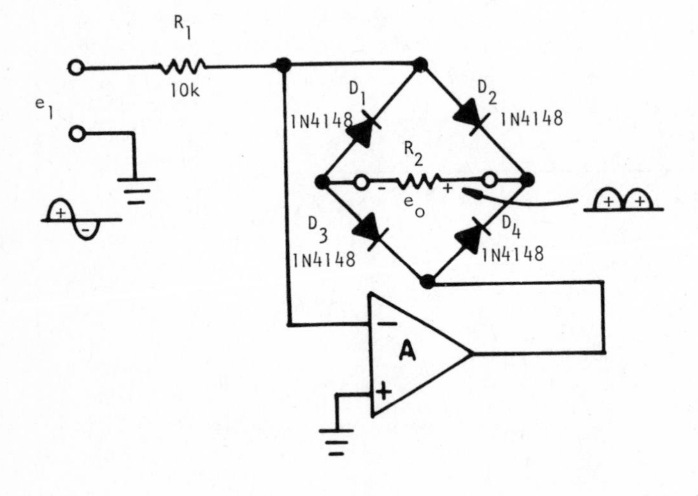

AC-to-DC Converter (Full Wave)

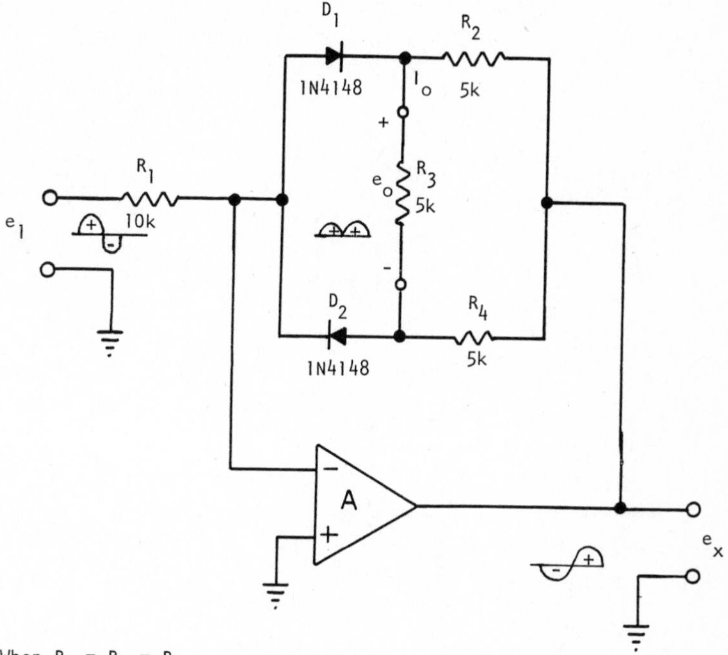

When $R_2 = R_3 = R_4$

$e_o = \frac{1}{2} |e_x|$

neglecting diode drop errors,

$e_o = \frac{1}{3} \frac{R_3}{R_1} |e_1|$

$|e_x| = \frac{R_2 || (R_3 + R_4)}{R_1} e_1 = \frac{2}{3} \frac{R_3}{R_1} |e_1|$

$I_o = \frac{1}{3} \left| \frac{e_1}{R_1} \right|$

A DC ammeter may be placed in series with resistor R_3.

$$I_o = \frac{e_1}{3(10k\Omega)} = \frac{e_1}{30} \text{ ma}$$

AC-to-DC Converter (Full Wave)

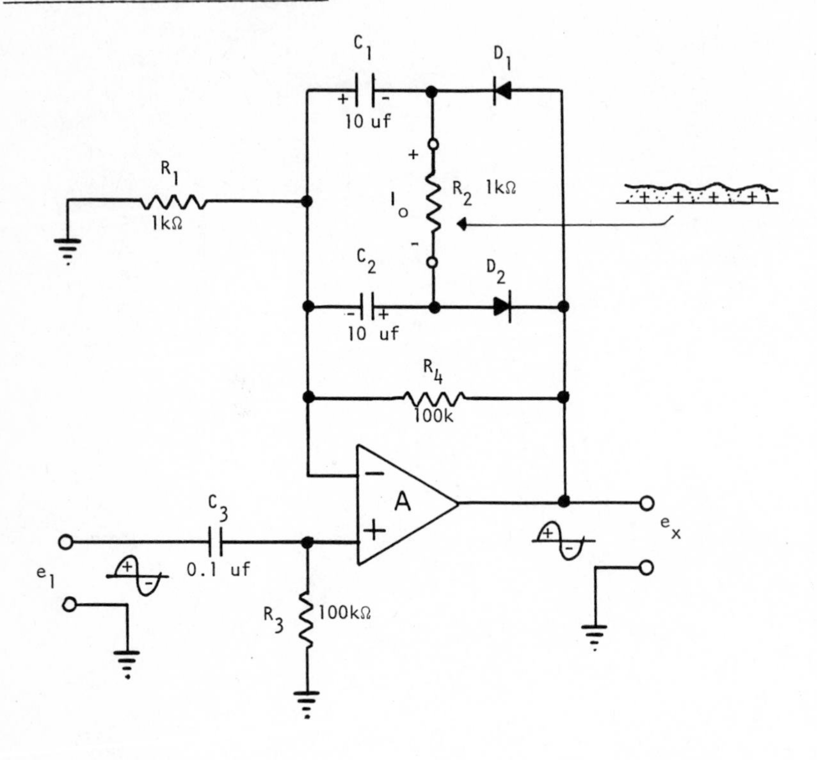

$$I_o = \left|\frac{e_1}{R_1}\right|$$

$$= \frac{+|e_1|}{1k\Omega} = +|e_1| \text{ ma}$$

The low frequency response is 16 Hz (R_2C_2, R_3C_3). High frequency response is limited by the slew rate of Op Amp.

COMPARATORS

Introduction

There are two basic categories of comparators: (1) voltage input and (2) current input.

The voltage input type comparator will switch when the input voltage signal passes through the reference threshold voltage. It should be remembered that this reference voltage never exceeds the input common mode voltage limit of the amplifier.

The current input type comparator will switch when the input current signal passes through the reference threshold current. The reference current is usually obtained by connecting the desired resistor from a reference voltage source to the summing junction of the Op Amp. Input common mode voltage is not of concern in this type comparator.

In all cases, the in input signal must pass through the threshold voltage by an amount equal to the output voltage divided by the open loop gain of the Op Amp.

$$e_{error} = \frac{e_o}{A}$$

Simple Voltage Type Comparator

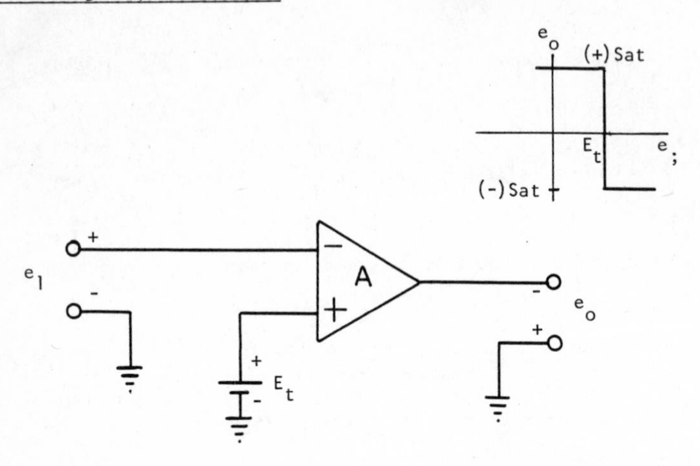

Threshold Voltage = E_t

e_o = (-) saturation for $e_1 > E_t$

e_o = (+) saturation for $e_1 < E_t$

This circuit offers the main advantage of simplicity. The disadvantages are: no input protection and slow speed. Also the input voltage range is limited to the common mode voltage of the Op Amp.

Input protection can be achieved by the addition of a $1k\Omega$ - $10k\Omega$ resistor in series with each input terminal. The speed can be improved by clamping the output to avoid saturation.

The output switching speed will then be limited only by the slew rate of the Op Amp.

Simple Current Type Comparator

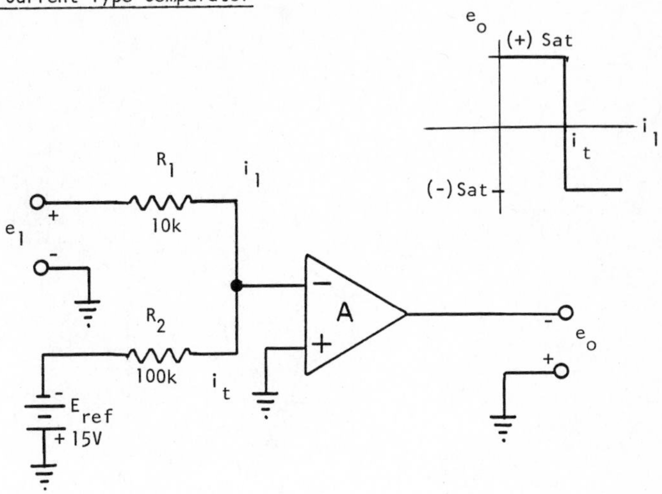

Threshold Current $I_t = \frac{E_{ref}}{R_2} = \frac{15V}{100k} = -0.15$ ma

Resulting input switching threshold voltage $= +i_t R_1$

$$e_{in(t)} = + (0.15 \text{ ma})(10k) = +1.5V$$

e_o = (-) saturation for $e_1 > +1.5V$

e_o = (+) saturation for $e_1 < +1.5V$

The circuit offers the advantage of a large input voltage range. Speed can be improved by clamping the output to avoid saturation.

The output switching speed will then be limited only by the slew rate of the Op Amp.

Zero-Crossing Detector

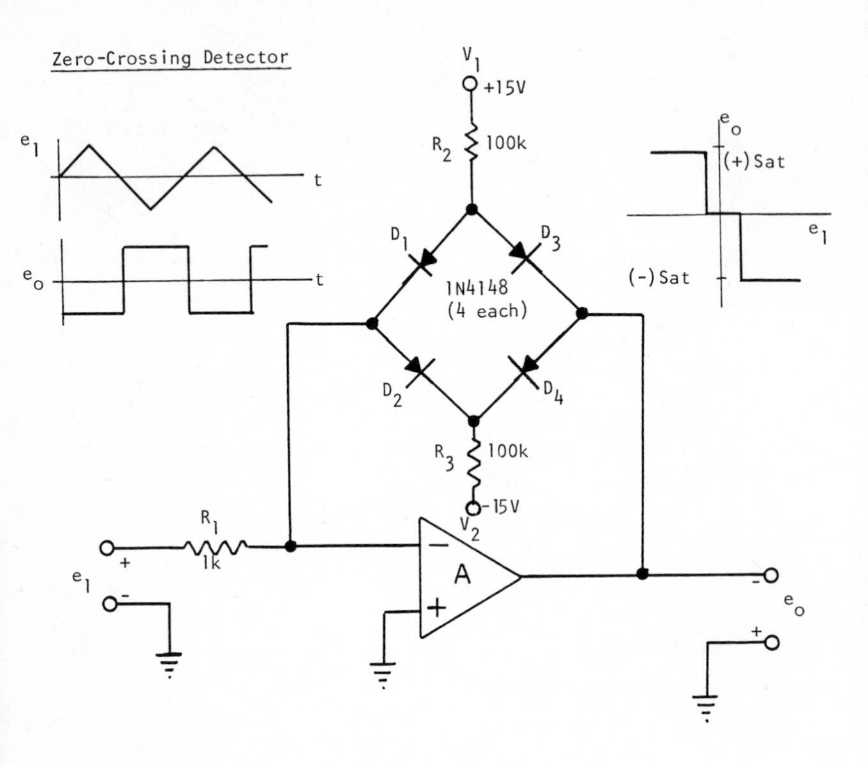

e_o = (-) saturation for $e_1 > +150\text{mV}$

e_o = (+) saturation for $e_1 < -150\text{mV}$

$$I_{THRES} = \frac{V_1}{R_2} = \frac{V_2}{R_3} \quad , \quad \text{therefore } e_{THRES} = I_{THRES} R_1 = 150 \text{ mV}$$

$$= 150 \ \mu\text{A}$$

At precisely zero input, the output will be zero, and all diodes will conduct equally. Any input current caused by e_1 flowing through R_1 will unbalance D_1 and D_2 until one of these diodes (depending on polarity) cuts off. At the output, the corresponding (diagonal) diode will follow

Zero-Crossing Detector (continued)

the same pattern. For example, if e_1 is positive D_2 and D_3 will conduct, while D_1 and D_4 will cut off. For high accuracy requirements, D_1 and D_2 should be matched.

This network is actually a special case of the "dead zone comparator" discussed later in this section.

Latching Comparators

These circuits offer noise immunity using hysteresis.

Assuming output saturation voltage is ±10V, the input hysteresis voltage V_H will be

$$V_H = \pm 10V \frac{R_1}{R_1+R_2} = \pm \frac{(10)(1k)}{101k} \approx \pm 100 \text{ mV}$$

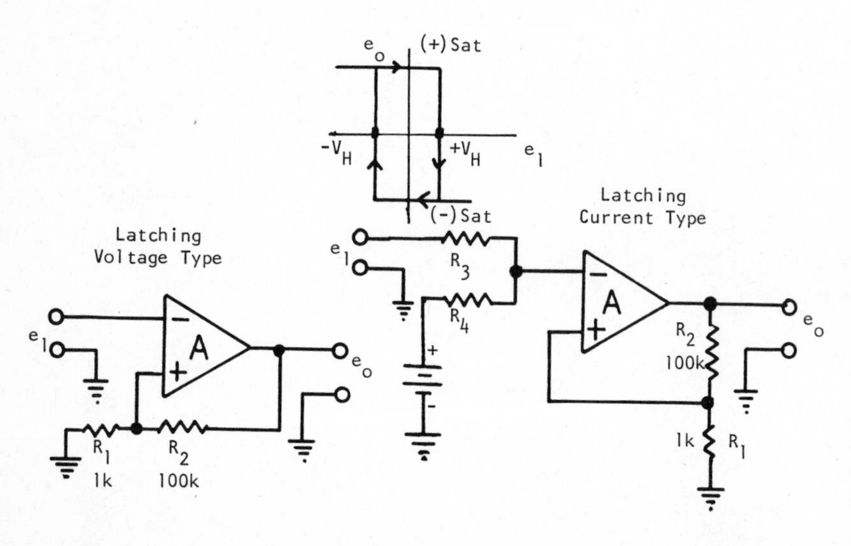

This means the input signal must overcome the hysteresis 100 mV voltage to switch the output. Once the output has switched, however, it cannot erroneously be switched back again by a random noise spike unless the noise spike is greater than 200 mV in the opposite direction from the signal.

Latching Comparators (continued)

$$e_o = (-) \text{ saturation for } e_1 > +100 \text{ mV}$$

$$e_o = (+) \text{ saturation for } e_o < -100 \text{ mV}$$

Resistor R_1 need not always be grounded. It can be connected to a reference voltage for the purpose of changing the reference threshold level as shown earlier in the simple voltage type comparator.

Simple Analog Switch (Uni-polar)

When the logic input (e_L) is zero volts, the network acts as an ordinary inverter for positive input signals (e_1). When the logic switches to negative 15 volts, the output (e_o) goes to zero.

The "on" resistance of Q_1 should be low as compared to R_1 and R_2 to insure that the output (e_o) will come as close to zero as possible. (Q_1 is an N-channel junction FET.)

$$e_o = 0, \qquad \text{for } e_L = -15V$$

$$e_o = -\frac{R_2}{R_1}, \qquad \text{for } e_L = 0$$

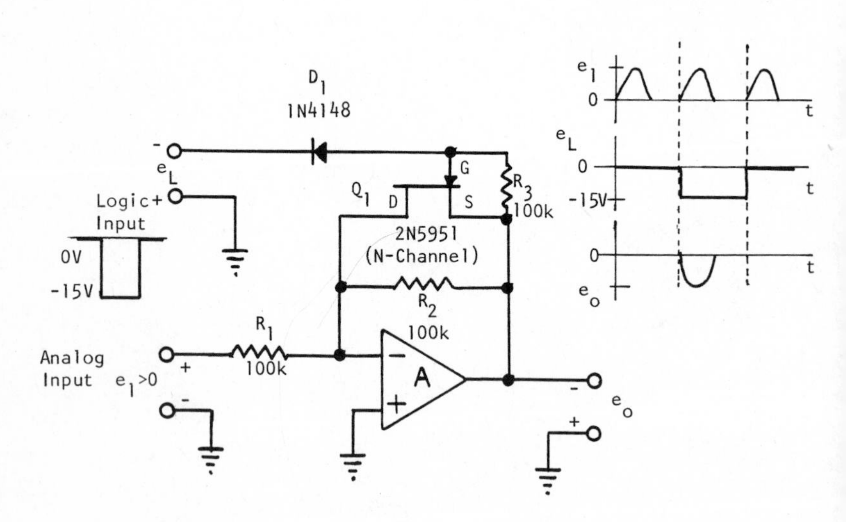

Analog Switch (Uni-polar)

When the logic input (e_L) is positive the analog Op Amp network operates as a normal inverter (for negative input signals only). When the logic level goes to zero, the analog output (e_o) will be inhibited and forced to zero.

The polarity of this analog switch can be reversed if all diodes and input signals are reversed in polarity.

This network is useful for driving and resetting integrator circuits.

$$e_o = 0 \,, \qquad \text{for } e_L = +5V$$

$$e_o = -\frac{R_2}{R_1} e_1 \,, \quad \text{for } e_L = 0V$$

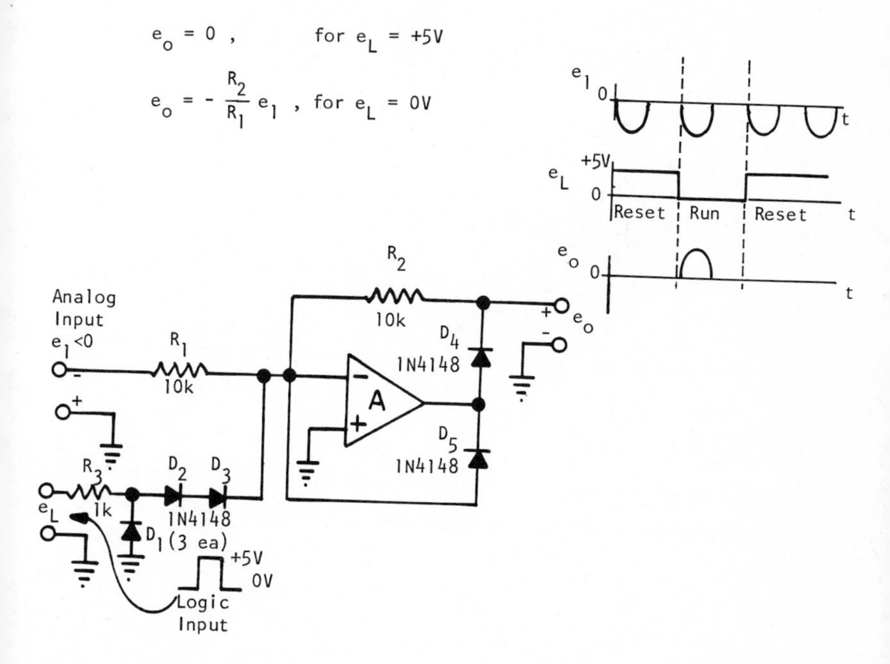

Analog Switch (Bi-polar)

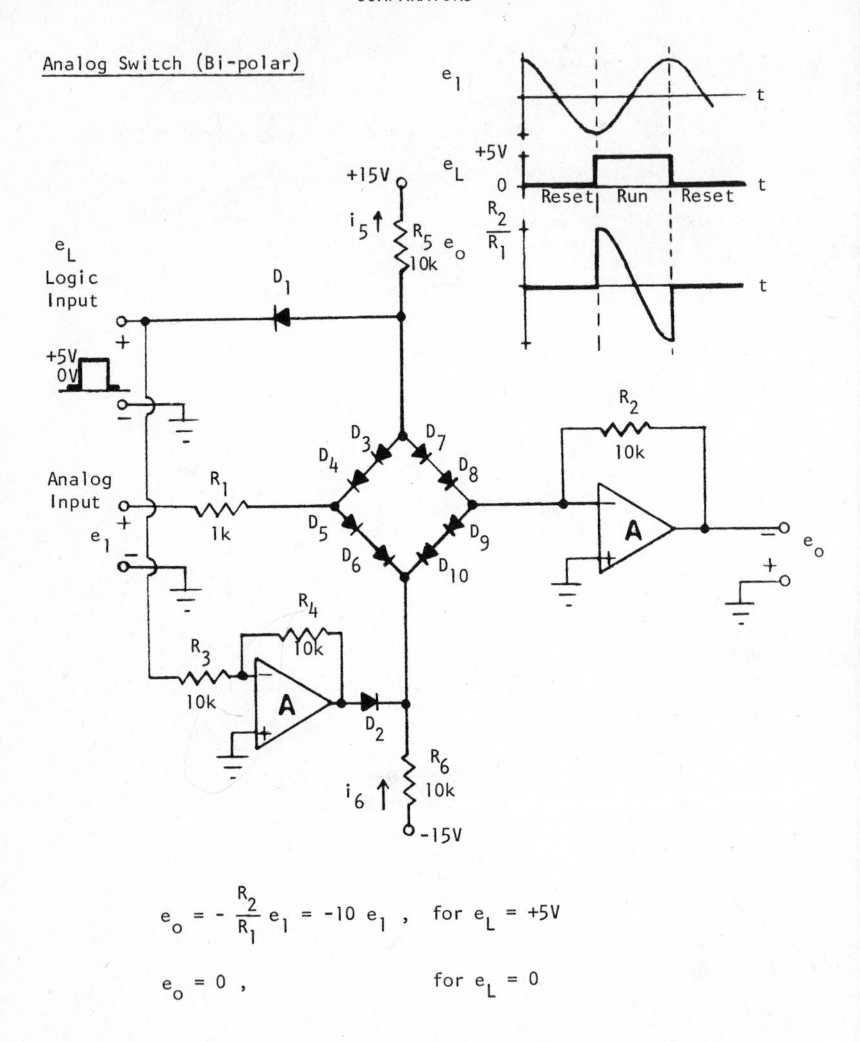

$$e_o = -\frac{R_2}{R_1} e_1 = -10\ e_1 , \quad \text{for } e_L = +5V$$

$$e_o = 0 , \quad \text{for } e_L = 0$$

When the logic input is positive (+5V), the diode bridge conducts. The feedback resistor R_2 will carry a current equal to the input current through R_1. However, it must be remembered that this input current is transferred to R_2 by way of the bridge network; therefore in designing

Analog Switch (Bi-polar) (continued)

the bridge circuit make sure that the quiescent current (i_5, i_6) is sufficient enough to carry the maximum current required.

Design note for establishing the value of i_5 and i_6:

$$e_o = 10V \quad \text{full scale}$$

$$\text{for} \quad e_1 = 1V$$

$$i_1 = \frac{1V}{1k} = 1 \text{ ma}$$

$$\text{therefore} \qquad i_5 = i_6 \geq 1 \text{ ma}$$

$$\text{and} \quad i_5 = i_6 = \frac{30V}{R_5+R_6}$$

$$= 1.5 \text{ ma}$$

The circuit has the advantage of being able to operate under both positive and negative input signals.

The network is useful for driving and resetting integrator circuits.

Signal Selector

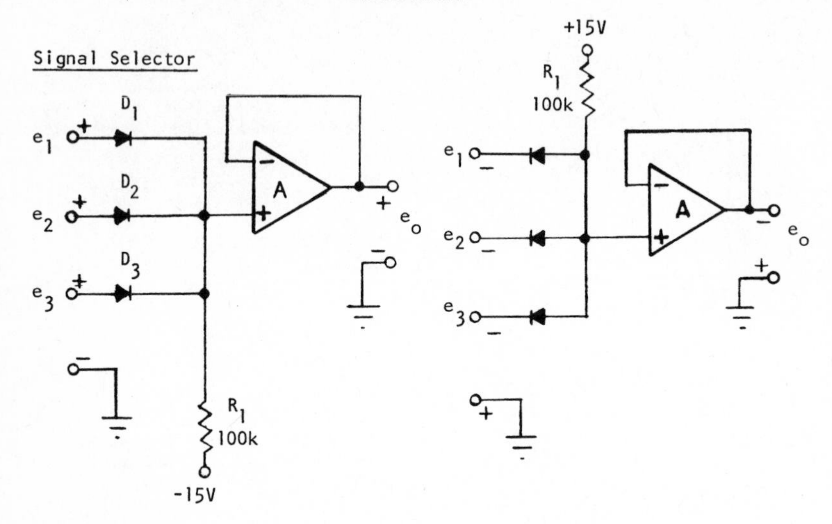

Figure 1

Most Positive
Signal Selector

Figure 2

Most Negative
Signal Selector

The output e_o in Figure 1 will always track the input that has the highest positive value. The output magnitude will be equal to the most positive signal magnitude minus the diode drop.

The identical process takes place in Figure 2 except it is the largest negative signal that will be tracked.

Current sources may be used in place of R_1 to minimize the variation in diode voltage drops as a result of large input voltage variations. (See Appendix G for current source design techniques.)

Precision Signal Selector

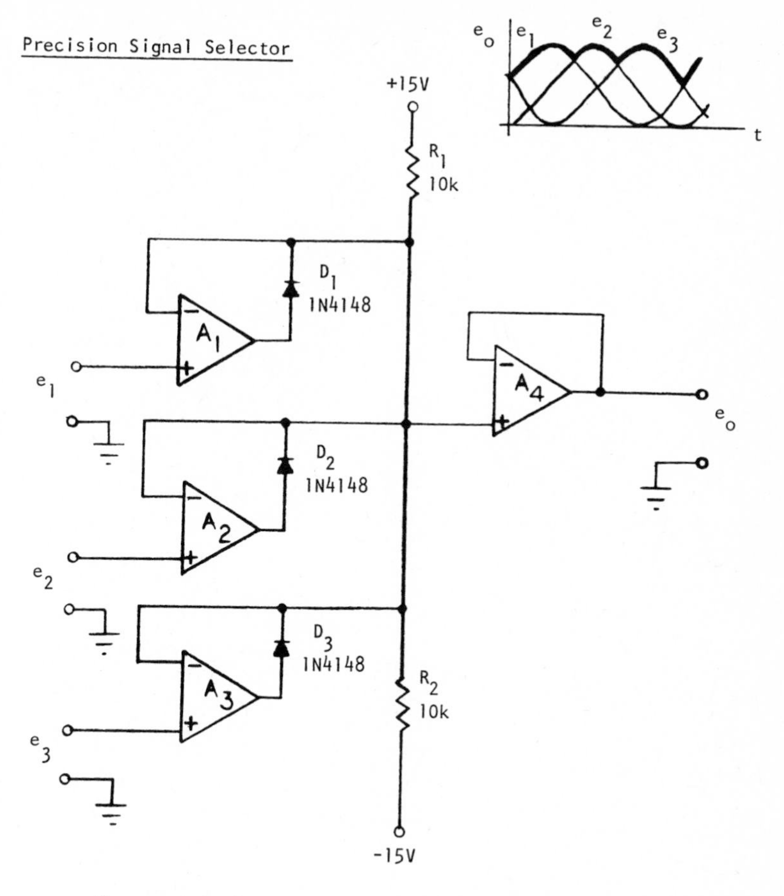

The network shown is a "Most-Positive Signal Selector."

This selector has the advantage of eliminating both the diode voltage drop and diode temperature drift errors. No current sources are needed in place of R_2 and R_3 for large signal applications.

To obtain a "Most-Negative Signal Selector," simply reverse the diodes.

Polarity Separator

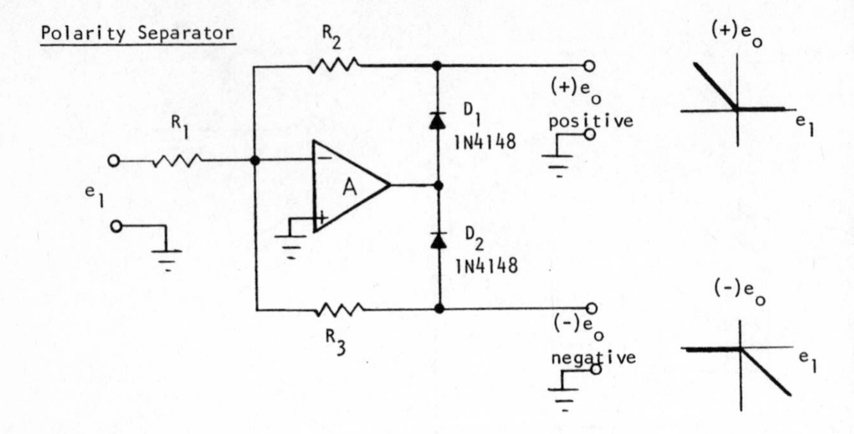

This network separates the positive and negative input signals. Each output is inverted from the input and only one polarity is recognized. The gains are adjusted independently by R_2 and R_3.

$$e_o(\text{pos}) = + \left| \frac{R_2}{R_1} e_1 \right| \quad : \quad e_o(\text{neg}) = - \left| \frac{R_3}{R_1} e_1 \right|$$

This is another variation of the precision half-wave rectifier network.

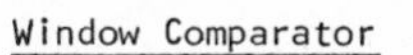

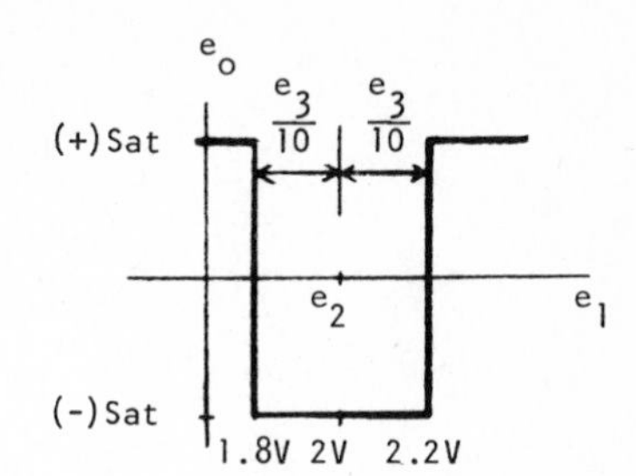

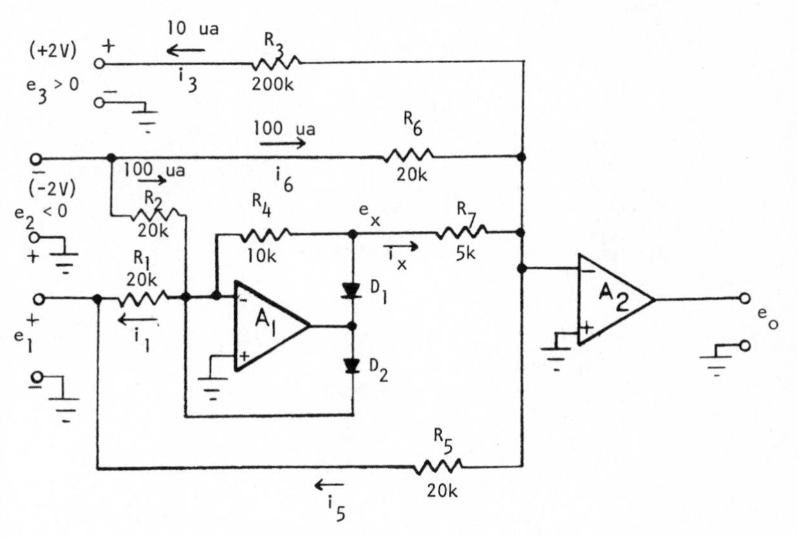

$$e_x = 0 , \qquad \text{for } e_1 < e_2$$

$$e_x = - \frac{e_1 - e_2}{2} , \quad \text{for } e_1 > e_2$$

$$= - \frac{e_1 - 2}{2}$$

This network is also sometimes referred to as a "Go No-Go Comparator."

The circuit is best understood by following the sequence of events starting from a zero input signal:

<u>Window Comparator</u> (continued)

1. At $e_1 = 0$, e_2 establishes the (+) saturation state at the output e_o. It also biases the rectifier (A_1 network) at $e_x = 0$ volts (until such a time as e_1 overcomes e_2).

2. As e_1 increases slightly from zero, a current i_5 now begins to oppose the current through R_6. Remember, this is happening while e_x is still at zero, because e_1 is still less than e_2.

3. The next thing that will happen is the sum of i_5 and the current through R_3 will become sufficient enough to overcome the current through R_6, and the output comparator (e_o) will switch states.

$$i_5 + i_3 = i_6$$

$$i_5 + 10 \text{ ua} = 100 \text{ ua} \quad ,$$

therefore $$i_5 = 90 \text{ ua} \quad .$$

The input voltage level that makes this happen is

$$e_1 = i_5 R_5 = (90 \text{ ua})(20\text{k}) = 1.8 \text{ volts}$$

The output comparator (e_o) switches to (-) saturation at this time.

4. As e_1 further increases toward the value of e_2, the output e_o remains at (-) saturation. However, when e_1 reaches e_2, the rectifier (A_1) is now ready to yield a negative output (e_x).

Window Comparator (continued)

5. As e_1 passes slightly beyond e_2 the output e_x generates a current (i_x) which is in the direction that will cause the output comparator to switch back to (+) saturation.

6. Although i_5 and i_x are opposing currents, i_x will override i_5. The net result for reaching the switching point is

$$\frac{e_1 - e_2}{20k} = i_3 \quad : \quad \frac{e_1 - 2V}{20k} = 10 \text{ ua}$$

$$e_1 = 2.2V$$

At this point the output comparator will switch to the (+) saturation state.

By further examination, it can be seen that the "window" location is determined by the value of e_2, and that the "window size" is proportional to the value of e_3.

Since the value of R_3 is ten times larger than R_5 or R_6, the window size will be proportional to one tenth of e_3.

$$\text{window size} = \frac{e_3}{10} = \frac{2V}{10} = 0.2V$$

The window is symmetrical around the value of e_2, and therefore the total width is twice the value shown above.

$$\text{total window width} = (2)\left(\frac{e_3}{10}\right) = 0.4 \text{ V}$$

The output comparator may be improved by the clamping techniques shown in the section "Clips and Clamps."

Dead Zone Amplifier

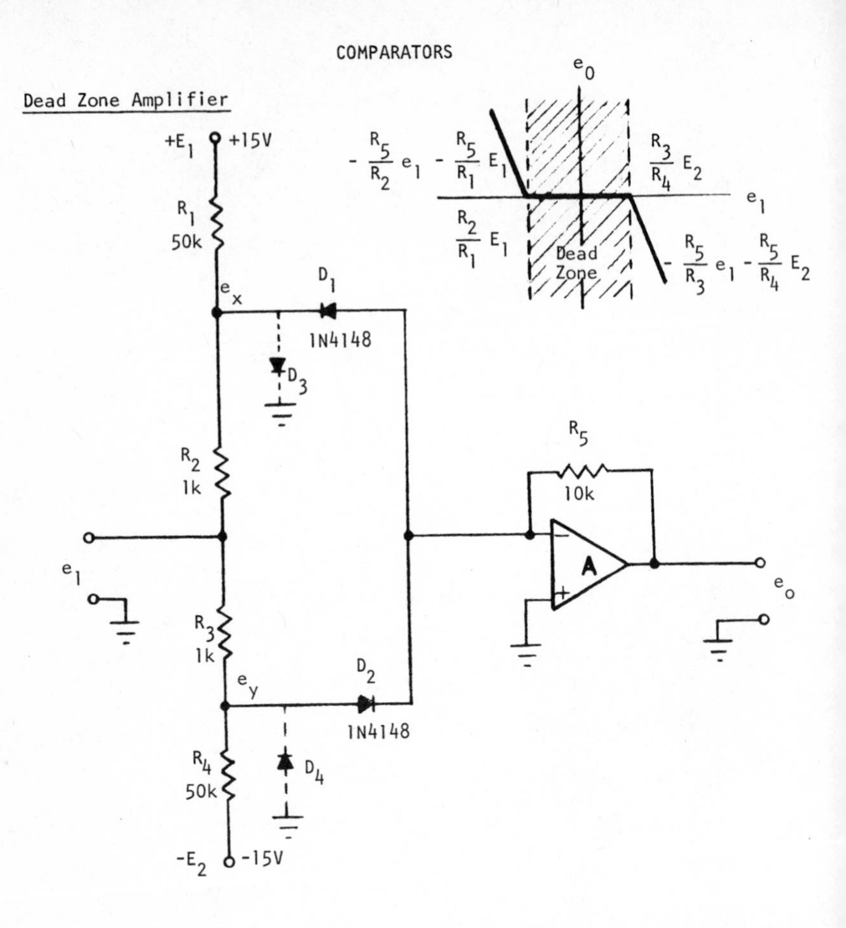

Diodes D_1 and D_2 are reversed biased in the "dead zone" region (for small signals). The thresholds are reached when either e_x or e_y approach zero volts.

The dead zone threshold location is established at the point where the currents flowing into the nodes (e_x or e_y) are equal and opposite.

Dead Zone Amplifier (continued)

For e_1 positive threshold ($e_y = 0$):

$$i_3 = i_4 \quad : \quad \frac{e_1}{R_3} = \frac{E_2}{R_4} \quad ,$$

$$\text{therefore } e_1 = \frac{R_3}{R_4} E = \left(\frac{1k}{50k}\right)(15V) = 0.3V \quad .$$

For e_1 negative threshold ($e_x = 0$):

$$i_1 = i_2 \quad : \quad -\frac{e_1}{R_2} = \frac{E_1}{R_1} \quad ,$$

$$\text{therefore } e_1 = -\frac{R_2}{R_1} E_1 = \left(\frac{1k}{50k}\right)(15V) = -0.3V \quad .$$

After threshold is reached, one of the diodes (depending on input polarity) will forward bias. This connects the input (e_1) and reference (E), through their corresponding resistors, to the Op Amp summing junction. The inputs are now free from the dead zone, and the network will act as an inverting summing amplifier (neglecting diode drops). The voltages add algebraically.

$$e_o = -\frac{R_5}{R_3} e_1 - \frac{R_5}{R_4} E_2 \quad , \qquad \text{for } e_1 \text{ positive region}$$

$$= -\left(\frac{10k}{1k}\right) e_1 - \left(\frac{10k}{50k}\right)(-15V)$$

$$= -10\, e_1 + 3$$

Dead Zone Amplifier (continued)

$$e_o = -\frac{R_5}{R_2}e_1 - \frac{R_5}{R_1}E_1\,, \quad \text{for } e_1 \text{ negative region}$$

$$= -\left(\frac{10k}{1k}\right)e_1 - \left(\frac{10k}{50k}\right)(15V)$$

$$= -10\,e_1 - 3$$

The advantage of this circuit is its component simplicity. The disadvantages are the diode offset and drift errors.

Diodes D_3 and D_4 may be added to minimize reverse bias leakage of D_1 and D_2. Also the switching time will be improved.

Hysteresis Loop Amplifier

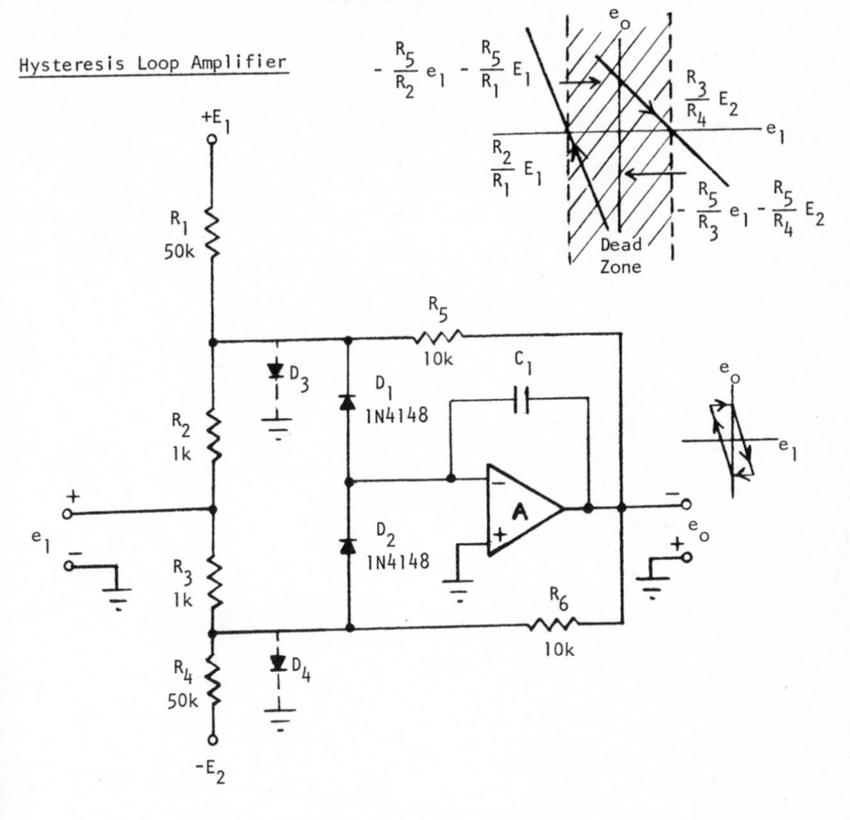

This network simulates the well known "Hysteresis Loop." The output will lag behind the input in time except when the input is passing through the dead zone region. During this time the output will "hold," or retain, its amplitude (the value just before the input entered the dead zone). Once the input signal passes to the other side of the dead zone, the ouput voltage will again begin to track.

The time lag between input and output is determined by the feedback RC time constant (R_5, R_6 and C_1).

Hysteresis Loop Amplifier (continued)

While the amplifier is "holding" drift errors will be generated by the reverse leakage of D_1 and D_2. The addition of D_3 and D_4 (dotted lines) will minimize this effect. Also the Op Amp input leakage can cause a similar error. FET input Op Amps will minimize this error.

Dead Zone Comparator

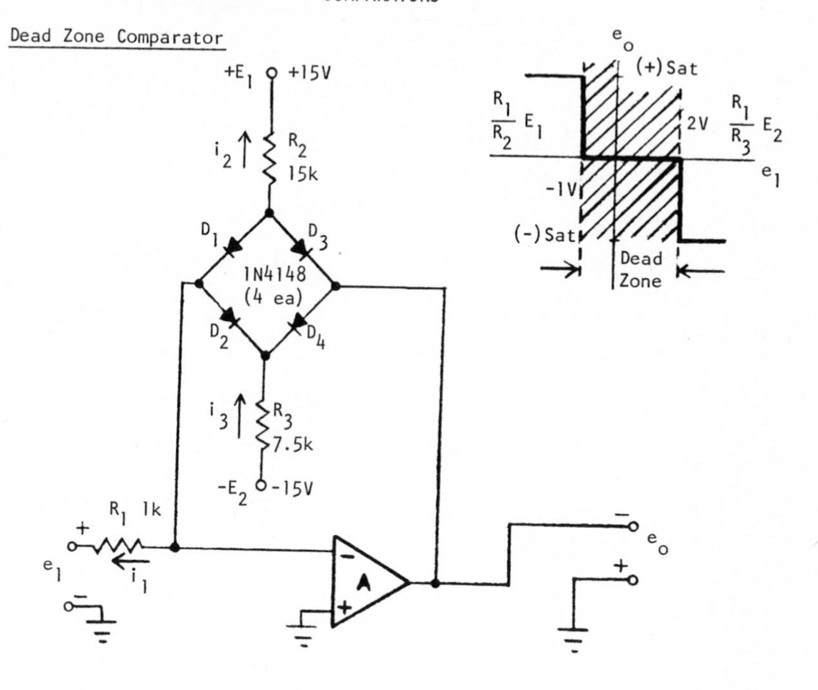

The Dead Zone is established by the currents i_2 and i_3:

$$i_2 = \frac{E_1}{R_2} = \frac{+15V}{15k} = +1 \text{ ma} \quad : \quad i_3 = \frac{E_2}{R_3} = \frac{15V}{7.5k} = 2 \text{ ma}$$

At negative threshold, $i_1 = i_2$, $e_1 = i_1 R_1$

e_1 negative threshold $= i_2 R_1 = (-1 \text{ ma})(1k) = -1V$

At positive threshold, $i_1 = i_3$, $e_1 = i_1 R_1$

e_1 positive threshold $= i_3 R_1 = (2 \text{ ma})(1k) = 2V$

The Dead Zone region in this network is 3 volts wide and not symmetrical around zero.

Dead Zone Comparator (continued)

When the input voltage e_1 is increased, the resulting current i_1 will affect the diode bridge currents in such a manner as to cut-off D_3 or D_4 (depending on input polarity). When one of these output diodes cuts off, the Op Amp output will immediately switch to saturation in a direction opposite to that of the input polarity. The resulting output can be used for sorting devices into 3 separate categories. The standard hysteresis and clamping circuits can be applied to this network for improved noise immunity and higher speed if desired.

Designing symmetry close to zero volts for the Dead Zone region results in an excellent "Zero Crossing Detector" network, as was discussed earlier in this section.

Dead Zone Amplifier

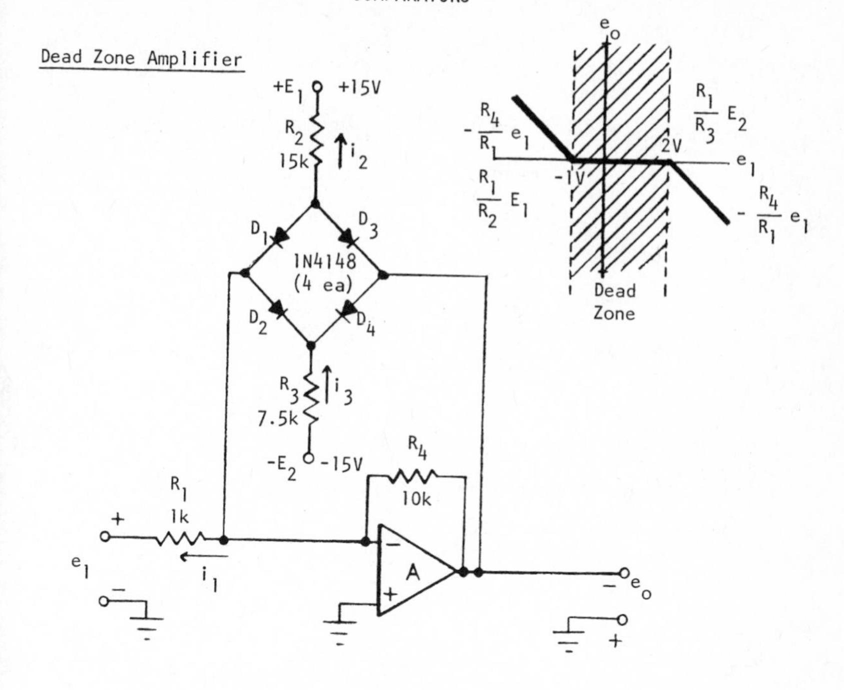

The dead zone is established by the currents i_2 and i_3. The positive input signal reaches its threshold when $i_1 = i_3$,

$$\frac{e_1}{R_1} = \frac{E_2}{R_3} \qquad \text{for positive input threshold}$$

$$e_1 = \frac{R_1}{R_3}E_2 = \left(\frac{1k}{7.5k}\right)(15V) = 2V$$

Similarly, the negative input signal reaches its threshold when $i_1 = i_2$.

$$-\frac{e_1}{R_1} = \frac{E_1}{R_2} \qquad \text{for negative input threshold}$$

$$e_1 = -\frac{R_1}{R_2}E_1 = -\left(\frac{1k}{15k}\right)(15V) = -1V$$

Dead Zone Amplifier (continued)

After the input passes through threshold, the network acts as an inverter with the gain determined by R_4.

$$e_o = -\frac{R_4}{R_1} e_1$$

$$= -\left(\frac{10k}{1k}\right) e_1$$

$$= -10\, e_1$$

This circuit offers the advantage of excellent accuracy. The disadvantage is that the output gains for both positive and negative signals cannot be controlled independently.

Two Mode Inverting Amplifier

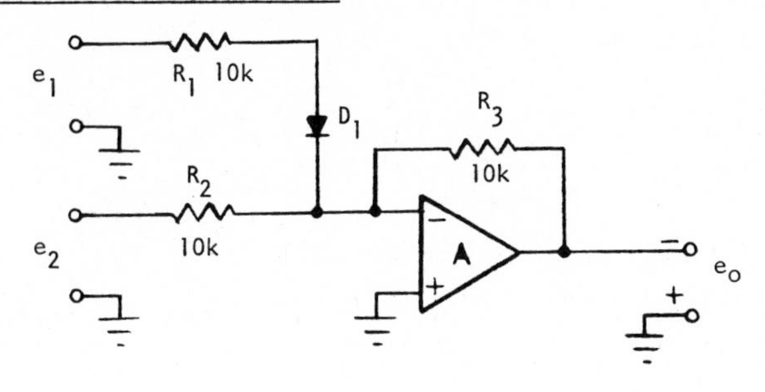

Mode I $\quad e_o = -\left(\frac{R_3}{R_1} e_1 + \frac{R_3}{R_2} e_2\right), \qquad \text{for } e_1 > 0$

$$= -(e_1 + e_2)$$

Mode II $\quad e_o = -\frac{R_3}{R_2} e_2, \qquad \text{for } e_1 < 0$

$$= -e_2$$

In mode I, the network acts as an inverting summing amplifier. When the polarity of e_1 goes negative, the network then switches to mode II, a simple inverting amplifier for the e_2 signal only. The diode drop will cause errors and must be taken into consideration where high accuracies are important.

Two Mode Non-Inverting Amplifier

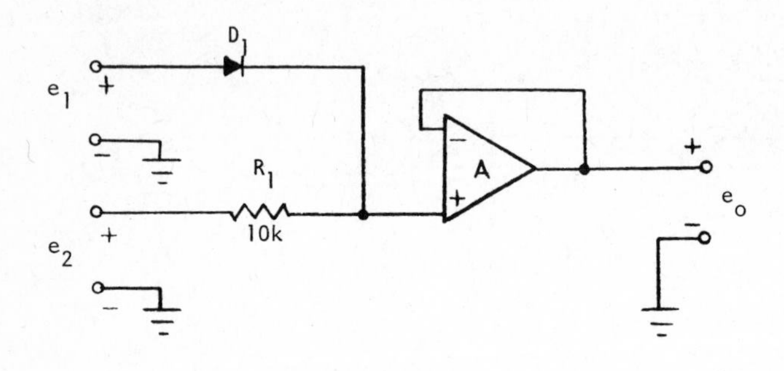

$$e_o = e_1\ , \qquad \text{for } e_1 > e_2$$

$$e_o = e_2\ , \qquad \text{for } e_1 < e_2$$

When e_1 is more positive than e_2, the unity gain buffer network will track e_1 through the diode drop. If e_1 falls below e_2, the diode will reverse bias, and the buffer amplifier will track e_2.

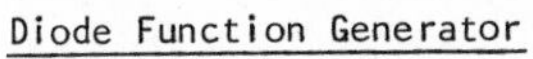

Diode Function Generator

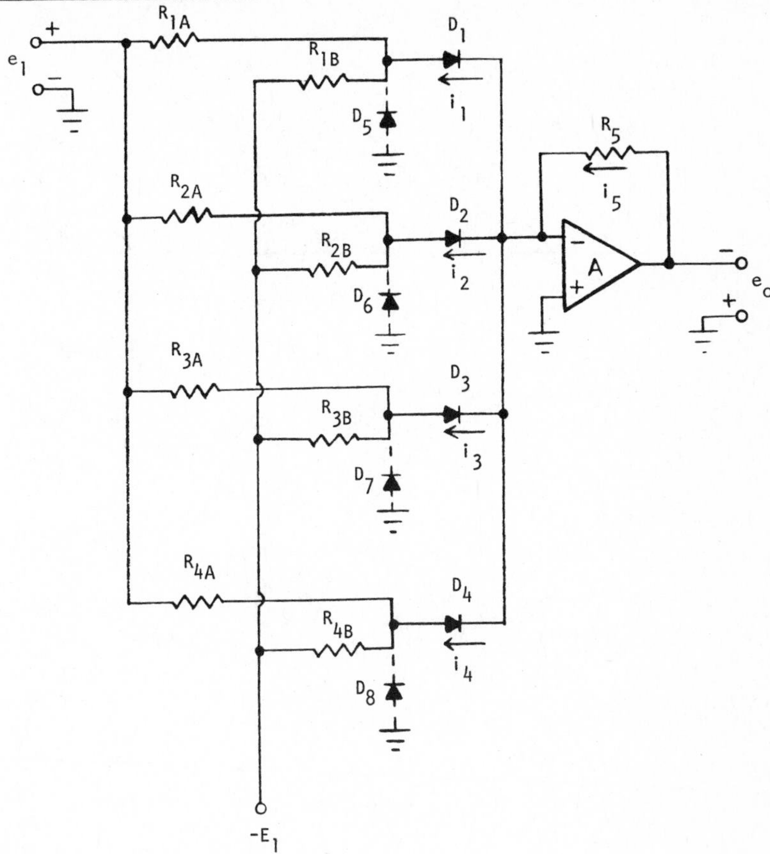

Initially diodes D_1 - D_4 are reversed biased for e_1 equal zero. As e_1 increases, each diode will "kick in" at the point where the current due to e_1 equals the current due to E_1.

For the (first) breakpoint, $i_{1A} = i_{1B}$

or,
$$\frac{e_1}{R_{1A}} = \frac{E_1}{R_{1B}}$$

Diode Function Generator (continued)

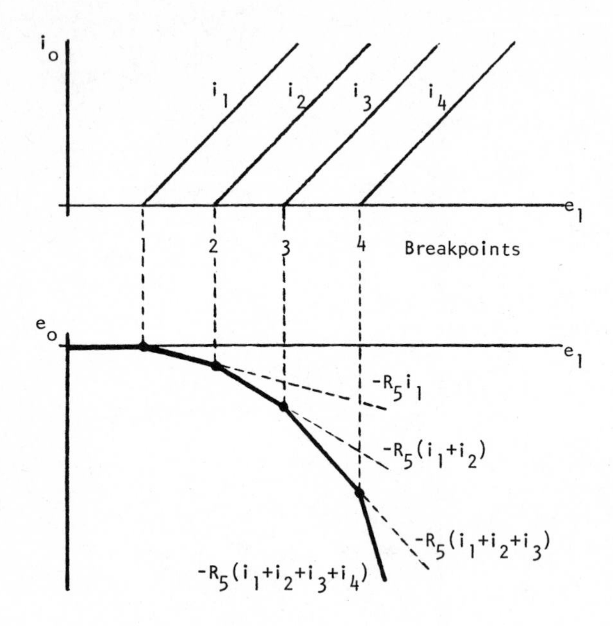

First breakpoint $\quad e_1 = \dfrac{R_{1A}}{R_{1B}} E_1$

Second breakpoint $\quad e_1 = \dfrac{R_{2A}}{R_{2B}} E_1$, and so on.

The output slope of e_o is determined by the resultant currents through the diodes into the feedback resistor R_5

$$i_1 = 0 \ , \quad \text{before breakpoint}$$

$$i_1 = \frac{e_1}{R_{1A}} - \frac{E_1}{R_{1B}} \ , \quad \text{after first breakpoint}$$

$$\frac{e_o}{R_5} = - \frac{e_1}{R_{1A}} - \frac{E_1}{R_{1B}}$$

Diode Function Generator (continued)

$$e_o = -\frac{R_5}{R_{1A}} e_1 - \frac{R_5}{R_{1B}} E_1 \text{, where } \frac{R_5}{R_{1B}} E_1 \text{ is a constant.}$$

The process continues through each breakpoint until the final diode yields

$$e_o = -\left(\frac{R_5}{R_{1A}} + \frac{R_5}{R_{2A}} + \frac{R_5}{R_{3A}} + \frac{R_5}{R_{4A}}\right) e_1 - \left(\frac{R_5}{R_{1B}} + \frac{R_5}{R_{2B}} + \frac{R_5}{R_{3B}} + \frac{R_5}{R_{4B}}\right) E_1$$

All the above equations have neglected diode drops and drift errors. These errors cannot be neglected when high accuracies are required. Extraneous drift and offset compensation circuitry is generally required. (Diodes D_5 - D_8 maybe used to minimize reverse leakage of D_1 - D_4; speed is

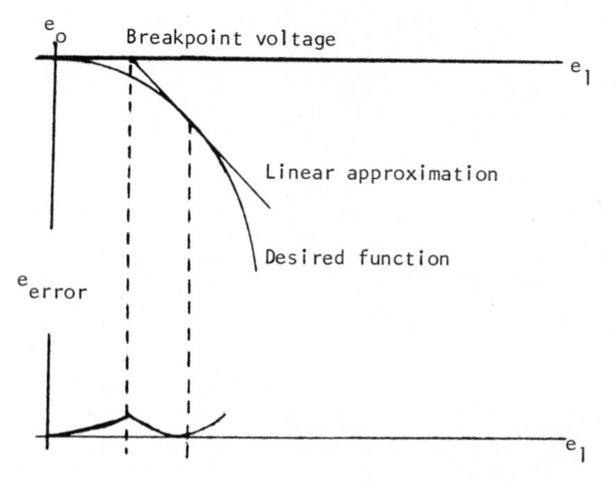

also increased due to less capacitance.) The theoretical errors inherent in the linear approximation technique are shown in the above diagram.

It can be seen that more breakpoints will minimize the theoretical error. However, adding more breakpoints increases drift problems and

Diode Function Generator (continued)

diode voltage offsets. The practical trade-off occurs somewhere in the vicinity of 8 breakpoints. This generally yields accuracies of better than 1% for most function modules.

CLIPS/CLAMPS

Introduction

In the world of Op Amp applications clips are generally used for protection against input overvoltage damage. No great effort is spent in trying to achieve accuracy since the purpose is only to protect the Op Amp input circuitry.

Clamps, on the other hand, require somewhat more attention because of their broad spectrum of applications. Basic comparator networks were chosen to demonstrate the clamping techniques. The clamping techniques illustrated in this section can be applied to the output of practically any Op Amp network.

Input Overvoltage Protection Clipper

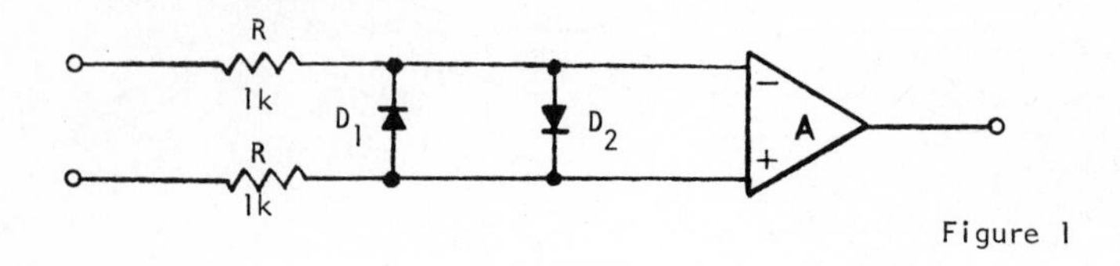

Figure 1

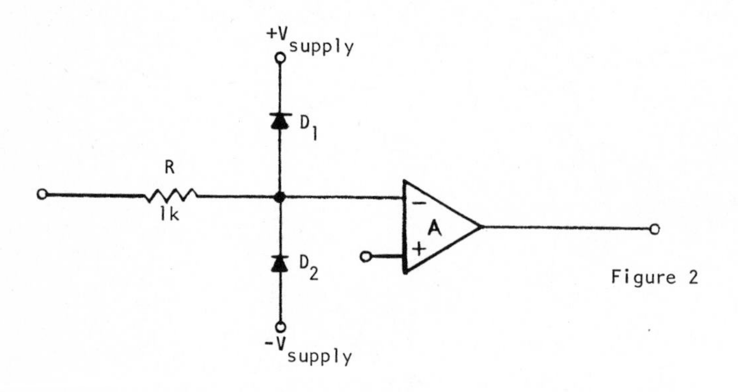

Figure 2

The circuit in Fig. 1 protects the Op Amp from an excessive voltage differential between the input terminals.

The technique in Fig. 2 prevents the input from exceeding the Op Amp power supply voltage. The circuit can be used on one, or both, input terminals. Low leakage diodes should be used where high impedance applications prevail.

Input Clipper

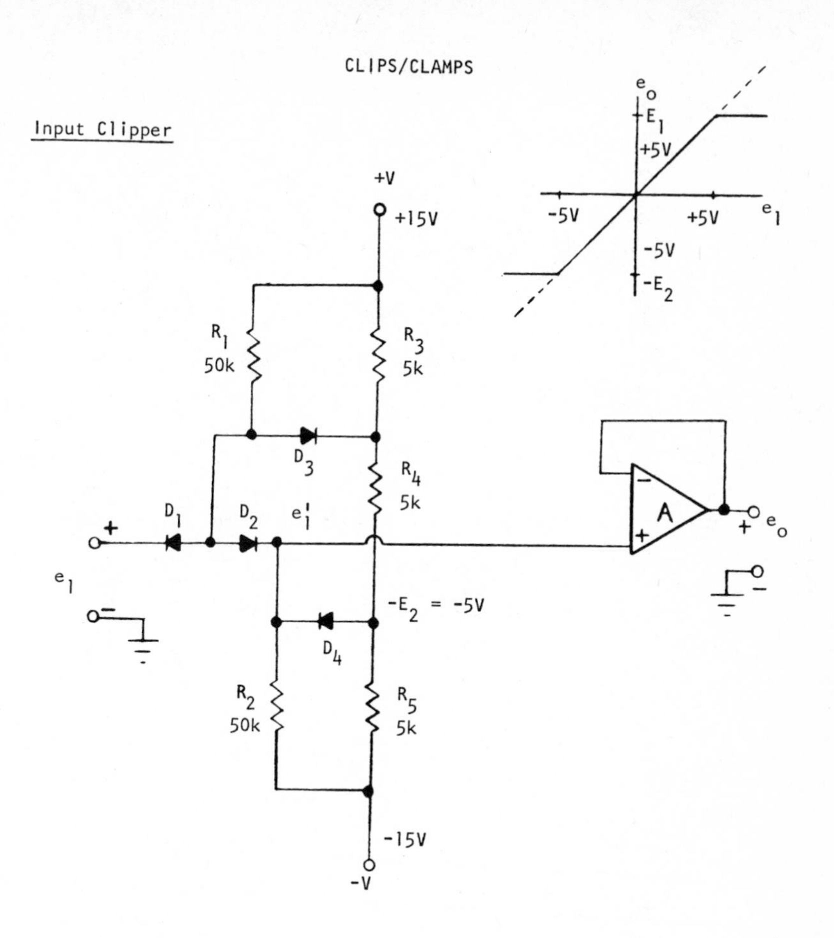

$e_o = e_1$, for e_1 between the limits of E_1 and E_2

$e_o = E_1$, for $e_1 > E_1$

$e_o = E_2$, for $-e_1 < -E_2$

Within the clipper limits, the output voltage e_o is equal to the input voltage e_1 of the network, provided that the diodes D_1 and D_2 are ideally matched. The voltages e_1 and e_1' are identical until either D_3

Input Clipper (continued)

or D_4 (depending on input polarity) is forward biased which is caused by the input exceeding the clipping limits.

When all the diodes are matched properly, this circuit is automatically temperature compensated except for $-E_2$ (D_4). This can be accomplished by connecting a diode in series with R_5, which will offset the drift from D_4.

If more precision is required, the resistors R_1 and R_2 can be replaced by current sources (see Appendix G). This is especially advisable if the input swing approaches +V or -V.

Simple Diode Clamp (Uni-polar)

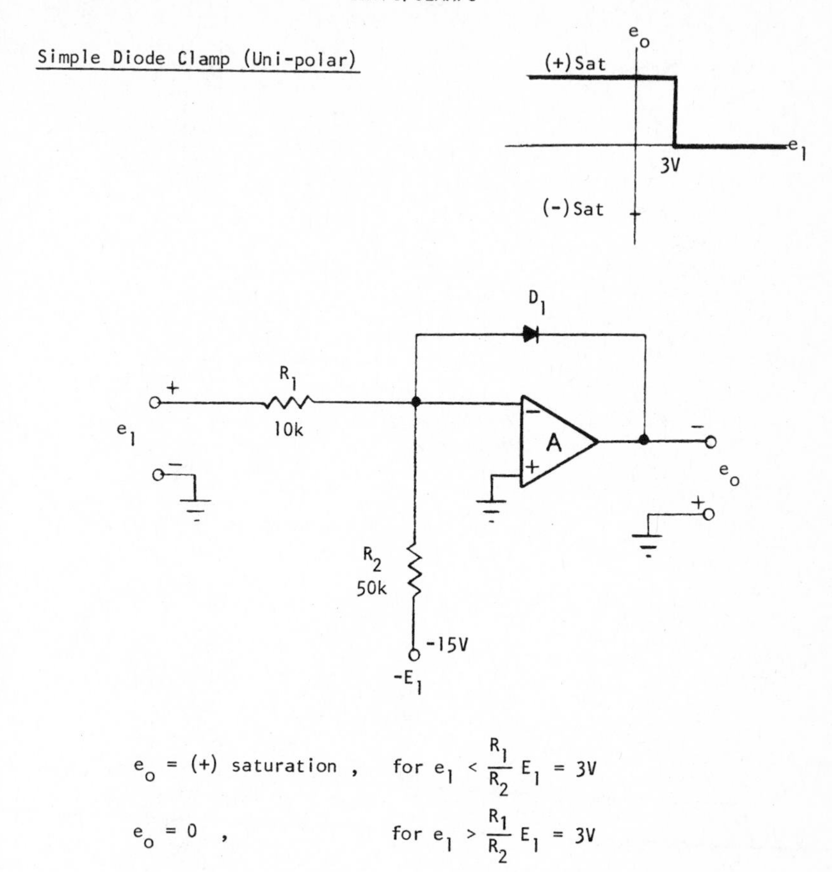

$$e_o = (+)\ \text{saturation}, \qquad \text{for } e_1 < \frac{R_1}{R_2} E_1 = 3V$$

$$e_o = 0, \qquad \text{for } e_1 > \frac{R_1}{R_2} E_1 = 3V$$

The diode D_1 is reversed biased for a positive output swing. When the output goes negative, D_1 becomes forward biased, clamping e_o essentially to zero.

This network has the advantage of simplicity. The major disadvantage is the diode forward voltage drop is seen directly at the output during the clamped mode.

Series Zener Diode Clamp (Bi-polar)

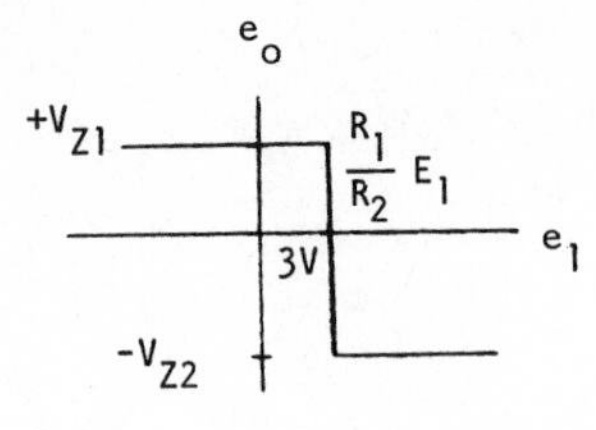

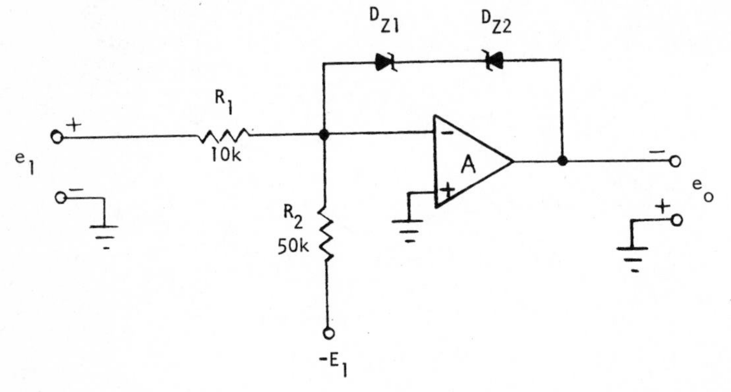

$$e_o = (+)\ V_{Z1}\ , \qquad \text{for } e_1 < \frac{R_1}{R_2} E_1 = 3V$$

$$e_o = (-)\ V_{Z2}\ , \qquad \text{for } e_1 > \frac{R_1}{R_2} E_1 = 3V$$

The Zener diodes in the feedback circuit work in an alternate fashion depending upon output voltage polarity. When e_o is positive D_{Z1} yields the Zener voltage while D_{Z2} acts as a simple forward biased diode. The output clamped voltage V_{Z1} is equal to the sum of the Zener voltage of D_{Z1} and the forward diode drop of D_{Z2}. For negative e_o the diodes reverse roles.

The circuit has the advantage of simplicity and versatility. The disadvantage is the drift characteristics of the diodes which directly affect the output voltage. If commercial temperature compensated diodes are to be used the Shunt Zener Diode Clamp network will be required.

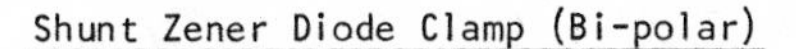

Shunt Zener Diode Clamp (Bi-polar)

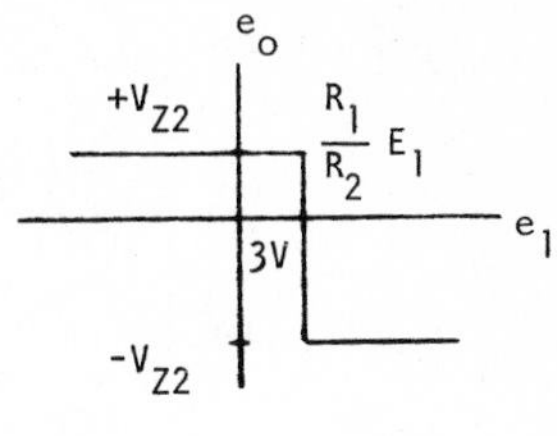

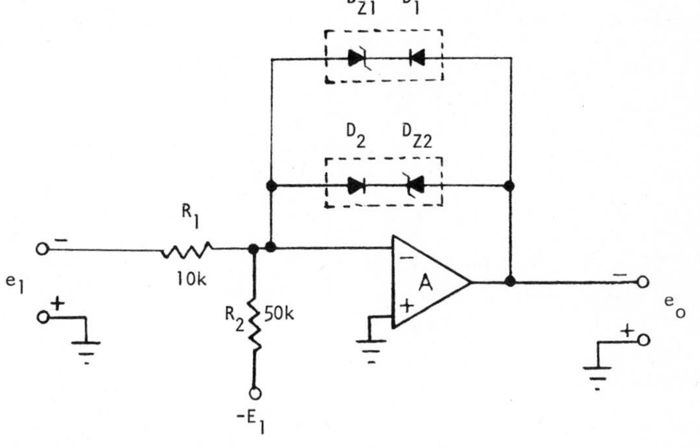

$$e_o = (+)\ V_{Z1}\ , \qquad \text{for } e_1 < \frac{R_1}{R_2} E_1 = 3V$$

$$e_o = (-)\ V_{Z2}\ , \qquad \text{for } e_1 > \frac{R_1}{R_2} E_1 = 3V$$

This network is useful when temperature compensated Zener diodes are used (represented by a Zener diode in series with a standard diode).

The diode feedback circuits alternate conduction depending on the polarity of the output voltages. The clamped voltage (V_{Z1}, V_{Z2}) is equal to the sum of the Zener voltage and diode forward voltage drop.

Shunt Zener Diode Clamp (continued)

This network has the advantage of flexibility since the clamped voltages for positive and negative signals are independent from each other. The major disadvantage is the output clamping voltages are dependent on the temperature characteristics of both the switching diode and Zener diode.

Low Leakage Diode Clamp (Bi-polar)

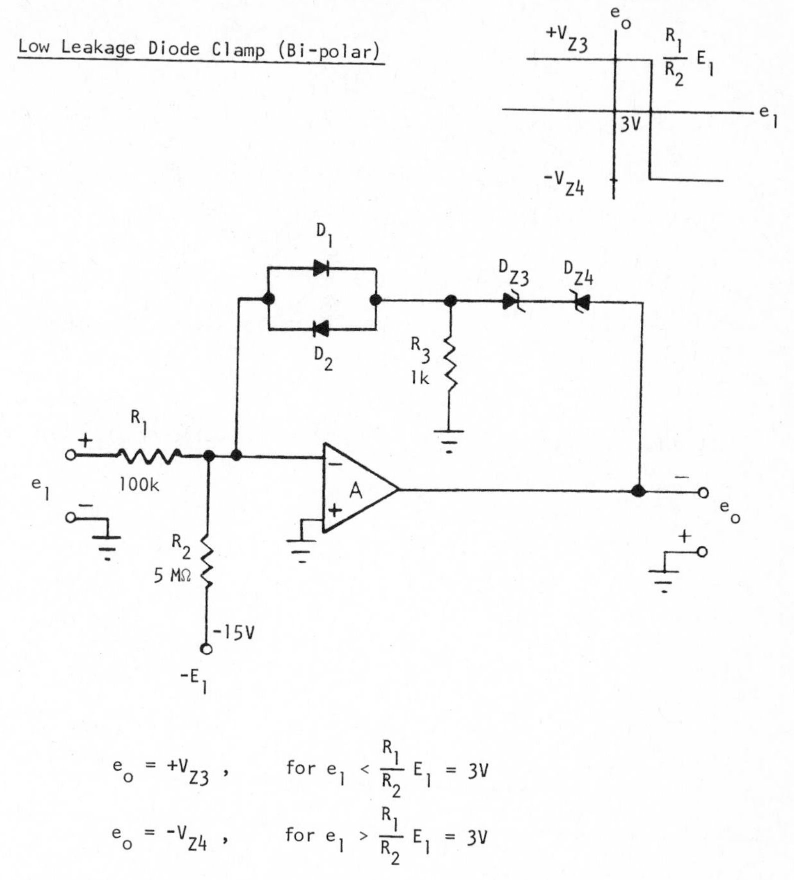

$$e_o = +V_{Z3}, \qquad \text{for } e_1 < \frac{R_1}{R_2} E_1 = 3V$$

$$e_o = -V_{Z4}, \qquad \text{for } e_1 > \frac{R_1}{R_2} E_1 = 3V$$

This technique is useful when it is necessary to work with high input impedances. In this situation, the slightest extraneous leakage current can cause a large error in the threshold point. It is assumed that either a FET Op Amp or Chopper Stabilized Op Amp is used in the above network and any extraneous leakage is therefore due to the network diodes only.

Low Leakage Diode Clamp (continued)

The diode leakage error is minimized in the above circuit by the action of R_3. The leakage of the Zener diodes are shunted through R_3 to ground rather than into the critical summing junction.

The diodes D_1 and D_2 limit their own reverse bias voltage drop across each other; this also greatly reduces the leakage to the summing junction.

This circuit functions well in μA input current range applications.

Diode-Diode Clamp

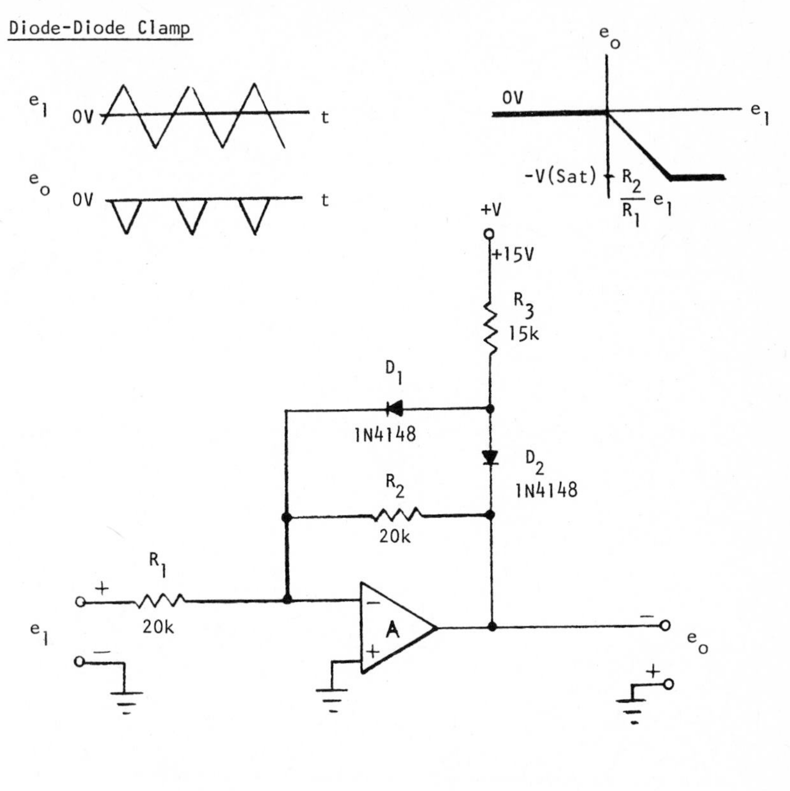

$$(+)\ e_o = 0\ V, \qquad \text{for negative } e_1 \text{ (clamped by } D_1, D_2\text{)}$$

$$(-)\ e_o = \frac{R_2}{R_1} e_1, \qquad \text{for positive } e_1$$

$$= e_1$$

The direction of current flow through R_1 determine the clamping action of D_1. A negative input signal e_1 will make both diodes D_1, D_2 conduct. This causes the output voltage e_o to clamp to zero. A positive input signal e_1 makes the output e_o go negative which, in turn, reverse

Diode-Diode Clamp (continued)

biases the diode D_1. The Op Amp is then free to operate as a normal inverter for positive input signals.

The maximum negative input current I_1 must not exceed the clamping current I_3.

$$I_3 = \frac{V}{R_3} = 1 \text{ ma}$$

If this current is exceeded, the clamping action is lost. The output voltage will go positive, and D_2 will reverse bias.

Resistor-Diode Clamp (Uni-polar)

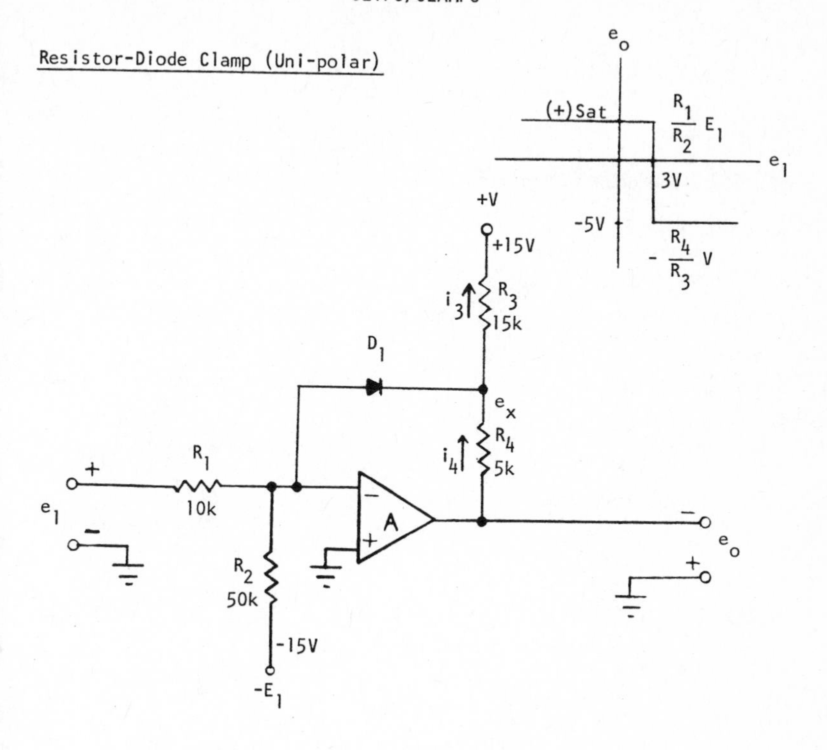

$$e_o = (+) \text{ saturation}, \qquad \text{for } e_1 < 3V$$

$$e_o = -\frac{R_4}{R_3} V, \qquad \text{for } e_1 > 3V$$

The switch point for clamping occurs when e_x reaches zero volts (neglecting diode drop). At this point

$$i_3 = i_4 \qquad \text{clamping threshold point}$$

$$\frac{V}{R_3} = -\frac{e_o}{R_4}$$

and $$e_o = -\frac{R_4}{R_3} V = -\frac{5k}{15k}(15V) = -5V$$

Resistor-Diode Clamp (continued)

This circuit has two weak points: (1) at very low input currents (high input impedance applications) the leakage of D_1 can cause a significant error, and (2) for large input currents (very high input voltages) the clamping voltage will increase due to the high current through D_1. This additional current flows through R_4, thus increasing e_o.

The circuit has optimum application in the 1 ma input current range.

Resistor-Diode Clamp (Bi-polar)

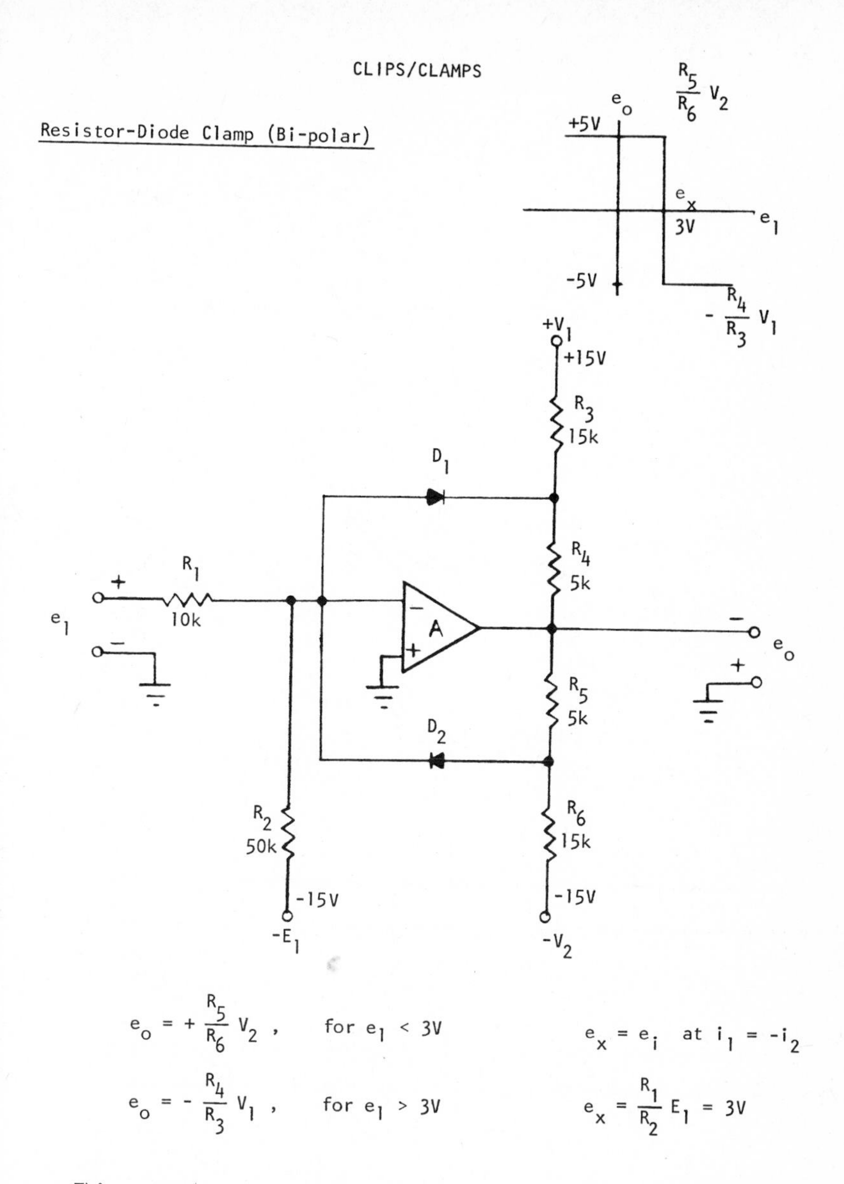

$$e_o = +\frac{R_5}{R_6} V_2 \,, \qquad \text{for } e_1 < 3V \qquad\qquad e_x = e_i \quad \text{at } i_1 = -i_2$$

$$e_o = -\frac{R_4}{R_3} V_1 \,, \qquad \text{for } e_1 > 3V \qquad\qquad e_x = \frac{R_1}{R_2} E_1 = 3V$$

This network is identical to the uni-polar Resistor-Diode Clamp except the technique is applied to both output polarities. This is a very useful and economical network.

Transistor-Resistor Clamp (Uni-polar)

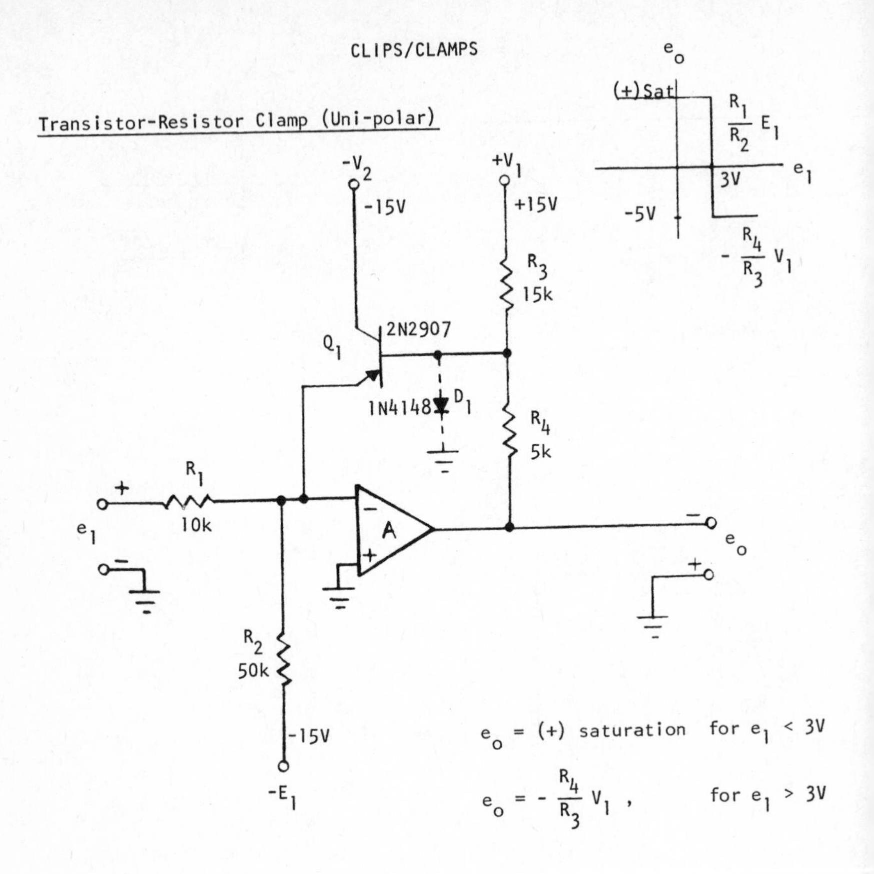

$$e_o = (+)\ \text{saturation} \qquad \text{for } e_1 < 3V$$

$$e_o = -\frac{R_4}{R_3} V_1 , \qquad \text{for } e_1 > 3V$$

This network overcomes the major disadvantages of the Diode-Resistor Clamp. The transistor (Q_1) configuration drastically reduces any leakage currents to the summing junction. This offers a solution to the high input impedance application.

The output clamping voltage also remains virtually constant even for large input currents. Transistor Q_1 diverts excess current directly the the power supply ($-V_2$). Diode D_1 is used to protect the Q_1 emitter-base junction from excessive reverse bias voltage.

CLIPS/CLAMPS

Transistor Resistor Clamp (Bi-polar)

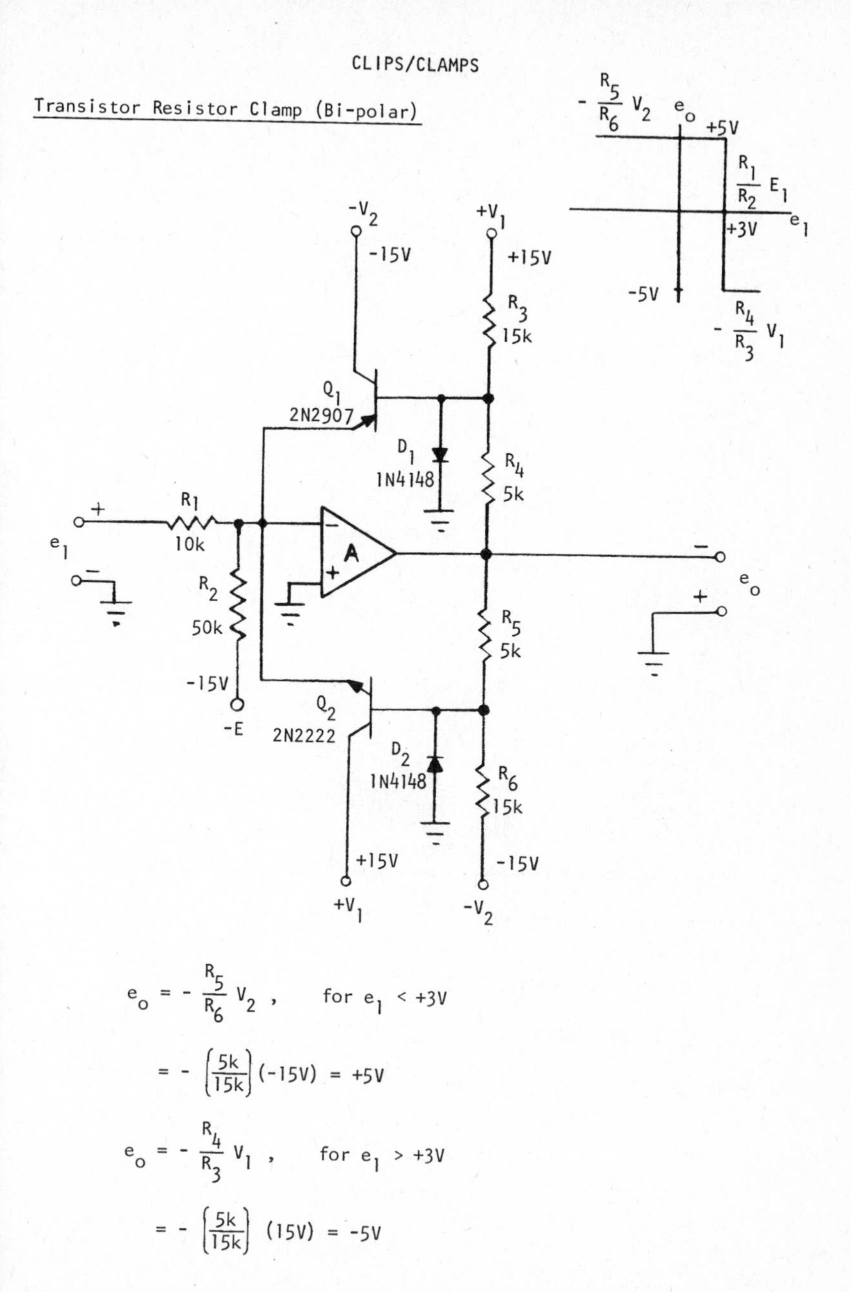

$$e_o = -\frac{R_5}{R_6} V_2 , \qquad \text{for } e_1 < +3V$$

$$= -\left(\frac{5k}{15k}\right)(-15V) = +5V$$

$$e_o = -\frac{R_4}{R_3} V_1 , \qquad \text{for } e_1 > +3V$$

$$= -\left(\frac{5k}{15k}\right)(15V) = -5V$$

Transistor-Resistor Clamp (continued)

This is an excellent clamping circuit. Although it requires more components, the network responds well to a wide range of inputs without sacrificing output accuracy.

The theory of operation is similar to that of the Resistor-Diode Clamps.

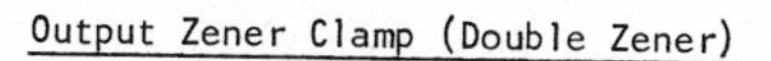

Output Zener Clamp (Double Zener)

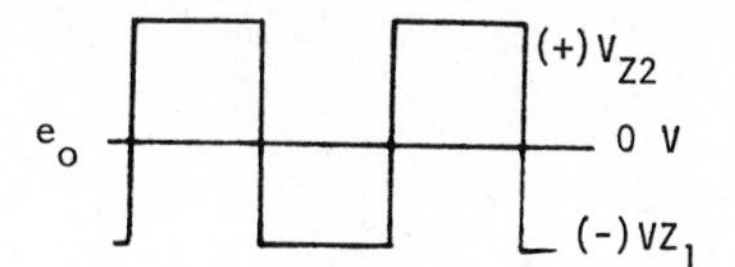

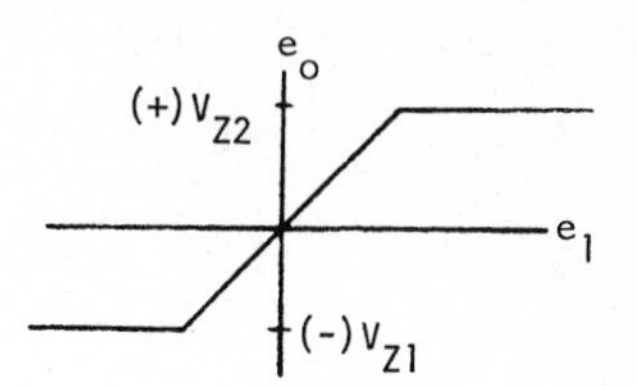

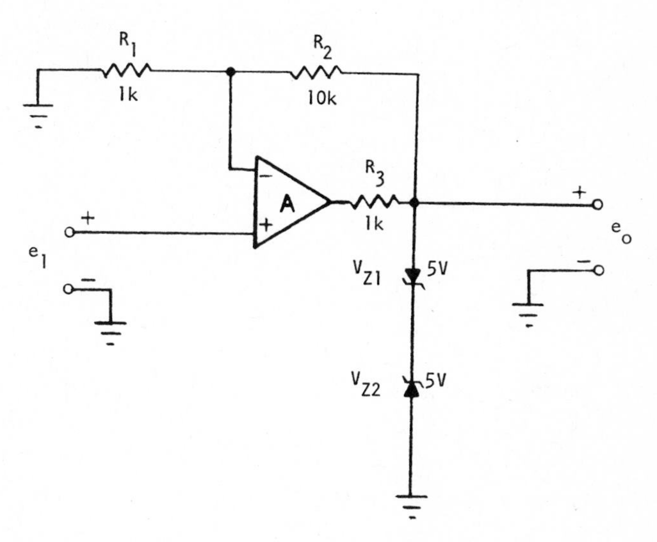

$$(+)\ e_o(\text{max}) = (+)\ V_{Z1}\ , \qquad \text{for } \left(1 + \frac{R_2}{R_1}\right) e_1 > (+)\ V_{Z1}$$

$$(-)\ e_o(\text{max}) = (-)\ V_{Z2}\ , \qquad \text{for } \left(1 + \frac{R_2}{R_1}\right) e_1 > (-)\ V_{Z2}$$

$$e_o = \left(1 + \frac{R_2}{R_1}\right) e_1\ , \qquad \text{for } \left(1 + \frac{R_2}{R_1}\right) e_1 < V_{Z1},\ V_{Z2}$$

$$= 11\ e_1$$

Output Zener Clamp (continued)

This technique will clamp a positive output signal to the value of the reversed bias Zener voltage V_{Z2} (5V) plus the forward biased voltage drop (0.6V) of the non-functioning series Zener V_{Z1}. For negative signals the Zeners change roles and V_{Z1} (5V) determines the clamping voltage.

Resistor R_3 is used to limit the output current of the Op Amp.

$$I_{max} = \frac{V_{sat} - (V_Z + 0.6V)}{R_3}$$

$$\approx \frac{12V - 5.6}{1k} = 6.4 \text{ ma}, \qquad \text{for } V_{sat} = 12V$$

This approach is particularly useful for non-inverting networks when feedback diode clamping cannot be applied.

The basic disadvantage to this design is that the Op Amp output is allowed to saturate. This will usually cause long settling times (recovery time) when the network moves back into the linear region.

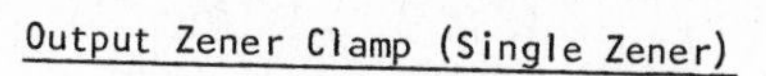

Output Zener Clamp (Single Zener)

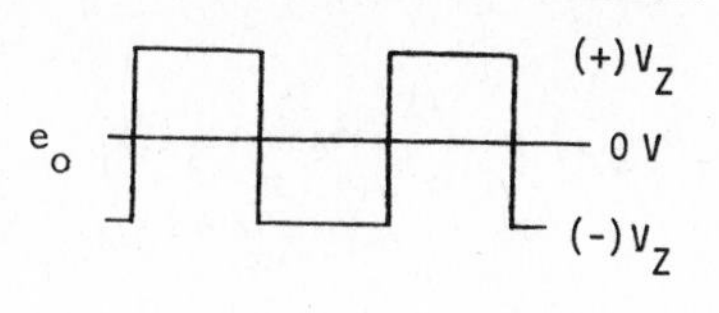

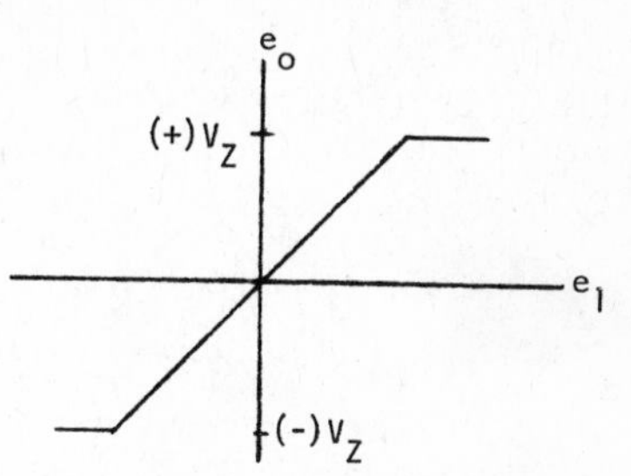

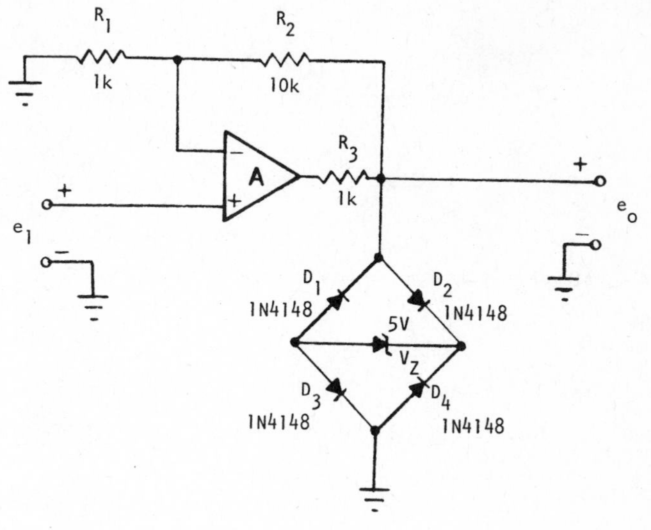

$$e_o(\text{max}) = (\;)\ V_Z , \qquad \text{for } \left(1 + \frac{R_2}{R_1}\right) e_1 \quad V_Z$$

$$e_o = \left(1 + \frac{R_2}{R_1}\right) e_1 , \qquad \text{for } \left(1 + \frac{R_2}{R_1}\right) e_1 \quad V_Z$$

$$= 11\ e_1$$

Output Zener Clamp (continued)

The output voltage will be clamped to the Zener voltage V_Z (5V) plus two diode forward voltage drops (1.2V). The diode bridge provides a current path to the Zener for clamping signals of both polarities.

Resistor R_3 serves as a current limit to protect the Op Amp.

$$I_{max} = \frac{V_{(Sat)} - (V_Z + 1.2V)}{R_3}$$

$$= \frac{12V - 6.2}{1k} = 5.8 \text{ ma}, \quad \text{for } V_{(Sat)} = 12V$$

This clamping technique is useful for non-inverting Op-Amp networks.

The basic disadvantage is that the output of the Op Amp is allowed to saturate. This action will usually cause excessive recovery times when the network goes back to linear operation.

INTEGRATORS

Introduction

In general, integrators deal in the areas of slow output signal changes and long steady time periods. The major difficulties encountered in this type of situation are generally DC drift and leakage type parameters.

The particular error parameters that must be confronted in Op Amp integrator design applications are:

(1) Op Amp Input Bias Currents

(2) Op Amp Input Offset Voltage

(3) Capacitor Dielectric Leakage

(4) Capacitor Dielectric Absorption

(5) Set/Reset Switching Times

The fundamental operation of an Op Amp integrator is similar to the inverting amplifier except that the input current is transferred to a feedback capacitor rather than a resistor.

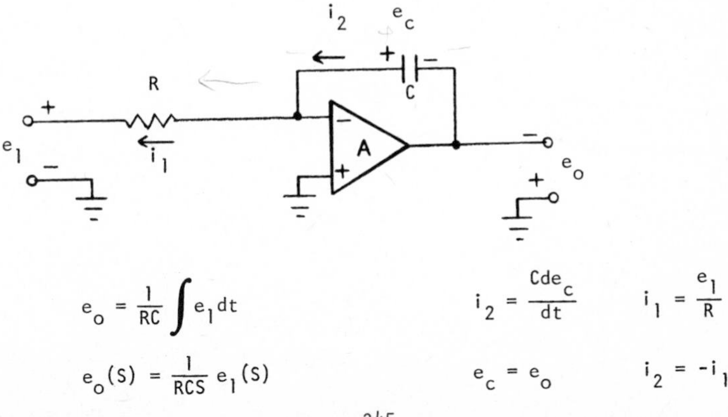

$$e_o = \frac{1}{RC}\int e_1 dt \qquad i_2 = \frac{Cde_c}{dt} \qquad i_1 = \frac{e_1}{R}$$

$$e_o(S) = \frac{1}{RCS}e_1(S) \qquad e_c = e_o \qquad i_2 = -i_1$$

Introduction (continued)

A fixed input voltage e_1 will cause a constant current i_1 to flow into the feedback circuit C. This current i_2 (since $i_1 = i_2$) will force the capacitor voltage e_c to climb at a specific rate of change $\frac{\Delta e_c}{\Delta t}$.

$$i_2 = C \frac{\Delta e_c}{\Delta t}$$

Since the summing junction is a virtual ground, the voltage across the capacitor is equal to the output voltage e_o.

Therefore $$i_2 = C \frac{\Delta e_o}{\Delta t} \quad : \quad i_1 = -i_2$$

and $$i_1 = -C \frac{\Delta e_o}{\Delta t} \quad : \quad i_1 = \frac{e_1}{R}$$

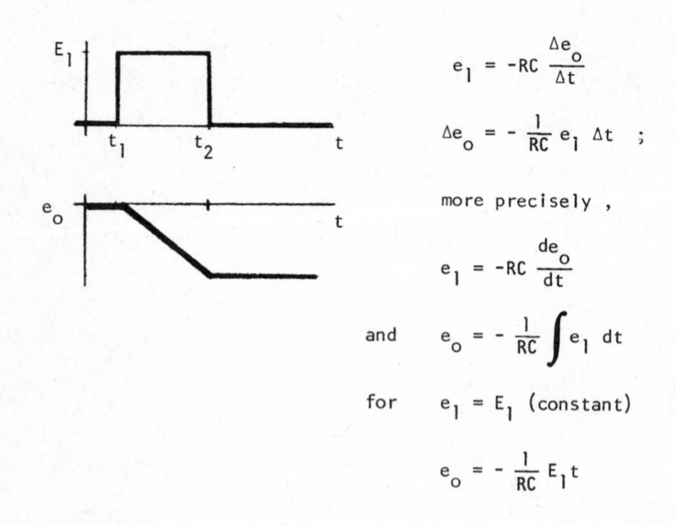

$$e_1 = -RC \frac{\Delta e_o}{\Delta t}$$

$$\Delta e_o = -\frac{1}{RC} e_1 \Delta t \quad ;$$

more precisely ,

$$e_1 = -RC \frac{de_o}{dt}$$

and $$e_o = -\frac{1}{RC} \int e_1 \, dt$$

for $e_1 = E_1$ (constant)

$$e_o = -\frac{1}{RC} E_1 t$$

Introduction (continued)

Notice that when E_1 returns to zero, the output stays constant and the capacitor remains charged, provided there are no extraneous leakage paths.

At this point, the typical errors encountered with integrators can be more clearly examined.

Op Amp Bias Current. It can be seen that any "extraneous" current entering the feedback capacitor circuit from the summing junction will cause an erroneous charge rate on the capacitor. The bias current from the input transistors in the Op Amp will cause such an effect. By leaving the input open i_1 will be zero. Therefore if the output voltage drifts away from zero with time, this will be due to the Op Amp's (-) input bias current charging the capacitor.

$$I_{BIAS} = C\frac{\Delta e_o}{\Delta t}$$

This effect can be minimized by using a Chopper Stabilized Op Amp or Low Leakage FET Input Op Amp. Another method would be to connect an input (resistor) to the summing junction which is calibrated to carry an equal and opposite current that will cancel the bias current. However, since the bias current usually varies with temperature, the cancelling network will also need temperature compensation circuitry.

Op Amp Offset Voltage. Another error current can occur if i_1 is not exactly equal to $\frac{e_1}{R_1}$. This error can be caused by the input offset voltage. This voltage will move the summing junction away from ground potential;

Introduction (continued)

therefore the actual current flowing through R_1 will be

$$i_1 = \frac{e_1 - V_{off}}{R_1}$$

and

$$i_{error} = \frac{V_{off}}{R_1}$$

With the input e_1 grounded, this error current (i_{error}) will charge the feedback capacitor

$$i_{error} = C \frac{\Delta e_o}{\Delta t}$$

This error can be minimized by adjusting the Op Amp Offset Voltage to zero externally. Since the offset voltage varies with temperature, usually the most accurate solution to the problem is to use a low drift Chopper Stabilized Op Amp.

Capacitor Dielectric Leakage. The error due to capacitor dielectric will cause the output voltage to "sag," which is equivalent to connecting a resistor directly across the capacitor terminals. The result is that the capacitor cannot hold its charge indefinitely and therefore will gradually discharge itself through its own shunt resistance.

The best capacitors for use in minimizing this effect are the polysytrene type. These devices are physically somewhat larger than other types. However, if physical size is the major factor, then polycarbonate and mylar can be used with a slight trade-off in the dielectric leakage specifications, respectively. In general, these capacitors should not be larger than 1 to 10 μf in order to optimize leakage trade-offs.

Introduction (continued)

Capacitor Dielectric Absorption. The output error due to this effect is usually very subtle and often quite illusive. This error is generally of importance where accuracies of better than 0.1% are required.

The concept behind the effect of dielectric absorption is somewhat analogous to that of residual magnetism. All dielectric materials have different electrostatic "residual" properties. The dielectric molecular "dipoles" must align themselves with the externally applied electrostatic field. This is how the capacitor holds its charge. However, these dipoles, like most other physical phenomena, tend to resist change.

The resulting outward effect shows itself like this: Charge a neutral capacitor to one volt. Now remove the voltage source and accurately measure the voltage across the isolated capacitor. It will be observed that when the voltage source is removed, the capacitor voltage will quickly drop to slightly below a volt. This is due to a few dipoles "snapping" back to their "original" positions after the external field was removed. By applying the one volt source again to the capacitor terminals and removing as before, it will be observed that the capacitor will retain a voltage closer to one volt. Repetition of the above procedure will result in the capacitor retaining the one volt charge to within the accuracy required.

If the capacitor was now short circuited in order to neutralize it again, this same effect would occur near zero volts. The device would retain some of the original one volt charge until the shorting procedure was repeated several times and all the dipole positions were neutralized.

Introduction (continued)

The best capacitors to use in minimizing this effect are, again, the polystyrene type. Similarly, if size is an important limitation, then polycarbonate is the next best trade-off. Sharper degradation becomes more apparent with mylar and other common lower-quality dielectrics, such as ceramic and mica, respectively.

Set/Reset Switching Times. Generally, the major switching errors encountered with Integrators are caused by the basic "RC Time Constants" associated with circuit component combinations. These RC Time Constants interchange when switching from one mode of operation to another.

For example, setting the initial condition E_i of integration requires time for the feedback capacitor to charge up to this point before the network is switched into the integrate mode.

$$e_o = \underbrace{\frac{1}{R_1 C_1} \int_o^t e_1 \, dt}_{\text{integrate (RUN)}} - \underbrace{\frac{R_B}{R_A} E_i}_{\text{initial condition (SET)}}$$

The corresponding circuit for the above equation is shown on the following page.

Introduction (continued)

In the network below it can be seen that the output will require a finite amount of time to "acquire" the initial condition voltage during the "SET" mode.

$$e_o = \frac{R_B}{R_A} E_i + \left(e_s - \frac{R_B}{R_A} E_i\right) \varepsilon^{-t/R_B C_1}$$

where e_s is the value of e_o at the start of the set interval.

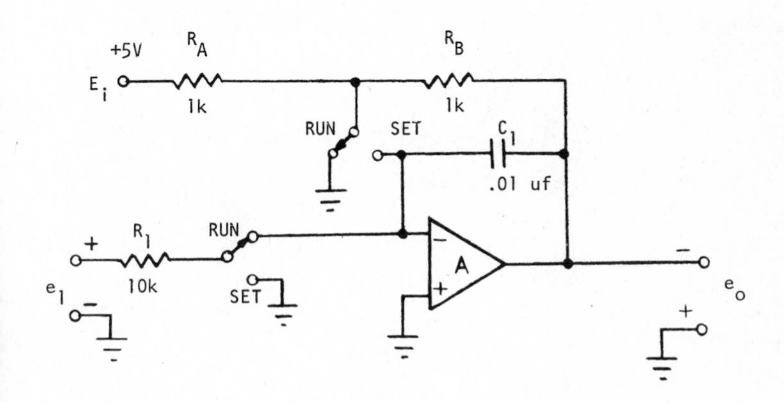

$$e_o = \frac{10^6}{(10k)(0.01)} \int_o^t e_1 \, dt - \left(\frac{1k}{1k}\right) (5V)$$

$$= (10^{-6}) \int_o^t e_i \, dt - 5V$$

Introduction (continued)

The "SET" interval must be long enough to allow the output to reach the initial condition within a prescribed accuracy (i.e., for 0.1% accuracy, the time allowed should be at least 7 times the value of $R_B C_1$). In this case, the design limitation is found in R_B. The smaller the value of R_B, the higher the accuracy will be for a given acquisition time. However, the value of R_B will be ultimately limited by the output current capability of the Op Amp. Often an output "booster" amplifier (unity gain driver module, see Appendix M) is connected within the loop to give the additional current capability. This booster amplifer is available commercially from most Op Amp manufacturers.

The remainder of this section will show various network configurations used for integrating applications.

Basic Integrator (Bias and Offset Compensated)

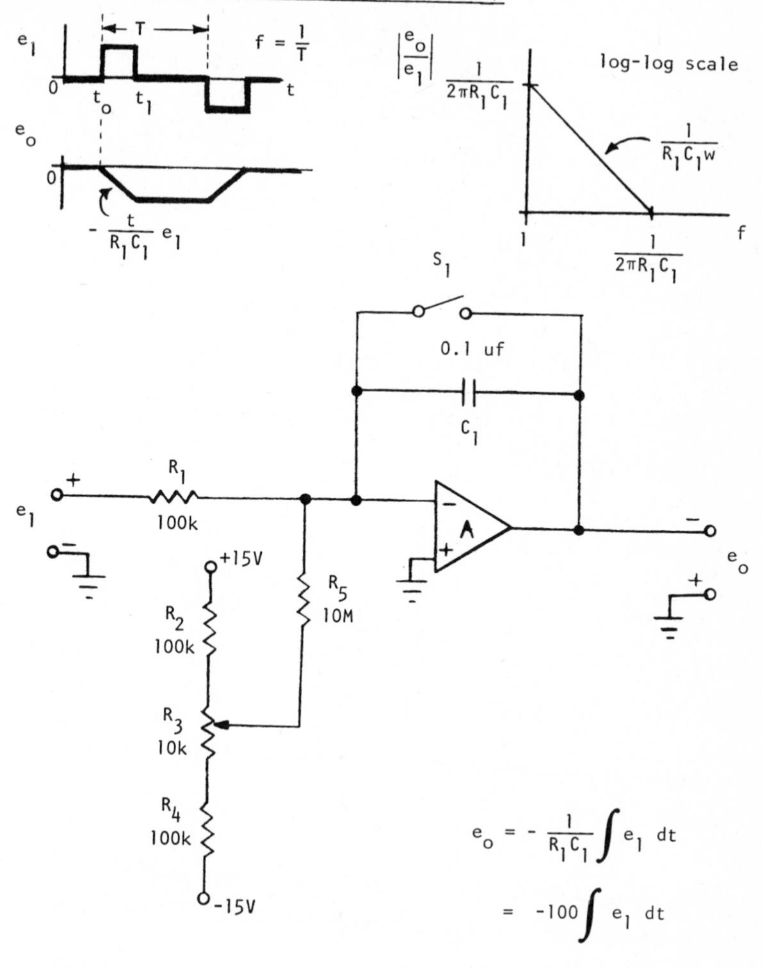

$$e_o = -\frac{1}{R_1C_1}\int e_1\,dt$$

$$= -100\int e_1\,dt$$

Components R_2, R_3, R_4, and R_5 make up error current compensation circuit. By properly adjusting R_3 both offset voltage and bias current errors can be cancelled simultaneously for a given operating temperature.

For a square wave input, the output will show a linear climb proportional to time.

Basic Integrator (continued)

The output will ramp to a maximum value determined by the time duration and amplitude of the input pulse.

$$-e_o(\text{max}) = \frac{(t_1 - t_o)}{R_1 C_1} e_1(\text{max})$$

When the input drops back to zero, the output will remain constant. However, the output may be instantaneously reset to zero by closing the switch S_1.

The general expression for the frequency response of this network is

$$e_o(S) = - \frac{1}{R_1 C_1 S} e_1(S)$$

Relay Reset Circuit

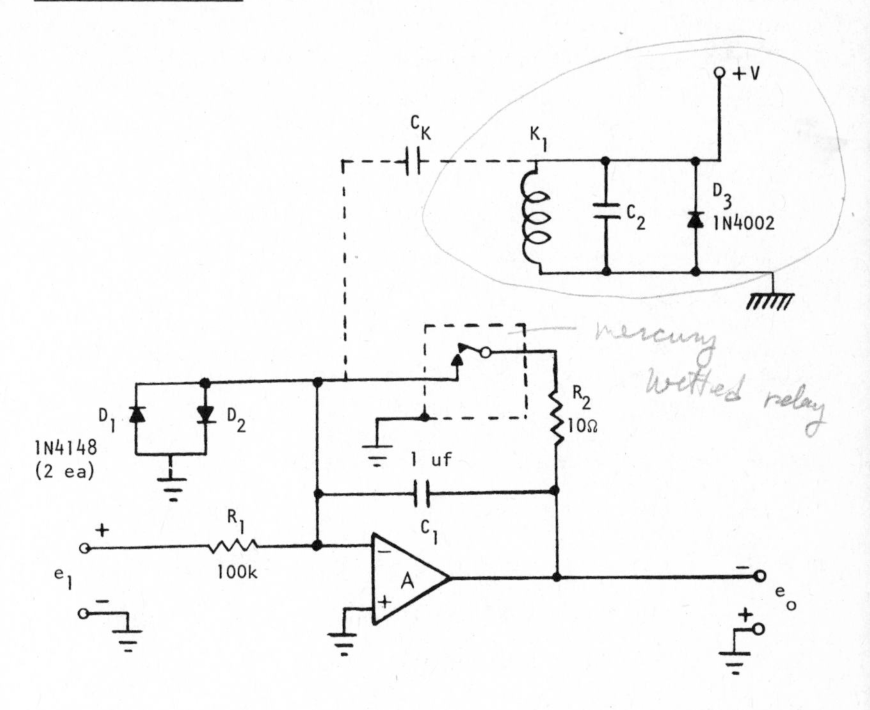

The following points should be examined when designing reset circuitry using relays:

(1) Relay contact maximum current capability.

(2) Relay coil inductive kick in switching.

(3) Leakage current across relay contacts.

(4) Leakage capacitance from relay source voltage to the relay contact connected closest to the summing junction (C_K).

(5) Relay contact bounce after switch is closed.

Relay Reset Circuit (continued)

Relay Contacts. Resistor R_2 serves as a current limiter for discharging C_1.

$$i_{max} = \frac{E_{C1}}{R_2}$$

The trade-off usually occurs when using "fast" switching reed relays. These devices switch in milliseconds; however, the relay contacts are often quite fragile and usually cannot carry over an ampere (or less) without damage.

When using sensitive relays, a design compromise must be made between the relay switching time and the time constant (R_2C_1) decay time. The speed at which the output resets is dependent on the sum of these time intervals.

Relay Coil. A poorly chosen relay coil can induce an "inductive kick" due to transformer action into surrounding circuitry. This can be minimized by damping the coil with C_2 and short circuiting the reverse voltage inductive switching transient with a rectifier diode D_3. This phenomenon occurs at the worst possible time--just as the integration mode begins.

Leakage Current. There is always a finite resistance between the open relay contacts. This resistor will cause the network to approach the function of a RC lag network rather than an ideal integrator. Selection of a good quality relay will minimize this problem.

Relay Reset Circuit (continued)

Leakage Capacitance. The leakage capacitance can cause relatively large voltage spikes to be transferred to the summing junction when the relay is switched. This effect is minimized by shielding and grounding the relay. The diode clamps D_1 and D_2 will limit the peak amplitude of any remaining transient spikes to the summing junction.

Contact Bounce. The action occurs when the relay is initially closed. The contacts will vibrate upon impact and require a finite time to settle. In critical timing applications "mercury-wetted" relays should be used. These relays have their contacts coated with thin film of liquid mercury. When the switch is closed, this film sets up a conductive liquid surface tension between the two contacts, which virtually eliminates the effect of contact bounce.

The main disadvantage of these devices is they must be physically oriented in a specific position in order to prevent the "pool of mercury" from inadvertently short circuiting the contacts.

An additional point that should be noted is the relay ground path. Particularly with "voltage type" devices, relay energizing currents are relatively high. It is best to avoid passing these currents through any "signal" ground path. A separate ground path, leading directly to the power supply common, should be provided for the relays.

FET Reset Circuit

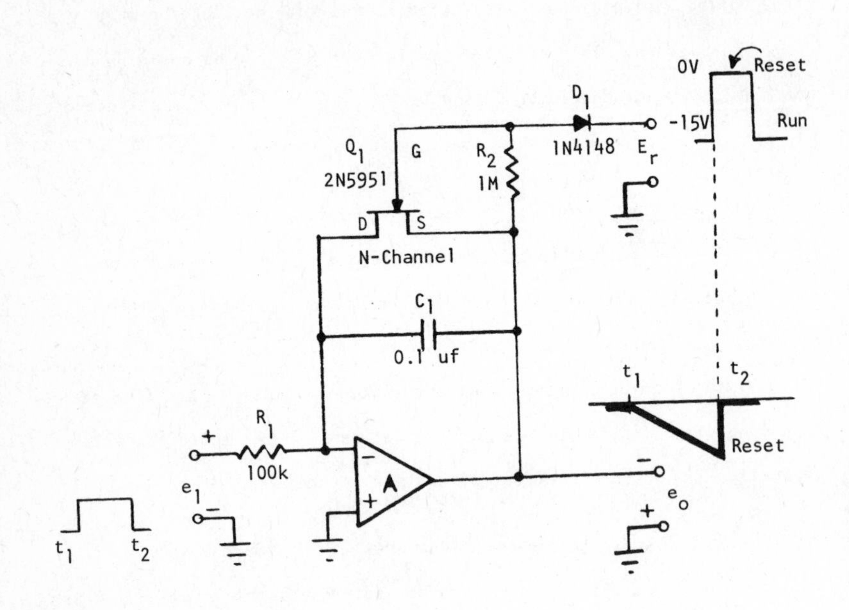

$$e_o = -\frac{1}{R_1C_1}\int e_1\,dt$$

$$= -100\int e_1\,dt$$

This is a simple, but very useful, integrator. Transistor Q_1 (FET) acts as a switch to reset the output to zero. Refer to Appendix H for +5V TTL level converter circuit (E_r). Better than 1% accuracy can be achieved with this network using a FET or Chopper Stabilized Op Amp.

FET Reset Circuit (continued)

The reset switching speed is limited by the "ON" resistance (R_{on}) of the FET (Q_1). This value usually ranges from 10Ω to 1000Ω depending on the type of FET. For the typical general purpose FET, as shown, the R_{on} is approximately 100Ω.

The reset accuracy is determined by the time constant of R_{on} and C_1.

$$\tau = R_{on} C_1 = (100\Omega)(0.1)(10^{-6}) = 10 \mu\text{sec}$$

$$e_o = e_{C_1} \varepsilon^{-t/\tau}$$

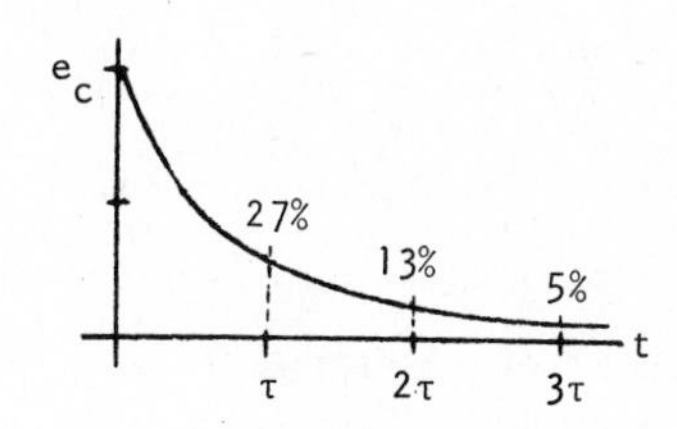

for $t = \tau$, the output will drop about 63% from the original value, or yield an accuracy of only 27%

for $t = 2\tau$, accuracy will increase to 13%

for $t = 3\tau$, accuracy will be about 5%

and for $t = 7\tau$, accuracy will reach 0.1%

FET Reset Circuit (continued)

Therefore, the reset time required for the above circuit to yield 0.1% accuracy must be at least 70 μsec.

Another source of error due to the switching FET (Q_1) is the leakage current from the gate (G) to the drain (D). This current will flow through R_1 causing a voltage offset that will charge C_1. However, with a good quality FET this current can be kept as low as 1 pA.

Care should be taken to keep all surfaces clean around the FET and near Op Amp summing junction. Dirty surfaces between terminals can cause excessive leakage currents across any voltage gradient. All integrators are very sensitive to this effect. Electrostatic shielding techniques are often necessary to minimize these gradients (see Appendix C).

One advantage of the FET switch is the low leakage capacitance. The capacitance from gate to drain is typically less than 2 pf. Together with low cost and simplicity, the FET makes an excellent switch for many integrator applications.

Summing Integrator

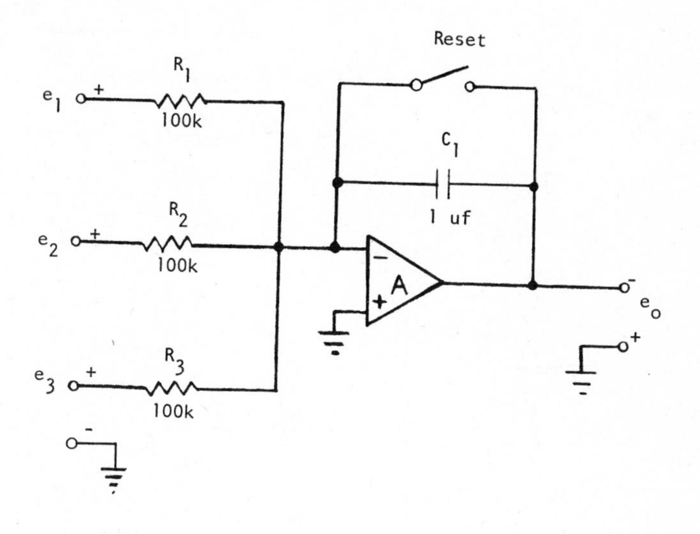

$$-e_o = \frac{1}{R_1C_1}\int e_1\,dt + \frac{1}{R_2C_1}\int e_2\,dt + \frac{1}{R_3C_1}\int e_3\,dt$$

$$= 10\int (e_1 + e_2 + e_3)\,dt\,, \qquad \text{for } R_1 = R_2 = R_3$$

All input currents are algebraically summed at the junction point. The capacitor C_1 integrates the total current and this is seen at the output as a voltage.

Initial Condition Reset Circuit

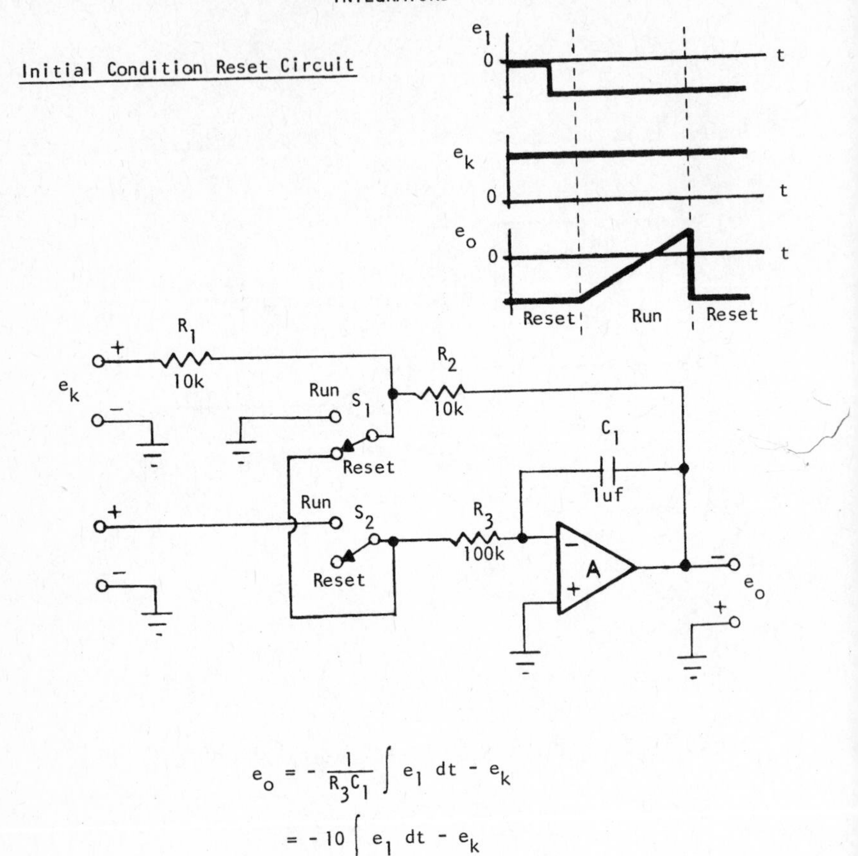

$$e_o = -\frac{1}{R_3C_1}\int e_1\,dt - e_k$$

$$= -10\int e_1\,dt - e_k$$

In this network, an initial reference state e_k shows up at the output as the constant of integration. The initial condition is established by charging the capacitor C_1 to the voltage $-e_k$; then the input e_1 is switched in to integrate. The integration will continue until the output e_o reaches the saturation level of the Op Amp or until it is reset or terminated. This method of switching minimizes the effect of switching transients at the summing junction.

Initial Condition Reset Circuit (continued)

The integrator drift may be measured by setting e_k at any known value, then switching to the integrate mode (Run) with $e_1 = 0$. Under ideal conditions, the output should not deviate from the value e_k. However, the Op Amp bias current will cause the integrator to drift.

$$\frac{i}{C_1} = \frac{\Delta e_o}{\Delta t}$$

The current i is the (-) input bias current of the Op Amp. In practical applications the integration times may vary from a few milli-seconds to several hours; therefore the choice of the type Op Amp as to input bias current will vary accordingly. Refer to the introduction at the beginning of this section and also see Offset Current and Bias Current in the section devoted to Op Amp Parameters for a more compre-hensive coverage of this subject.

In place of S_1 and S_2, electronic switching circuits may be used, such as shown in the comparator section. Excellent electronic switch modules are available commercially from Op Amp and Function Module manufacturers.

Difference Integrator (Single Op Amp)

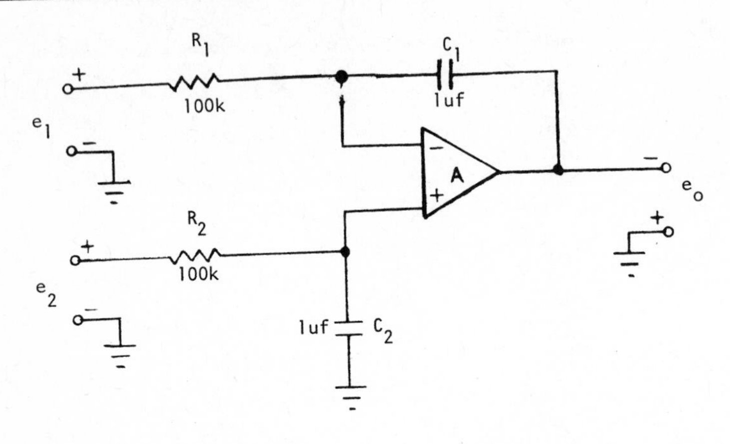

$$e_o = \frac{1}{R_1C_1}\int (e_2 - e_1)\, dt \,, \qquad \text{for } R_1 = R_2,\ C_1 = C_2$$

$$= 10 \int (e_2 - e_1)\, dt$$

Note that a common mode voltage appears at the (+) input terminal of the Op Amp. Many Chopper Stabilized Op Amps cannot be used in this configuration because internal design limitations require that the (+) input terminal be grounded for proper operation.

Difference Integrator (Inverter Type)

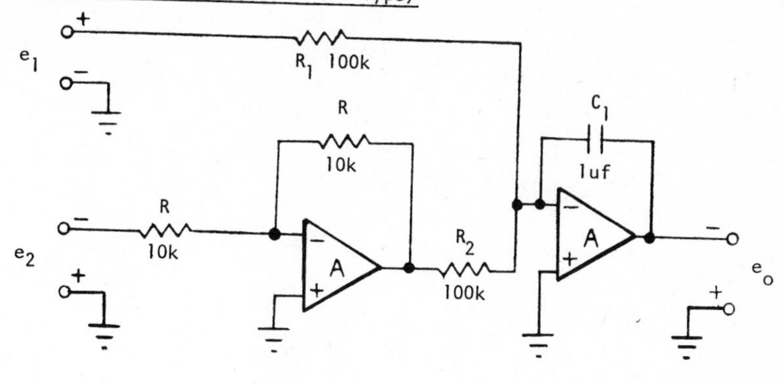

$$e_o = \frac{1}{R_1C_1}\int (e_2 - e_1)\,dt\,, \qquad \text{for } R_1 = R_2$$

$$= 10\int (e_2 - e_1)\,dt$$

The circuit offers an advantage in that high accuracy general purpoase Chopper Stabilized Op Amps may be used because all (+) input terminals are grounded.

Note that additional input resistors (R_1) may be added to yield algebraically weighted outputs.

$$-e_o = \frac{1}{R_1C_1}\int e_1\,dt + \frac{1}{R_kC_1}\int e_1\,dt + \frac{1}{R_nC_1}\int e_1\,dt$$

INTEGRATORS

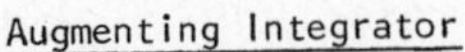

Augmenting Integrator

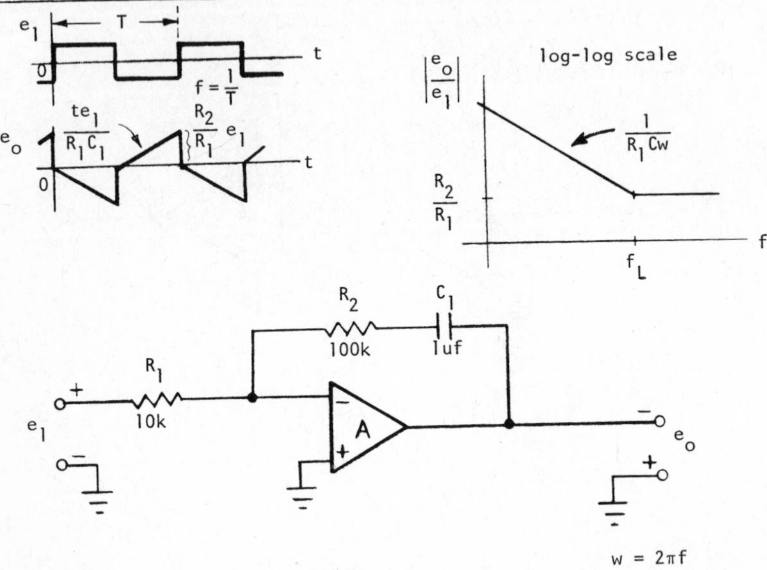

$$w = 2\pi f$$

$$-e_o = \frac{R_2}{R_1} e_1 + \frac{1}{R_1C_1} \int e_1 \, dt$$

$$= 10\ e_1 + 10 \int e_1 \, dt$$

For a square wave input, the output will respond first to the direct transition of the input signal ($e_o = - \frac{R_2}{R_1} e_1$) and then to the time integral of the input signal ($e_o = - \frac{t}{R_1C_1} e_1$).

In the frequency domain, this network acts as a dual-pass filter. The frequency breakpoint f_L occurs when R_2 equals X_{C1}.

$$f_L = \frac{1}{2\pi R_2 C_1}$$

$$= \frac{1}{2\pi(10^5)(10^{-6})} = 1.6\ \text{Hz}$$

Augmenting Integrator (continued)

The expression for the general response is

$$e_o(S) = -\frac{R_2}{R_1}\left(1 + \frac{1}{R_2 C_1 S}\right) e_1(S)$$

This circuit is adaptable to weighted inputs. Note that these inputs will affect only the output gains, and not the breakpoint.

INTEGRATORS

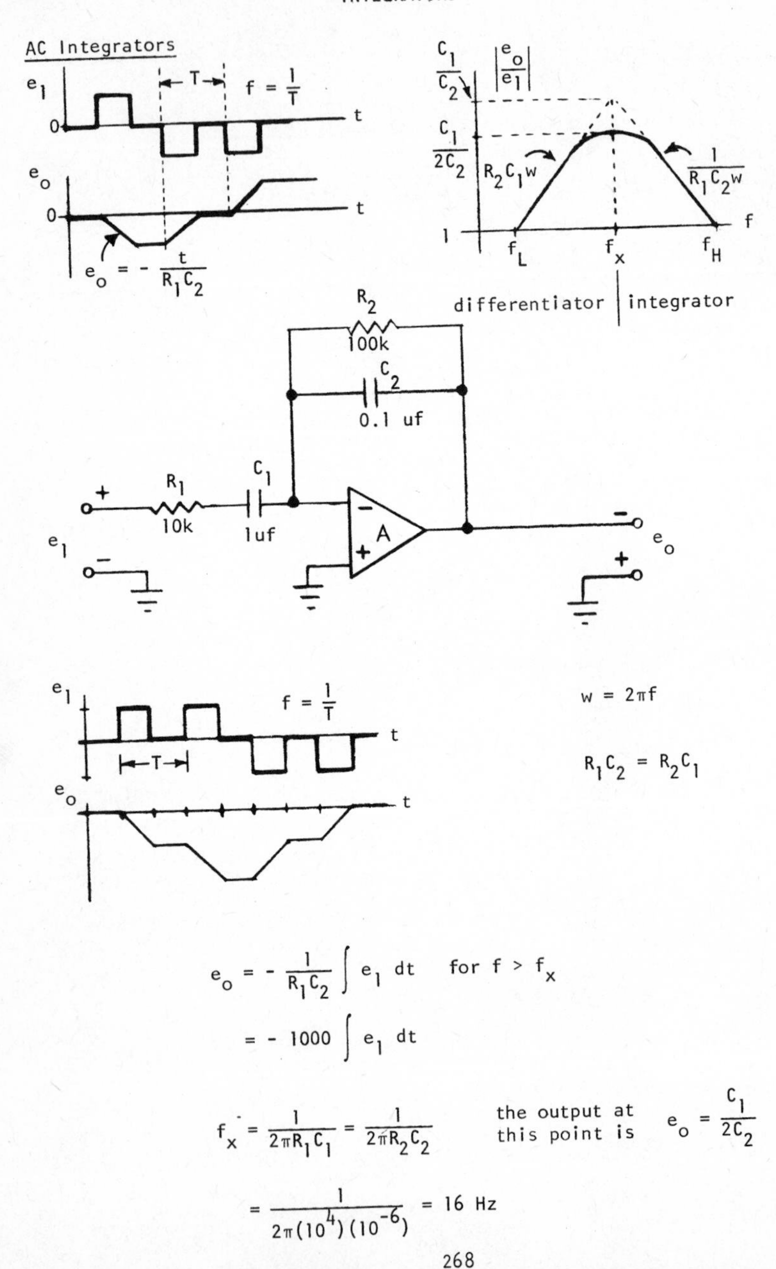

$$e_o = -\frac{1}{R_1C_2}\int e_1\,dt \quad \text{for } f > f_x$$

$$= -1000\int e_1\,dt$$

$$f_x = \frac{1}{2\pi R_1C_1} = \frac{1}{2\pi R_2C_2}$$

the output at this point is $e_o = \frac{C_1}{2C_2}$

$$= \frac{1}{2\pi(10^4)(10^{-6})} = 16\text{ Hz}$$

AC Integrators (continued)

$$f_L = \frac{1}{2\pi R_2 C_1}$$ output equals unity gain at low frequency point f_L

$$= \frac{1}{2\pi(10^5)(10^{-6})} = 1.6 \text{ Hz}$$, where $e_o = 1$

$$f_H = \frac{1}{2\pi R_1 C_2}$$ output equals unity gain at high frequency point f_H

$$= \frac{1}{2\pi(10^4)(10^{-7})} = 160 \text{ Hz}$$, where $e_o = 1$

The general expression for this network in the frequency domain is

$$e_o(S) = -\frac{1}{R_1 C_2 S}\left(\frac{\tau S}{1 + \tau S}\right)^2 e_1(S)$$

where $\tau = R_1 C_2 = R_2 C_1$

Since the network is a non-ideal integrator right at f_x, the actual operating frequency should extend beyond the center frequency to a value of at least twice f_x.

AC Integrator

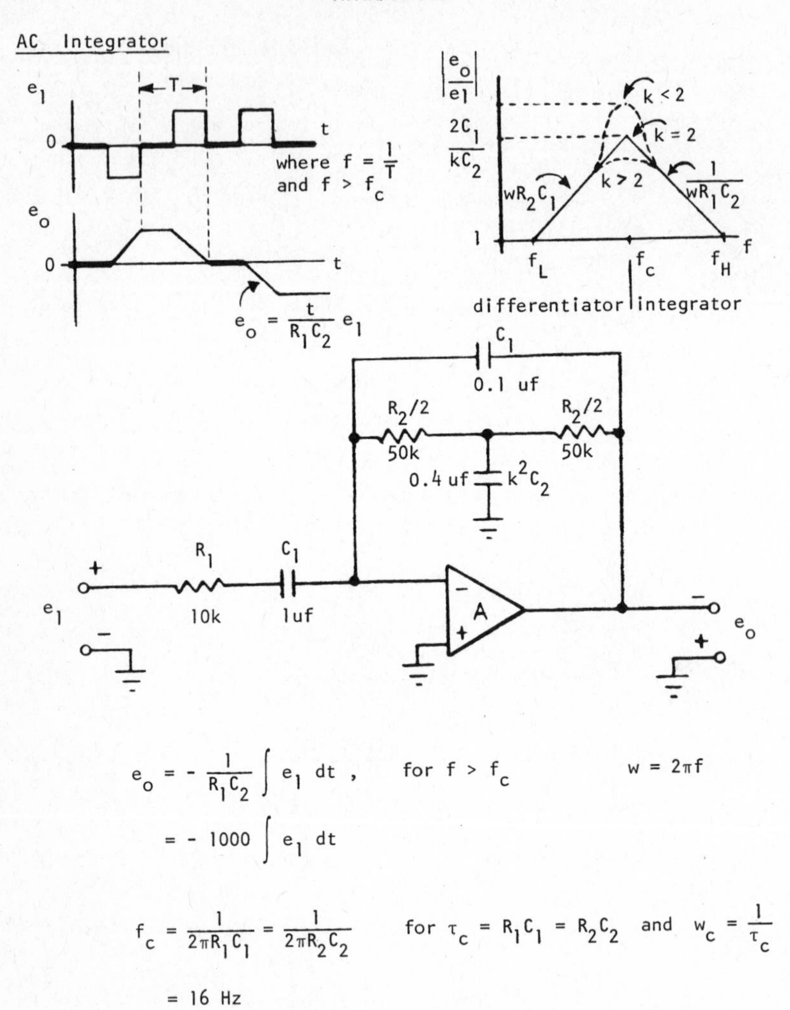

$$e_o = -\frac{1}{R_1C_2}\int e_1\,dt\,, \qquad \text{for } f > f_c \qquad w = 2\pi f$$

$$= -\,1000\int e_1\,dt$$

$$f_c = \frac{1}{2\pi R_1C_1} = \frac{1}{2\pi R_2C_2} \qquad \text{for } \tau_c = R_1C_1 = R_2C_2 \text{ and } w_c = \frac{1}{\tau_c}$$

$$= 16 \text{ Hz}$$

For optimum theoretical accuracy, the value for k is two (2), where

$$k = 2\sqrt{\frac{R_1C_1}{R_2C_2}}$$

AC Integrator (continued)

The limiting output voltage at f_c will then be

$$|e_o| = \frac{C_1}{C_2} |e_1|$$

$$f_L = \frac{1}{2\pi R_2 C_1}$$

$$= 1.6 \text{ Hz}$$

output equals unity gain at low frequency point f_L

$$e_o = 1$$

occurs where $R_2 = X_{C1}$

$$f_H = \frac{1}{2\pi R_1 C_2}$$

$$= 160 \text{ Hz}$$

output equals unity gain at high frequency point f_H

$$e_o = 1$$

occurs where $R_1 = X_{C2}$

The general expression for this network in the frequency domain is

$$e_o(s) = -\frac{e_1(S)}{\dfrac{1}{R_2 C_1 S} + \dfrac{C_2}{C_1} + R_1 C_2 S}, \qquad S = jw, \qquad k = 2\sqrt{\frac{R_1 C_1}{R_2 C_2}}$$

The basic advantage of this circuit is that the peak value of the output at frequency f_c can be optimized for nearly ideal performance ($k = 2$).

Double Integration

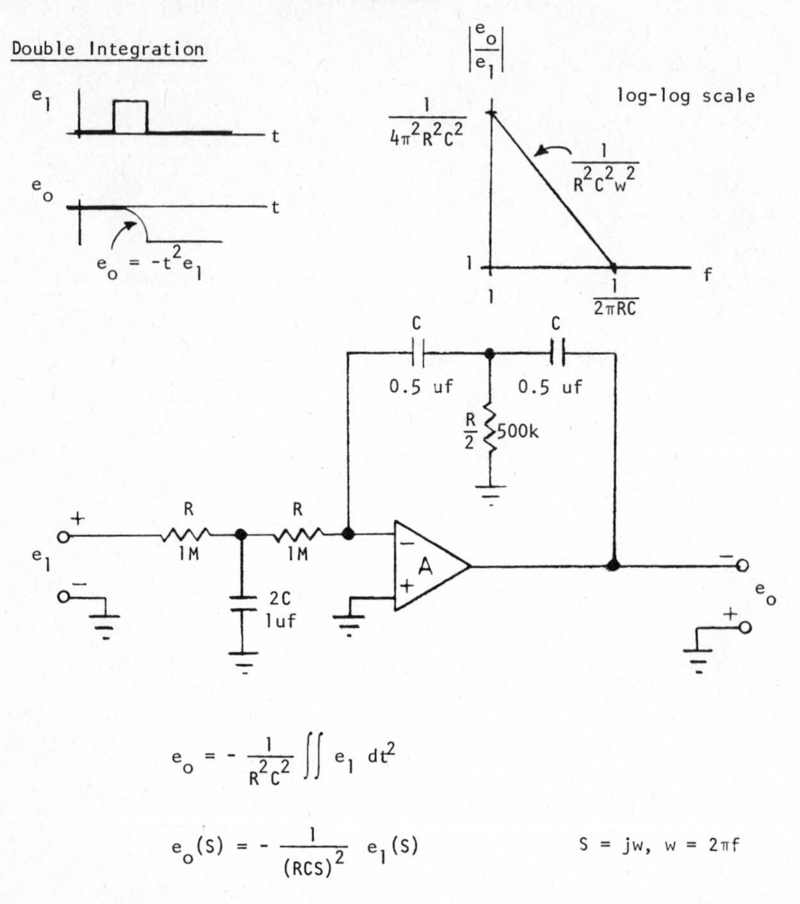

$$e_o = -\frac{1}{R^2C^2}\iint e_1\,dt^2$$

$$e_o(S) = -\frac{1}{(RCS)^2}\,e_1(S) \qquad S = jw,\ w = 2\pi f$$

This network integrates twice using only one Op Amp. For the derivation of the above equations refer to Appendix I.

Slew Rate Limiter

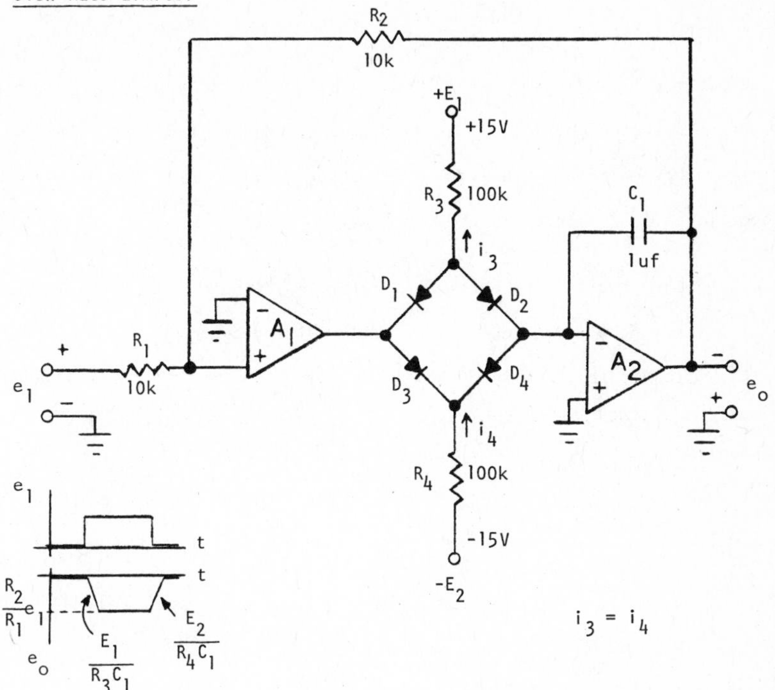

$$i_3 = i_4$$

$$e_o = -\frac{R_2}{R_1} e_1$$

$$(-)\text{ Slew Rate} = \frac{\Delta e_o}{\Delta t} = -\frac{E_1}{R_3 C_1}, \quad \text{for } e_1 \text{ positive going}$$

$$= \frac{-15}{(100k)(1\mu f)} = -1.5 \text{ V/sec}$$

$$(+)\text{ Slew Rate} = \frac{\Delta e_o}{\Delta t} = -\frac{E_2}{R_4 C_1}, \quad \text{for } e_1 \text{ negative going}$$

$$= \frac{-(-15)}{(100k)(1\mu f)} = +1.5 \text{ V/sec}$$

Slew Rate Limiter (continued)

When the network is in equilibrium, the output of A_1 is ideally zero, the diode bridge is balanced, and therefore no current flows into the summing junction of A_2. The resulting output is a constant voltage proportional to the input ($e_o = -\frac{R_2}{R_1} e_1$).

However, should the input make a relatively sudden "positive" change, the output of A_1 will go to (+) saturation, diverting the current in D_3 to flow into A_1, thus causing D_1 and D_4 to cutoff. This, in turn, makes the current from R_3 flow into the summing junction of A_2 causing C_1 to charge to a new level.

$$i_3 = -C_1 \frac{de_o}{dt}$$

$$i_3 = \frac{E_1}{R_3}$$

$$\text{Slew Rate} = \frac{de_o}{dt} = -\frac{E_1}{R_3 C_1}$$

The same phenomenon occurs for negative input charges except diodes D_3 and D_2 cutoff and i_4 flows into the summing junction.

For optimum operation the bridge network should be precisely matched at equilibrium.

Capacitance Multiplier (Miller Effect)

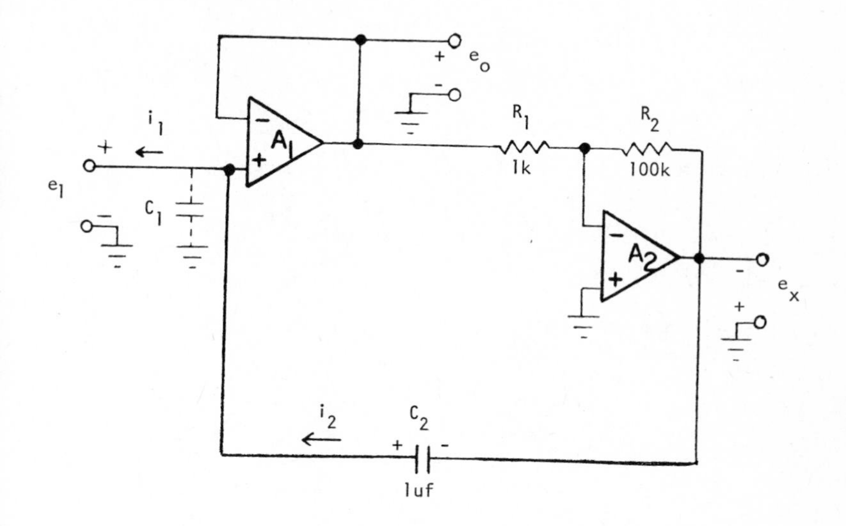

$$e_1 = e_o$$

$$C_1 = C_2 \left(1 + \frac{R_2}{R_1}\right)$$

$$\cong 100 \text{ uf}$$

$$e_x = -\frac{R_2}{R_1} e_1$$

$$= -100\ e_1$$

$$i_1 = C_i \frac{de_i}{dt}$$

$$i_2 = C_2 \frac{d(e_1 - e_x)}{dt}$$

$$i_1 = i_2$$

$$C_1 \frac{de_1}{dt} = C_2 \frac{de_1}{dt} - C_2 \frac{de_x}{dt}$$

$$de_x = -\frac{R_2}{R_1} de_1$$

$$C_1 = C_2 \left(1 + \frac{R_2}{R_2}\right)$$

This network synthesizes the feedback capacitor C_2 (or any other impedance) reflected back to the input e_1. Large capacitance can be derived from small capacitors by simply using high gain $(\frac{R_2}{R_1})$. The design

Capacitance Multiplier (continued)

trade-off occurs at the output e_x. High gain will produce a high output voltage which will quickly saturate the Op Amp A_2. In this situation A_2 should be a high voltage type Op Amp. These devices are presently available commercially with output over ±100V in low cost solid state modules.

The network is very adaptable to the high quality polystyrene capacitor. This type capacitor inherently has a high breakdown voltage; therefore, a very large high quality capacitor can be simulated at the input (e_1).

If the input terminals are connected to the output of an Op Amp Current Source (refer to section on DC Current Output), an excellent integrator network will be the result. The integrator output should be taken from the output of the buffer amplifier e_o, which offers excellent isolation.

$$i_1 = C_1 \frac{de_1}{dt}$$

where i_1 is the output of the Op Amp Current Source

$$e_1 = \frac{1}{C_1} \int i_1 \, dt$$

and since $e_1 = e_o$

$$e_o = \frac{1}{C_1} \int i_1 \, dt$$

Charge Amplifier

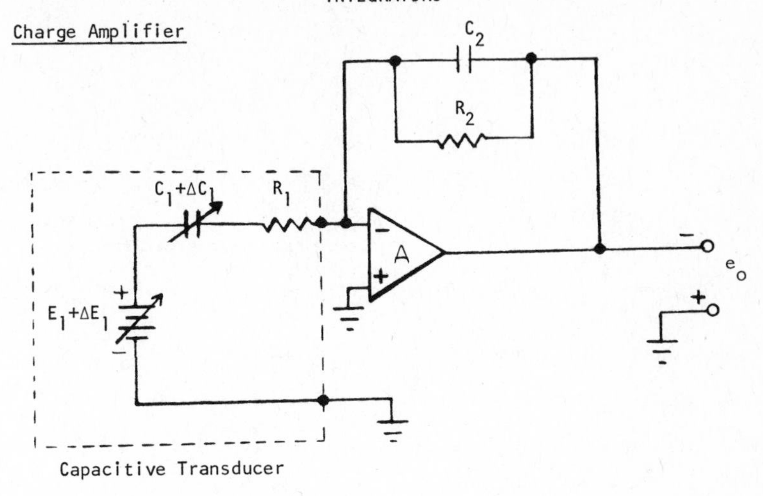

$$-e_o = \frac{E_1}{C_2}\int dC_1 + \frac{C_1}{C_2}\int dE_1$$

$$-\Delta e_o = \frac{E_1}{C_2}\Delta C_1 + \frac{C_1}{C_2}\Delta E_1$$

for

$$Q_2 = C_2\, e_o$$

$$dQ_2 = C_2\, de_o$$

$$Q_1 = C_1 E_1$$

$$dQ_1 = E_1\, dC_1 + C_1\, dE_1$$

Since

$$i_1 = -i_2$$

$$\frac{dQ_1}{dt} = -\frac{dQ_2}{dt}$$

or

$$dQ_1 = -\, dQ_2$$

$$E_1 dC_1 + C_1 dE_1 = -\, C_2 de_o$$

$$-de_o = \frac{E_1}{C_2} dC_1 + \frac{C_1}{C_2} dE_1$$

(derivation of above output e_o)

Charge Amplifier (continued)

This network is very useful in applications where a transducer energy conversion results in a varying electrical charge.

Examples of such devices are the common capacitance microphone (input circuit shown) and piezoelectric transducers (ΔE_1 omitted).

Note that either a change in C_1 or a change in E_1 will cause a change in output voltage. This phenomenon holds true because both reflect a change in charge ($\frac{dQ_1}{dt} = i_1$).

This network is virtually insensitive to shunt capacitance across the input. This capacitance is seen across the summing junction where the voltage is essentially zero, and therefore the change in charge is negligible. This inherent characteristic permits the use of long shielded cables between the Op Amp and the transducer without causing any significant output error.

A resistor may be required across capacitor C_2 in order to prevent the output from drifting into saturation due to the Op Amp bias current effects. The highest gain is achieved when capacitor C_2 is small; however the reactance of this capacitor must be small compared to R_2. at the lowest frequency of interest.

Because of the above design trade-offs it is usually best to use FET or Chopper Stabilized Op Amp for these applications.

Introduction

Differentiators generally deal with quick output signal changes for slow input time periods. The design problems encountered in the networks are therefore more likely to be susceptible to noise and high frequency instability parameters than most other general purpose Op Amp circuits. However, the other error parameters such as drift and leakage must also be dealt with just as in the case with integrators.

In fact, the most basic Op Amp differentiator circuit is dynamically unstable.

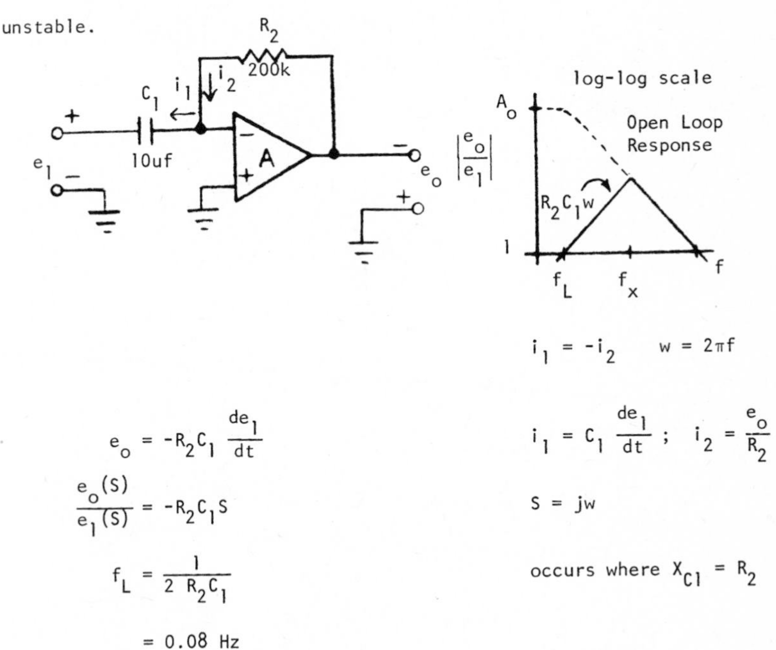

$i_1 = -i_2 \qquad w = 2\pi f$

$e_o = -R_2C_1 \dfrac{de_1}{dt}$

$i_1 = C_1 \dfrac{de_1}{dt} \; ; \quad i_2 = \dfrac{e_o}{R_2}$

$\dfrac{e_o(S)}{e_1(S)} = -R_2C_1S$

$S = jw$

$f_L = \dfrac{1}{2\ R_2C_1}$

$= 0.08$ Hz

occurs where $X_{C1} = R_2$

Introduction (continued)

Although this network is theoretically an ideal differentiator (for $f < f_x$), oscillations will occur due to the phase shift characteristics at frequency f_x. At this point, the summing junction sees a 360° phase shift: 90° from C_1; 180° from the DC inverting amplifier; and 90° from the standard Op Amp open loop frequency response. The total phase angle will cause the network to oscillate at f_x.

The purpose of showing the simple differentiator circuit is to demonstrate the overall fundamental design concepts. The differentiator characteristic time, R_2C_1, should be selected to give a full scale output when the input signal is producing its maximum rate of change.

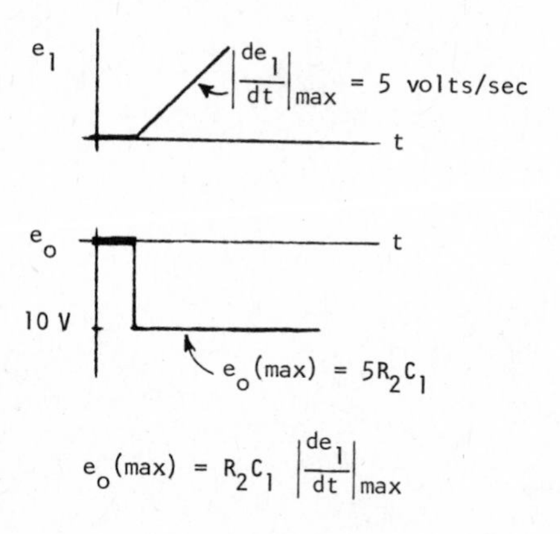

$$e_o(max) = R_2C_1 \left|\frac{de_1}{dt}\right|_{max}$$

Introduction (continued)

For a full scale ±10V output, and an input ramp of 5 volts/sec,

$$R_2C_1 = \frac{|e_o|_{max}}{\left|\frac{de_1}{dt}\right|_{max}}$$

$$= \frac{10V}{5V/sec} = 2 \text{ sec}$$

The design trade-off is now limited to the actual values of R_2 and C_1. As in the case with integrators (see Integrator Section, Capacitor Dielectric Leakage), the maximum capacitor value should not exceed 1 to 10 μf. This limits the resistor values to range from 200 kΩ to 2 MΩ. Using the lower resistance value will minimize any Op Amp bias current error at the output.

In general, it is wise to clamp the output of a differentiator to avoid off-scale saturation due to possible extraneous input transients.

The D.C. output errors due to the Op Amp input offset voltage V_{off} and bias current I_{BIAS} are seen directly.

$$e_o(\text{error}) = V_{off} + R_2 I_{BIAS}$$

The frequency stability problem is easily solved by placing an appropriate size resistor in series with C_1. This will cause a frequency breakpoint to occur before f_x is reached, thus greatly reducing the phase shift at f_x and eliminating oscillations.

Introduction (continued)

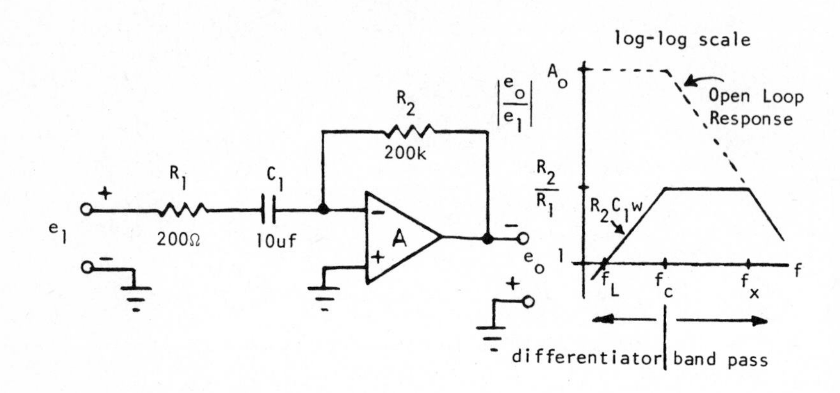

$$f_c = \frac{1}{2\pi R_1 C_1} \qquad \text{stabilizing breakpoint}$$

$$= 80 \text{ Hz}$$

Beyond the frequency f_c, the network no longer acts as a differentiator. The output gain at this point reaches its high.

$$\frac{e_o}{e_1} = \frac{R_2}{R_1}$$

$$= 1000$$

Although the network is stable, this circuit will contain a high degree of output noise. The high gain $(\frac{R_2}{R_1})$ band pass section of the response, beyond f_c, is the cause.

By decreasing the bandwidth of the band pass section, the noise can be reduced substantially. The addition of another frequency breakpoint between f_c and f_x will reduce the bandwidth. This is accomplished

Introduction (continued)

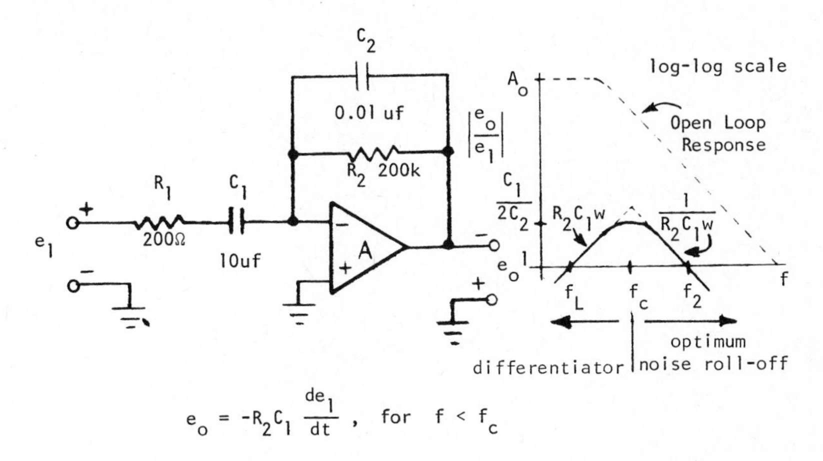

$$e_o = -R_2C_1 \frac{de_1}{dt}, \quad \text{for} \quad f < f_c$$

by connecting a capacitor C_2 across the feedback resistor R_2. The optimum breakpoint location for this purpose is as close to f_c as possible, or f_c itself ($R_2C_2 = R_1C_1$).

The general response equation for this network is

$$\frac{e_o(S)}{e_1(S)} = -\frac{R_2C_1S}{(1+R_1C_1S)(1+R_2C_2S)} \qquad S = jw$$

For $R_1C_1 = R_2C_2 = \tau_c$

$$\frac{e_o(S)}{e_1(S)} = -\frac{R_2C_1S}{(1+\tau_cS)^2}$$

This is the optimum response for achieving frequency stability with minimum noise output. (It should be realized that the design trade-off is output slew rate; if faster response is required, then the R_2C_2 breakpoint should be extended beyond f_c.)

Introduction (continued)

Further examination of the general response equation can be used to obtain a quantitative error analysis. Structuring the equation in terms of the radian frequency w (S = jw) will clearly show the phase and magnitude errors.

magnitude: $$-\left|\frac{e_o}{e_1}\right| = \underbrace{R_2 C_1 w}_{\text{ideal}} \underbrace{\left|\frac{1}{1+\left(\frac{w}{w_c}\right)^2}\right|}_{\text{error}}, \quad \text{where } w_c = \frac{1}{\tau_c}, \quad w_c = 2\pi f_c$$

phase: $$\phi = \underbrace{90°}_{\text{ideal}} - \underbrace{2\tan^{-1}\frac{w}{w_c}}_{\text{error}},$$ where ϕ is the output phase shift (lead) angle in degrees

At the point where $w = w_c$, the shift is essentially zero; this is good for stability purposes. However, the phase error is 90 degrees from the ideal value and the magnitude is one half the ideal value.

$$w = w_c = 2\pi f_c$$

$$w_c = \frac{1}{R_1 C_1} = \frac{1}{R_2 C_2}$$

$$\left|\frac{e_o}{e_1}\right| = \frac{R_2 C_1 w_c}{2}$$

$$\left|\frac{e_o}{e_1}\right| = \frac{R_2}{2R_1} = \frac{C_1}{2C_2},$$

at the point where $f = f_c$

Introduction (continued)

The following chart can be derived from the procedure to give basic design guidelines.

Differentiator Frequency vs. Error

Frequency f	$0.01\ f_c$	$0.1\ f_c$	f_c	$10\ f_c$	$100\ f_c$
Magnitude error*	-0.001%	-1%	-50%	-99%	-100%
Phase error*	-1.1°	-11.4°	-90°	-169°	-179°

*departure from ideal

The application of a RC "T network" in place of R_2 can be used to increase the accuracy in the vicinity of f_c. The network is shown later in circuit collections of this section.

DIFFERENTIATORS

Basic Differentiator

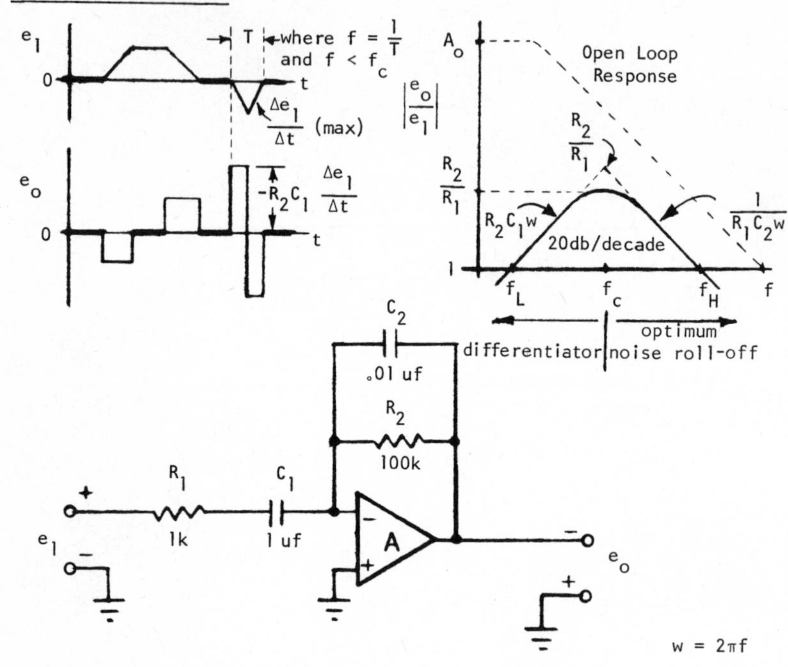

w = 2πf

$$e_o = -R_2C_1 \frac{de_1}{dt}, \qquad \text{for } f < f_c$$

$$= -\frac{1}{10} \frac{de_1}{dt}$$

$$f_L = \frac{1}{2\pi R_2C_1}, \qquad \text{occurs where } R_2 = X_{C1}$$

$$= 1.6 \text{ Hz}$$

$$f_c = \frac{1}{2\pi R_1C_1} = \frac{1}{2\pi R_2C_2},$$ occurs where $R_1 = X_{C1}$ and $R_2 = X_{C2}$ since $R_1C_1 = R_2C_2$

$$= 160 \text{ Hz}$$

Basic Differentiator (continued)

$$f_H = \frac{1}{2\pi R_1 C_2}, \qquad \text{occurs where } R_1 = X_{C2}$$

$$= 16 \text{ kHz}$$

The maximum rate of change of the input signal for full scale output ($e_o(max) = \pm 10V$) is

$$\left|\frac{de}{dt}\right|_{max} = \frac{e_o(max)}{R_2 C_1}$$

$$= 100 \text{ V/sec}$$

The general expression for the overall response of this network is

$$-\frac{e_o(S)}{e_1(S)} = \frac{R_2 C_1 S}{(1 + R_1 C_1 S)(1 + R_2 C_2 S)}, \qquad \text{where } S = jw \text{ and } w = 2\pi f$$

$$= \frac{R_2 C_1 S}{(1 + \tau_c S)^2} \qquad \text{where } \tau_c = R_1 C_1 = R_2 C_2 = \frac{1}{w_c}$$

Notice that the output is a square wave for a triangular, or ramp, input signal. In this situation, the rise and fall time of the output edges will be slew rate limited by the action of R_1 and C_2. These edges may be "sharpened" by separating one of the breakpoints from f_c. That is, let R_1, C_1 remain at f_c, but move R_2, C_2 to a higher frequency. This can be done simply by reducing the value of C_2. The design trade-off will be an increase in noise signal at the output.

For a more comprehensive explanation of the fundamentals behind this network refer to the Introduction of this section.

Resistor T Network Differentiator

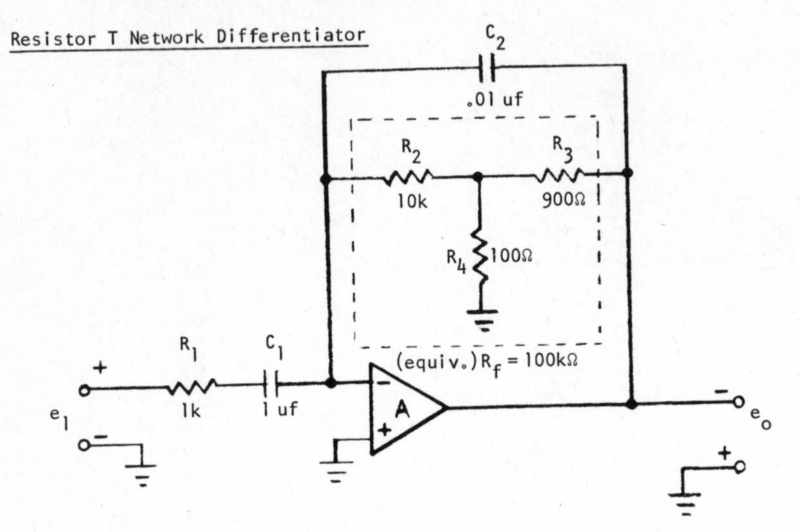

$$e_o = -R_fC_1 \frac{de_1}{dt}, \qquad \text{for } f < f_c \text{ where } R_f = R_2\left(1 + \frac{R_3}{R_4}\right)$$

$$= -\frac{1}{10} \frac{de_1}{dt}$$

$$f_L = \frac{1}{2\pi R_fC_1}, \qquad \text{occurs where } R_f = X_{C1}$$

$$= 1.6 \text{ Hz}$$

$$f_c = \frac{1}{2\pi R_1C_1} = \frac{1}{2\pi R_fC_2}, \qquad \text{occurs where } R_1 = X_{C1} \text{ and } R_f = X_{C2} \text{ since } R_1C_1 = R_2C_2$$

$$= 160 \text{ Hz}$$

$$f_H = \frac{1}{2\pi R_1C_2}, \qquad \text{occurs where } R_1 = X_{C2}$$

$$= 16 \text{ kHz}$$

Resistor T Network Differentiator (continued)

This circuit is identical in concept to the preceding Basic Differentiator. The T network (R_2, R_3, R_4) has replaced a single feedback resistor. (See DC Amplifier Resistor T Network and also Appendix F.) The equivalent single feedback resistance R_f is

$$R_f = R_2 \left(1 + \frac{R_3}{R_4}\right)$$

$$= 100\text{k}\Omega$$

The advantage of this technique is that smaller value precision resistors (less expensive) can be used to achieve the same output gain. This is especially useful when dealing with long-term differentiators that require large megohm resistors. It is best to keep R_4 at least 10 times smaller than R_2 in order to maintain the equation accuracy.

DIFFERENTIATORS

RC T-Network Differentiators

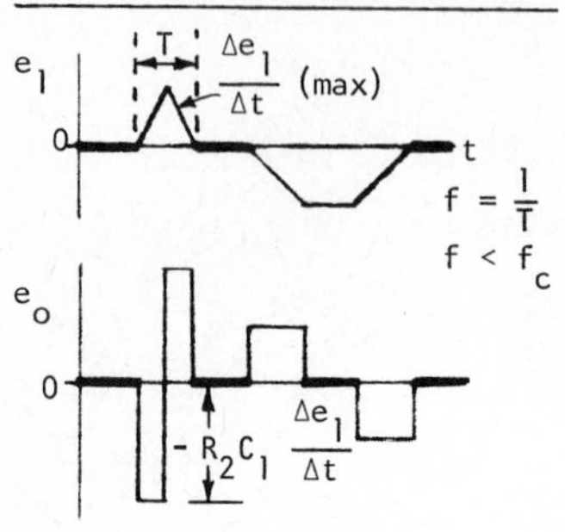

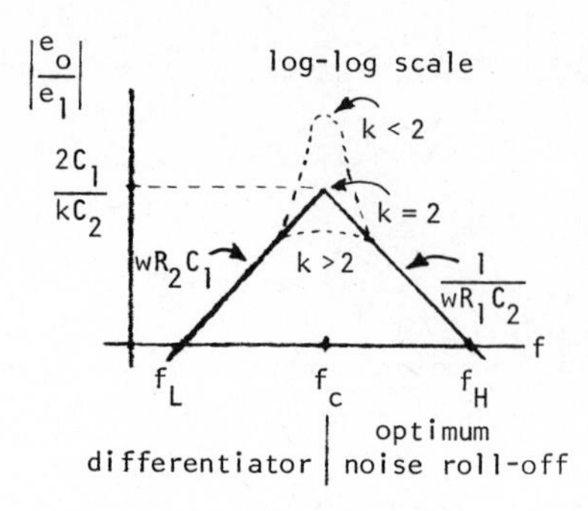

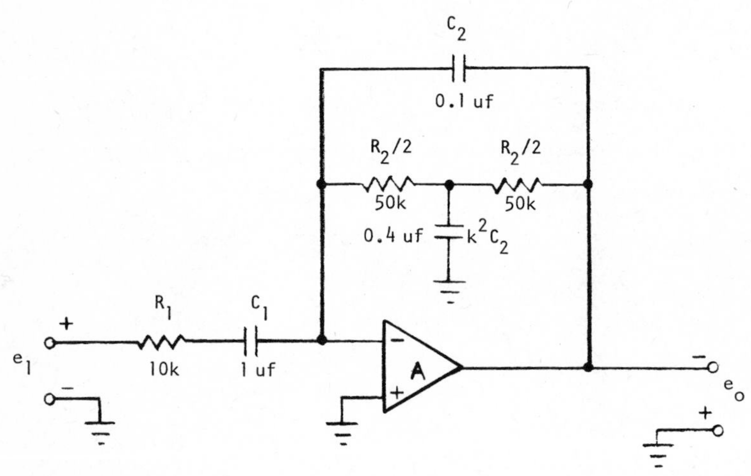

$w = 2\pi f$

$$e_o = -R_2C_1 \frac{de_1}{dt}, \qquad \text{for } f < f_c$$

$$= -\frac{1}{10} \frac{de_1}{dt}$$

$$f_c = \frac{1}{2\pi R_1C_1} = \frac{1}{2\pi R_2C_2}, \qquad \text{for } \tau_c = R_1C_1 = R_2C_2 \text{ and } w_c = 1/\tau_c$$

$$= 16 \text{ Hz}$$

RC T-Network Differentiator (continued)

$$f_L = \frac{1}{2\pi R_2 C_1}, \qquad \text{occurs where } R_2 = X_{C1}$$

$$= 1.6 \text{ Hz}$$

$$f_H = \frac{1}{2\pi R_1 C_2}, \qquad \text{occurs where } R_1 = X_{C2}$$

$$= 160 \text{ Hz}$$

For optimum theoretical accuracy, the value for k is two (2), where

$$k = 2\sqrt{\frac{R_1 C_1}{R_2 C_2}}$$

The limiting output voltage at f_c will then be

$$|e_o| = \frac{C_1}{C_2} |e_1|$$

The general expression for this network in the frequency domain is

$$\frac{e_o(S)}{e_1(S)} = \frac{1}{\frac{1}{R_2 C_1 S} + \frac{C_2}{C_1} + R_1 C_2 S}$$

This circuit has the advantage that it can attain optimum performance while using only one Op Amp.

DIFFERENTIATORS

DC Coupled Differentiator

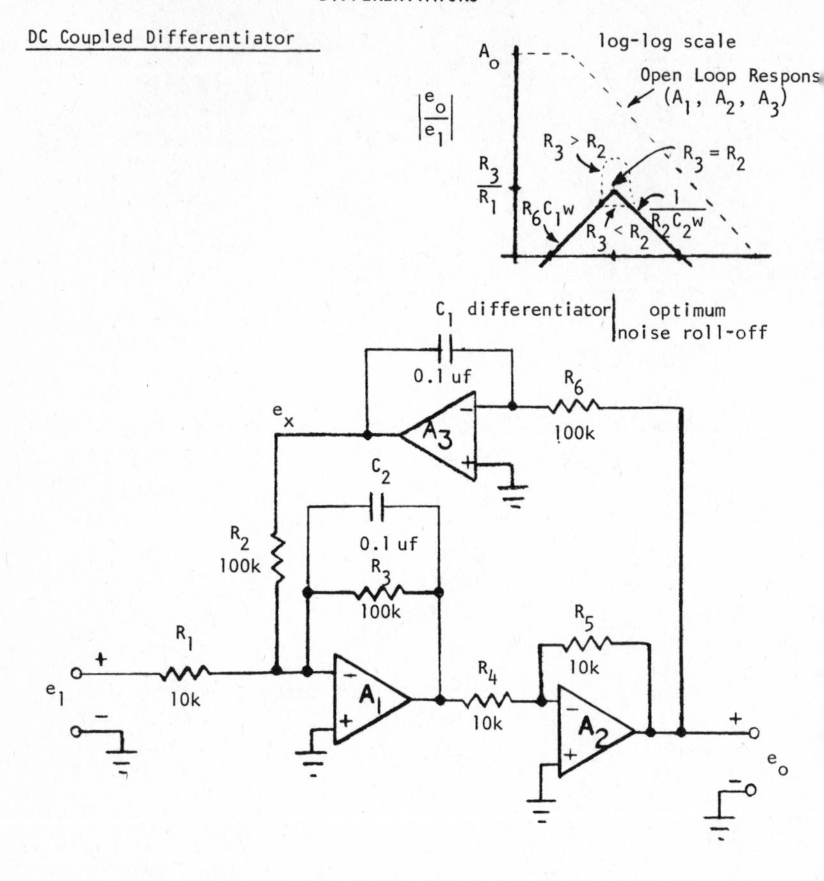

$$e_o = \left(\frac{R_2}{R_1}\right) R_6 C_1 \frac{de_1}{dt}, \qquad \text{for } f < f_c \text{ and } R_4 = R_5,\ w = 2\pi f$$

$$= \frac{1}{10}\frac{de_1}{dt}$$

$$f_c = \frac{1}{2\pi R_c C_1}$$

where $R_6 C_1 = R_2 C_2$
for optimum noise roll-off

$$= 16 \text{ Hz}$$

DC Coupled Differentiator (continued)

$$f_L = \left(\frac{R_1}{R_2}\right) \frac{1}{2\pi R_6 C_1} = \frac{R_1}{R_2} f_c$$

$$= 1.6 \text{ Hz}$$

$$f_H = \left(\frac{R_2}{R_1}\right) \frac{1}{2\pi R_2 C_2} = \frac{R_2}{R_1} f_c$$

$$= 160 \text{ Hz}$$

The value of R_3 can be adjusted for optimum peaking at f_c. The optimum theoretical value is

$$R_3(\text{opt}) = R_2$$

Note that the capacitor size can be reduced by increasing the ratio $\frac{R_2}{R_1}$. This makes an excellent long-term differentiator.

The general expression for the overall network is

$$\frac{e_o(S)}{e_1(S)} = \left(\frac{R_2}{R_1}\right) \frac{1}{\frac{1}{R_6 C_1 S} + \frac{R_2}{R_3} + R_2 C_2 S}$$

and for $\tau_c = R_6 C_1 = R_2 C_2$

$$\frac{e_o(S)}{e_1(S)} = \left(\frac{R_2}{R_1}\right) \frac{1}{\frac{1}{\tau_c S} + \frac{R_2}{R_3} + \tau_c S} \qquad \text{where } \tau_c = \frac{1}{w_c} \text{ and } w_c = 2\pi f_c$$

For the optimum theoretical response R_3 should equal R_2.

$$\frac{R_2}{R_3} = 1$$

DIFFERENTIATORS

Summing Differentiator

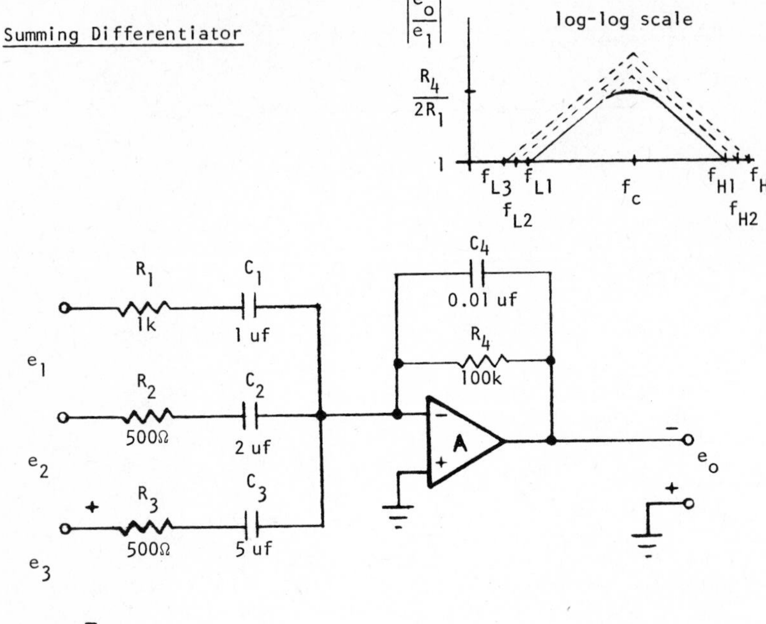

$$-e_o = R_4C_1 \frac{de_1}{dt} + R_4C_2 \frac{de_2}{dt} + R_4C_3 \frac{de_3}{dt}$$

$$-e_o(S) = R_2C_1S\ e_1(S) + R_4C_2S\ e_2(S) + R_4C_3S\ e_3(S)\ ,$$ where $S = jw$ and $w = 2\pi f$

$$f_c = \frac{1}{2\pi R_4C_4}\ ,$$ for $R_4C_4 = R_1C_1 = R_2C_2 = R_3C_3$

$$= 160 \text{ Hz}$$

$$f_{L1} = \frac{1}{2\pi R_4C_1}$$ occurs where $R_4 = X_{C1}$

$$= 1.6 \text{ Hz}$$

Summing Differentiator (continued)

$$f_{L2} = \frac{1}{2\pi R_4 C_2} \qquad \text{and} \qquad f_{L3} = \frac{1}{2\pi R_4 C_3}$$

$$= 0.8 \text{ Hz} \qquad\qquad = 0.3 \text{ Hz}$$

$$f_{H1} = \frac{1}{2\pi R_1 C_4}, \qquad \text{occurs where } R_1 = X_{C4}$$

$$= 16 \text{ Hz}$$

$$f_{H2} = \frac{1}{2\pi R_2 C_4} \qquad \text{and} \qquad f_{H3} = \frac{1}{2\pi R_3 C_4}$$

$$= 32 \text{ kHz} \qquad\qquad = 80 \text{ kHz}$$

The general expression for this network over its entire frequency range is

$$-\frac{e_o(S)}{e_1(S)} = \frac{1}{(1+\tau_c S)^2}\left[R_4C_1S\, e_1(S) + R_4C_2S\, e_2(S) + R_4C_3S\, e_3(S)\right]$$

where $\tau_c = R_4C_4$

and $R_4C_4 = R_1C_1 = R_2C_2 = R_3C_3$

DIFFERENTIATORS

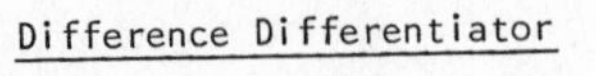

Difference Differentiator

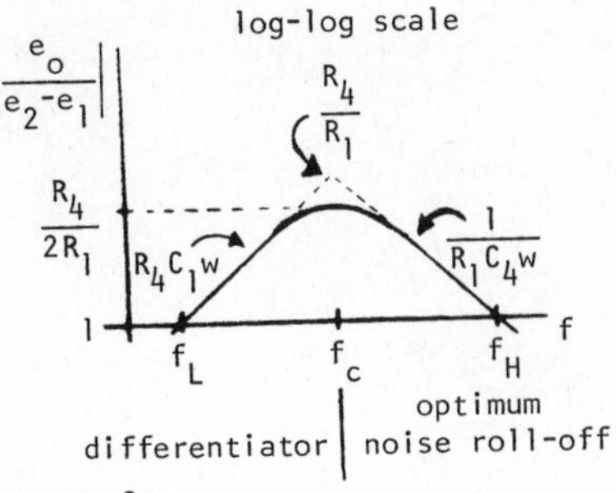

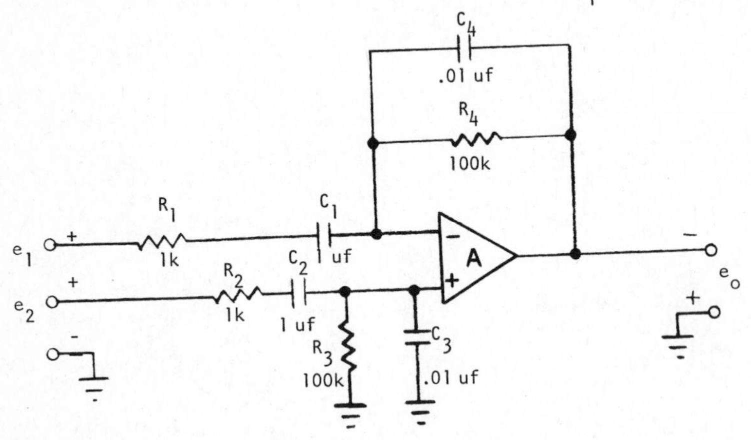

$$e_o = R_4C_1\left(\frac{de_2}{dt} - \frac{de_1}{dt}\right),$$

for $f < f_c$ and $R_1 = R_2$, $C_1 = C_2$, $R_3 = R_4$, $C_3 = C_4$, $\tau_c = R_4C_4 = R_1C_1 = \frac{1}{w}$

$w = 2\pi f$

$$e_o(S) = R_4C_1S\ [e_2(S) - e_1(S)] \quad S = jw$$

$$|e_o| = \frac{w}{10}(e_2 - e_1),$$

for $w < w_c$, or $f < f_c$

$$f_c = \frac{1}{2\pi R_4C_4}$$

$$= 160\ \text{Hz}$$

Difference Differentiator (continued)

$$f_L = \frac{1}{2\pi R_4 C_1}, \qquad \text{occurs where } R_4 = X_{C1}$$

$$= 1.6 \text{ Hz}$$

$$f_H = \frac{1}{2\pi R_1 C_4}, \qquad \text{occurs where } R_1 = X_{C4}$$

$$= 16 \text{ kHz}$$

The general expression for this network over the entire frequency range is

$$e_o(S) = \frac{R_4 C_1 W}{(1 + \tau_c S)^2} [e_2(S) - e_1(S)] \, , \qquad \text{where } {}_c = R_4 C_4 \quad \text{and } w_c = \frac{1}{\tau_c}$$

Care should be taken not to exceed the input common mode voltage limits of the Op Amp at the (+) input terminal.

DIFFERENTIATORS

Augmenting Differentiator

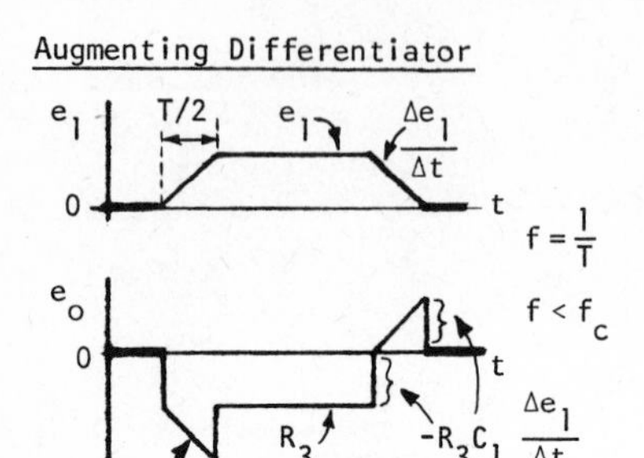

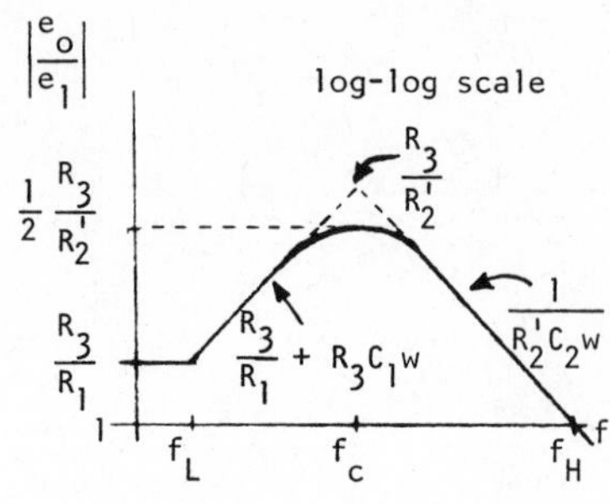

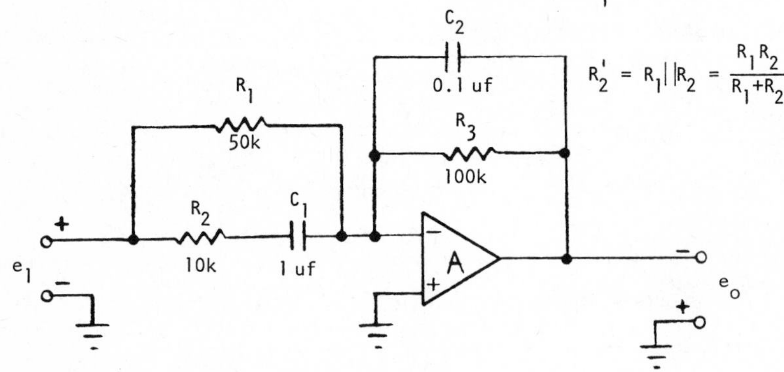

$$R_2' = R_1 || R_2 = \frac{R_1 R_2}{R_1 + R_2}$$

$$-e_o = \frac{R_3}{R_1} e_1 + R_3 C_1 \frac{de_1}{dt} \qquad \text{for } f < f_c$$

$$\left|\frac{e_o}{e_1}\right| = \frac{R_3}{R_1} + R_3 C_1 w \qquad w = 2\pi f$$

$$f_L = \frac{1}{2\pi R_1 C_1} = 3.2 \text{ Hz}$$

occurs where $R_1 = X_{C1}$

$$f_c = \frac{1}{2\pi R_2 C_1} = \frac{1}{2\pi R_3 C_2} = 16 \text{ Hz}$$

occurs where $R_2 = X_{C1}$ and $R_3 = X_{C2}$ for optimum noise roll-off

Augmenting Differentiator (continued)

$$f_H = \frac{1}{2\pi R_2 C_2} \qquad \text{occurs at } e_o = 1\text{, where } R_2 = X_{C2}$$

$$= 160 \text{ Hz}$$

The easiest method for understanding the network frequency response is to break it up into separate segments along the frequency spectrum.

First, at DC, the major contributing components are simply R_1 and R_3. Therefore, the output gain is

$$\left|\frac{e_o}{e_1}\right| = \frac{R_3}{R_1} \qquad 0 < f < f_L$$

Next, C_1 comes into play. The first breakpoint occurs where $R_1 = X_{C1}$. From this point, the output responds to both R_1 and C_1.

$$\left|\frac{e_o}{e_1}\right| = \frac{R_3}{R_1} + \frac{R_3}{X_{C1}} \qquad f_L < f < f_c$$

$$= \frac{R_2}{R_1} + R_3C_1w$$

Then, X_{C1} begins to fall out of the picture and R_2 comes into play. At f_c the two meet.

$$R_2 = X_{C2} \qquad \text{and} \qquad f_c = \frac{1}{2\pi R_2 C_2}$$

At this point the network begins to look like a pure resistive circuit (R_1, R_2, and R_3). The input is the parallel combination of R_1 and R_2. The output feedback impedance is supply R_3. Therefore, at this mid-frequency point (peak point at f_c), the theoretical gain is

Augmenting Differentiator (continued)

$$\left|\frac{e_o}{e_1}\right| = \frac{R_3}{R_2'}$$ at mid-frequency where $$R_2' = \frac{R_1 R_2}{R_1 + R_2}$$

The output does not actually reach this peak value at f_c since this is only the theoretical breakpoint. The actual output voltage magnitude at f_c is one half this value due to the 45 degree phase angle between R_2 and X_{C1}

$$\left|\frac{e_o}{e_1}\right| = \frac{1}{2}\frac{R_3}{R_2'} \quad \text{at } f_c$$

Finally, the last breakpoint (f_x) will appear when C_2 comes into play with R_3.

$$R_3 = X_{C2} \quad \text{and} \quad f_x = \frac{1}{2\pi R_3 C_2} \quad \text{(final breakpoint)}$$

The point can be located anywhere beyond f_c, or right at f_c for optimum noise roll-off ($f_c = f_x$).

Notice that beyond f_x, the network transforms into an integrator. The equivalent circuit becomes simply R_2' at the input, and C_2 (only) in the feedback.

$$\frac{e_o}{e_1} = \frac{X_{C2}}{R_2'} = \frac{1}{R_2' C_2 w}$$

This integrator will limit the output slew rate. Extending f_x further out (decreasing C_2) will permit a faster slew rate at the output at the expense of increased output noise.

HOLDING NETWORKS

Introduction

In the analog world, storage and memory are accomplished by "holding" a charge in a capacitor.

Designing these networks involves dealing with the common parameters associated with capacitors as well as Op Amp leakages. In addition, peripheral circuitry switching times (RC time constants) must be taken into account when high speed data sampling is the design objective.

In particular, the major error parameters that must be confronted in Op Amp memory storage circuits are:

(1) Op Amp Input Bias Current

(2) Op Amp Input Offset Voltage

(3) Capacitor Dielectric Leakage

(4) Capacitor Dielectric Absorption

(5) Acquisition and Aperture Times

Refer to the section on Network Parameters for a more detailed discussion of the Op Amp parameters.

Op Amp Input Bias Current. This current will continuously charge the holding capacitor at a rate directly proportional to its value. It can be seen that the bias current can cause a very serious error when long holding durations are required.

FET input Op Amps are usually recommended for these applications. These type Op Amps achieve input bias performances ranged from 100 pA down to less than 1 pA.

Introduction (continued)

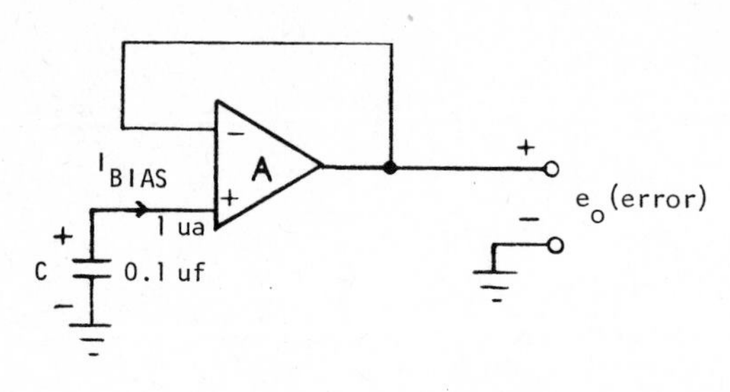

$$e_o(\text{error}) = \frac{I_{BIAS}}{C}\,t$$

$$= 10\,t\,, \quad \text{or 10 volts/sec}$$

Op Amp Input Offset Voltage. The offset voltage error is reflected to the output directly. This error can be eliminated completely at any given temperature by external offset adjustment circuitry (usually a simple pot). Refer to the Op Amp specifications sheet for the particular circuit configuration.

$$e_o(\text{error}) = V(\text{offset})$$

Capacitor Dielectric Leakage. The leakage resistance R_c across the holding capacitor will cause the output to "decay" exponentially (time constant equals RC). For E_c = 10V, the output will drop to less than 3V in ten seconds.

Introduction (continued)

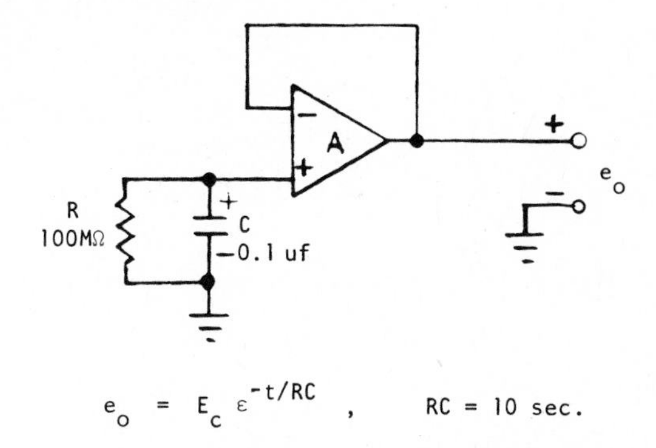

$$e_o = E_c\, \varepsilon^{-t/RC}, \qquad RC = 10 \text{ sec.}$$

The polystyrene and polycarbonate type capacitors, respectively, offer the best characteristics for minimizing this error. Generally it is not recommended to use capacitance values greater than 1 μf to 10 μf. The leakage trade-offs usually become excessive beyond these values.

When ultra-low leakage is required it is advisable to chose a "round" cylindrical polystyrene capacitor rather than an "oval" type. The mechanical stresses involved in manufacturing "oval" type capacitors are often very detrimental to ultra-low leakage performance.

Refer to Appendix C for low leakage circuit layout design considerations.

Capacitor Dielectric Absorption. This parameter creates an output error that is quite subtle. The effect is generally important when 0.1% storage accuracies are a design requirement.

Introduction (continued)

As explained earlier in the INTEGRATOR section of the text, the dielectric absorption of a capacitor can be analogous to residual magnetism. This phenomenon can adversely affect the ability of a capacitor to accurately store a precise voltage.

A neutral capacitor when connected to a known voltage source will not retain the exact value of that source voltage after the source voltage has been removed. The retained voltage value will be slightly less than the applied value due to the fact that a few of the dielectric molecular dipoles will "snap" back to their original neutral positions.

Conversely, when a capacitor is short circuited in order to reduce its voltage to zero, a residual charge remains in the dielectric material. This leaves a residual error voltage across the capacitor after the short is removed.

The accuracy of a memory storage network is highly dependent on the ability of the storage capacitor to retain a precise voltage. Capacitors exhibiting low dielectric absorption are recommended for designs of memory networks.

The best capacitors for minimizing this effect are the polystyrene type. However, they have the main disadvantage of being physically large. Smaller size can be attained by using a polycarbonate type with the disadvantage of a few percent degradation in dielectric absorption. Much sharper degradation comes with the smaller mylar types. The worst

Introduction (continued)

devices, from the author's experience, have been ceramic and silver mica types, respectively.

The designer must make the approximate trade-off between physical space and accuracy when it comes to the dielectric absorption parameter.

Acquisition and Aperture Times. A finite amount of time is required for the "hold" network to "acquire" and "track" the input signals accurately. This transition time is commonly called the Acquisition Time. It is primarily made up of the time in which the holding capacitor takes to charge. However, the delay time of the gating circuitry must also be included.

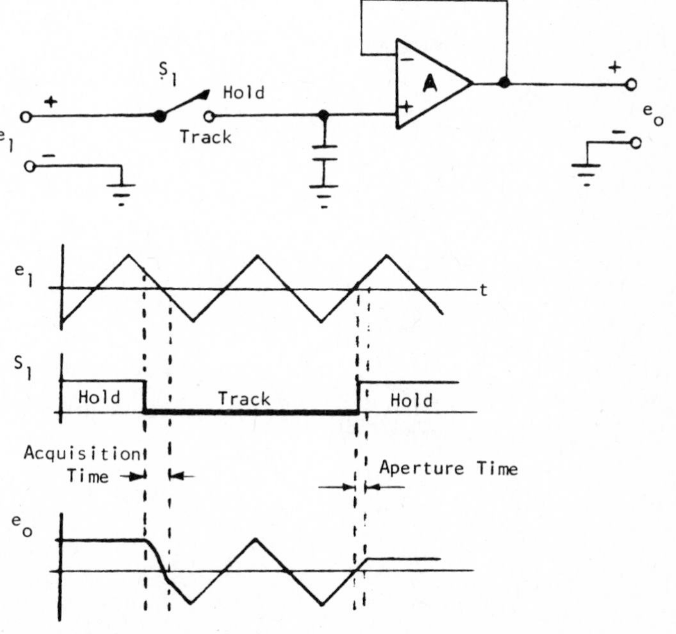

Introduction (continued)

Conversely, an incremental amount of time lapses when the network switches from "track" to "hold". This time delay is usually much smaller than the acquisition time and is referred to as the Aperture Time. It is the turn-off time of the control switch and therefore generally consists of only the delay time of gating circuitry.

It can be seen that fast acquisition can be a difficult design problem particularly when a large holding capacitor (long storage requirement) is used.

This problem is reduced substantially by using current amplifiers (gated buffer) in the switching circuitry in order to increase the charging rate of the capacitor (slew rate).

$$\frac{\Delta e}{\Delta t} = \frac{I}{C}$$

Basic Track/Hold Network

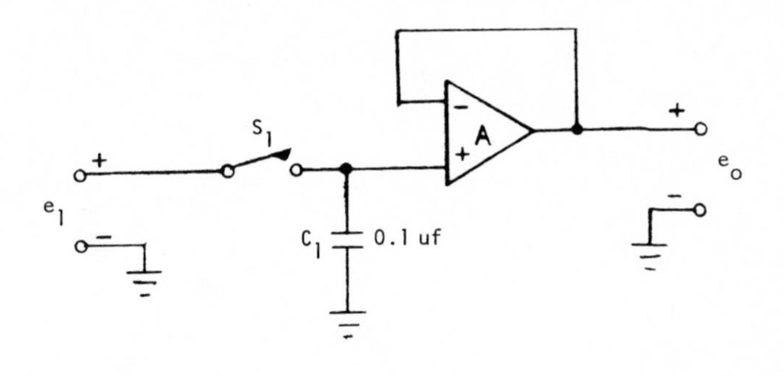

$e_o = e_1$,   for S_1 closed

e_o = Hold ,   for S_1 open

This circuit offers the basic advantage of simplicity. The major disadvantage is the input impedance is low.

Basically, the input e_1 instantaneously becomes a short circuit through capacitor C_1 when S_1 is closed. This loading action may be substantially reduced by placing a unity gain buffer network at the input.

Precision Track/Hold Network

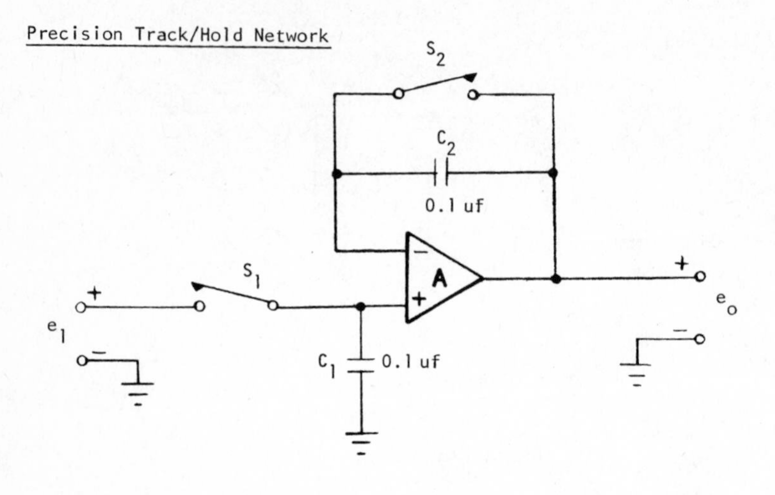

$e_o = e_1$, for S_1 and S_2 closed

e_o = Hold , for S_1 and S_2 open

If the bias current into C_2 is equal to the bias current into C_1, the corresponding error voltage generated across each capacitor will cancel and the output e_o will remain unaffected. The circuit limitation is reached when the voltage across C_1 (actual plus error at the (+) input) reaches the input common mode voltage limitation of the Op Amp. The network is extremely effective when used with FET input Op Amps.

When carefully designed, this network can store signals accurately for days at a time. Refer to Appendix C for low leakage design considerations.

Precision Sample/Hold (FET Switched)

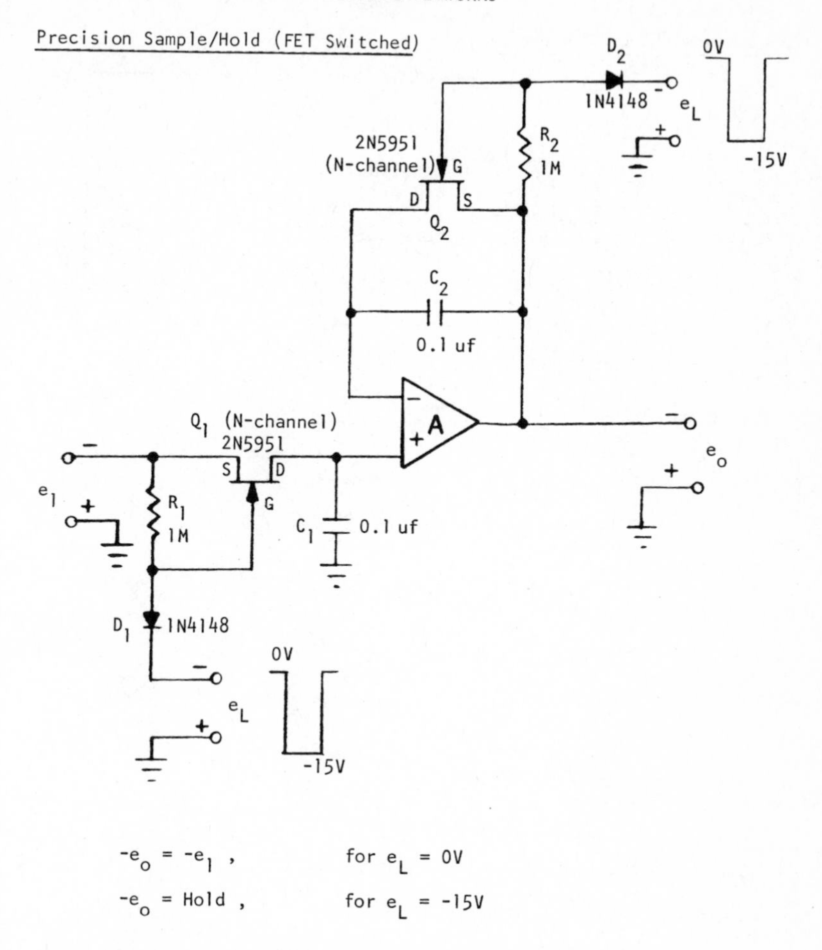

$-e_o = -e_1$, for $e_L = 0V$

$-e_o = \text{Hold}$, for $e_L = -15V$

This network operates only for negative input signals. (Refer to Appendix H for e_L level shifting circuit.) For positive signals Q_1, Q_2 must be P-channel FETs, diodes D_1, D_2 must be reversed, and switching level e_L polarity must be 0 to +15V.

HOLDING NETWORKS

Track/Hold Network (FET Switched)

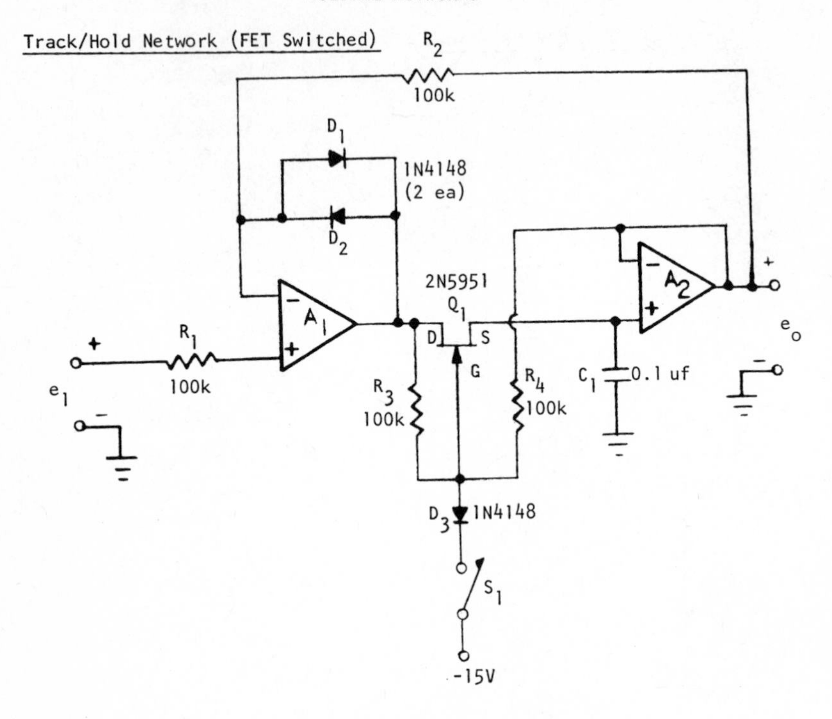

$e_o = e_1$, for S_1 open

e_o = Hold , for S_1 closed

Because of the symmetry of the N-channel junction FET, it is able to uniquely serve as a bilateral switch. Note that both R_3 and R_4 are tied to low output impedance points leaving the holding capacitor C_1 practically ideally isolated during the "Hold" mode.

Resistors R_1 and R_2 isolate the Op Amp A_1 during the non-linear region of operation (during switching and hold mode). Diodes D_1 and D_2 keep A_1 from saturating (thus improving switching times) during the non-linear "hold" mode. The switch S_1 may be replaced by transistor gating circuitry.

Track/Hold Network

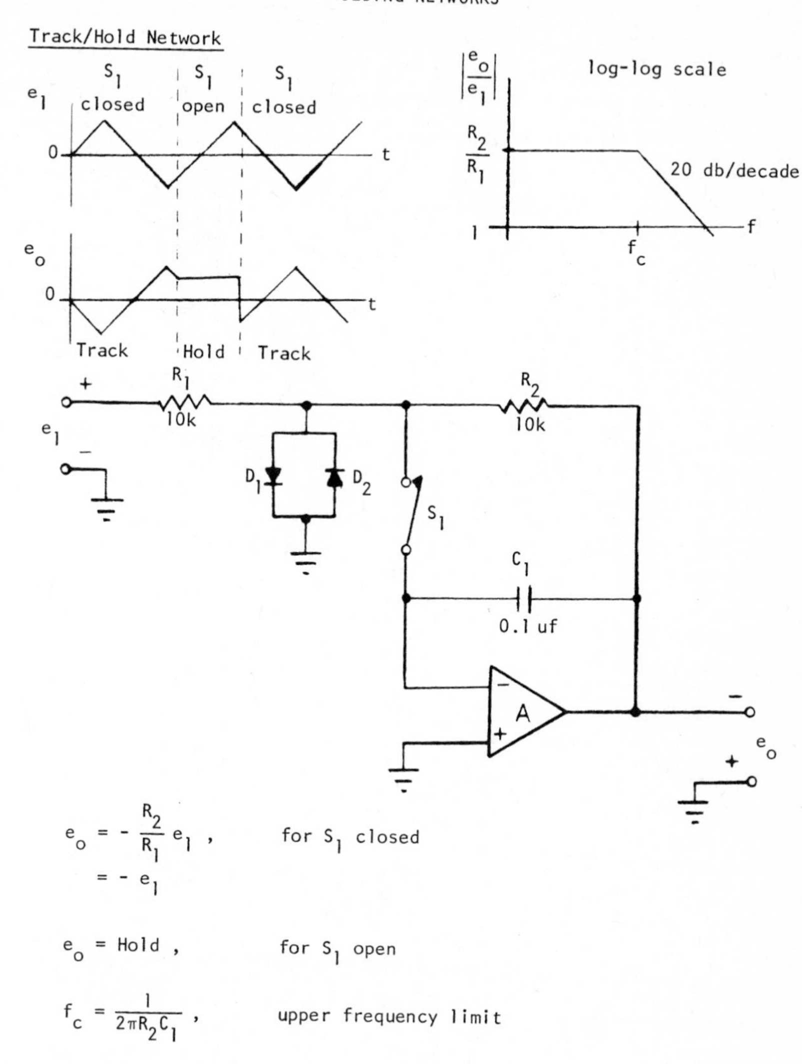

$$e_o = -\frac{R_2}{R_1} e_1 , \qquad \text{for } S_1 \text{ closed}$$

$$= - e_1$$

$$e_o = \text{Hold} , \qquad \text{for } S_1 \text{ open}$$

$$f_c = \frac{1}{2\pi R_2 C_1} , \qquad \text{upper frequency limit}$$

The upper frequency limit is caused by the time constant R_2C_1 which limits the acquisition time when switching to the track mode. Diodes D_1 and D_2 keep the summing point near ground to minimize leakage across S_1.

Sample/Hold Electronically Switched

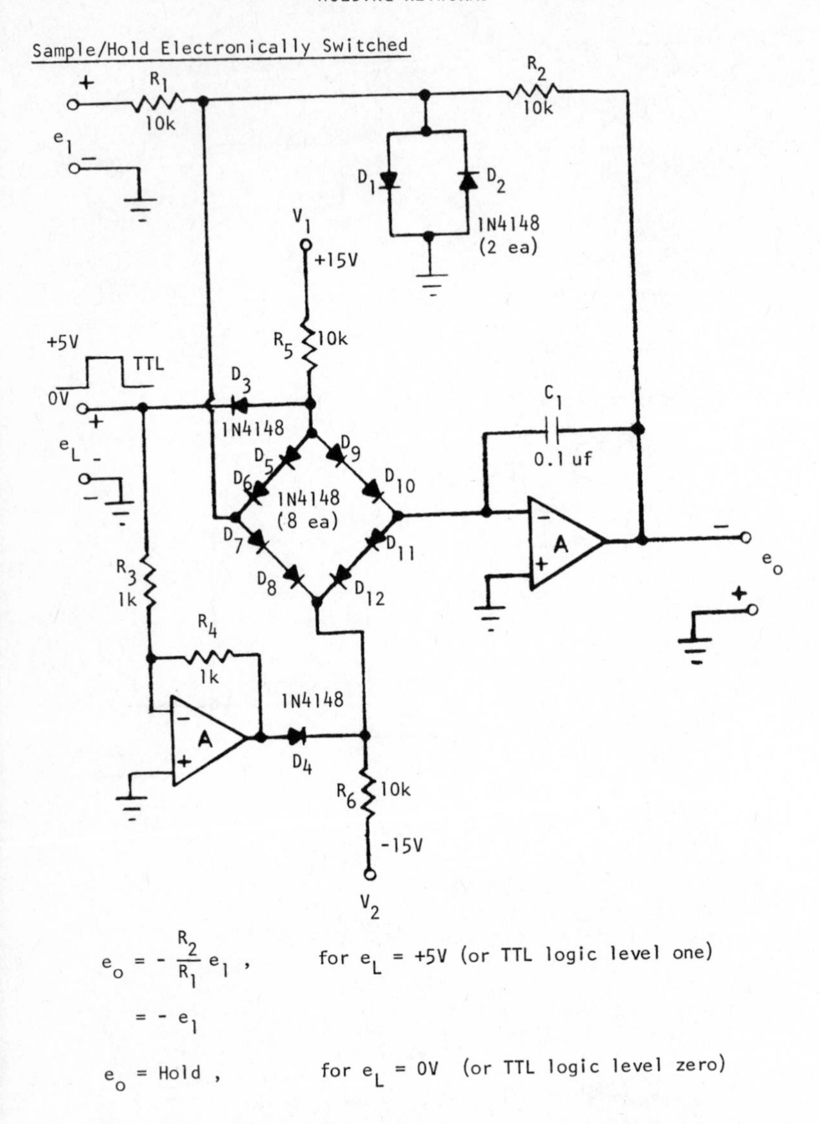

$$e_o = -\frac{R_2}{R_1} e_1 , \qquad \text{for } e_L = +5V \text{ (or TTL logic level one)}$$

$$= - e_1$$

$$e_o = \text{Hold} , \qquad \text{for } e_L = 0V \text{ (or TTL logic level zero)}$$

Diodes $D_6 - D_{12}$ should be matched for accurate operation of the switch. For more detail in designing the switching circuitry refer to the section on Comparators.

Sample/Hold Electronically Switched

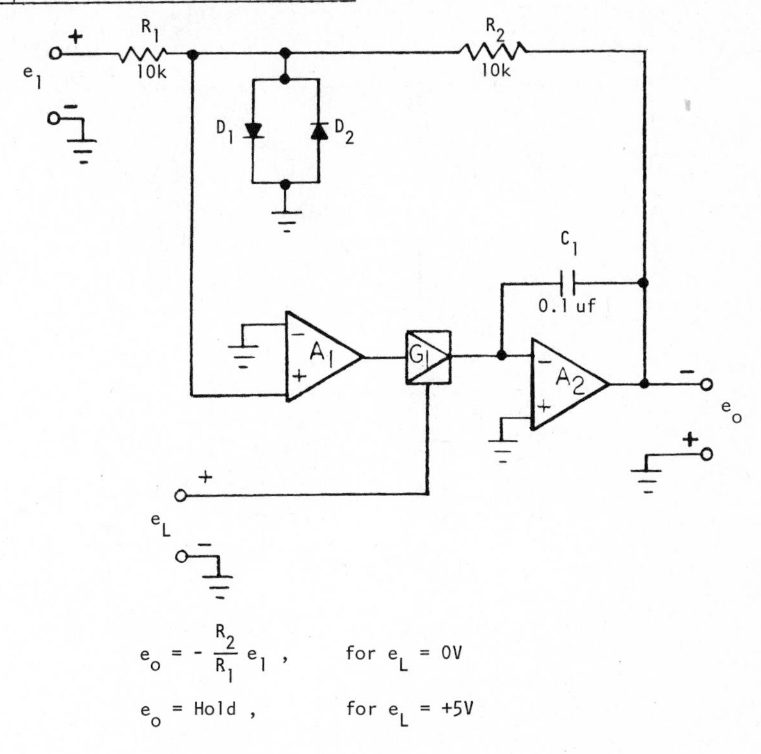

$$e_O = -\frac{R_2}{R_1} e_1 , \qquad \text{for } e_L = 0V$$

$$e_O = \text{Hold} , \qquad \text{for } e_L = +5V$$

Op Amp A_1 is followed by a gated buffer current amplifier G_1. This additional current improves acquisition time. (The fundamental circuit design of the gated buffer G_1 is shown in Appendix N.)

The integrated package of both the Op Amp A_1 and the gated buffer G_1 can be purchased as a single "electronic switch module" from most Op Amp manufacturers.

A slight variation of the above circuit allows the switch to operate within its own closed loop. This improves accuracy and output impedance.

Sample/Hold Electronically Switched

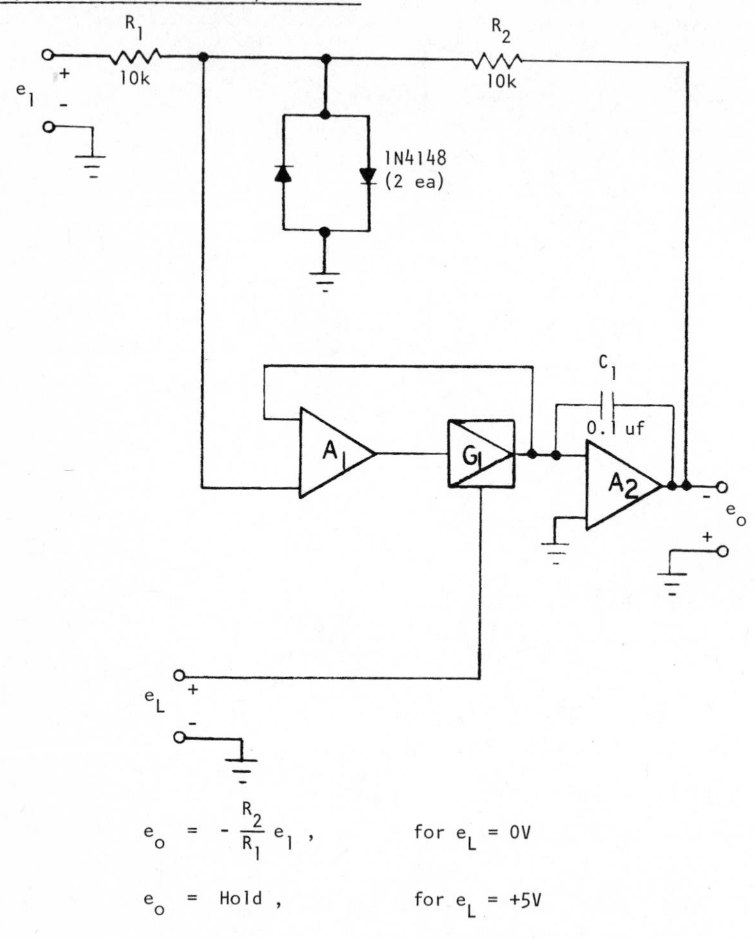

$$e_o = -\frac{R_2}{R_1} e_1 , \qquad \text{for } e_L = 0V$$

$$e_o = \text{Hold} , \qquad \text{for } e_L = +5V$$

This circuit is identical to the previous network except that the gated buffer G_1 is incorporated into the closed loop of A_1. This lowers the output impedance of G_1, which improves switching time and accuracy of the network.

Integrate/Hold Network

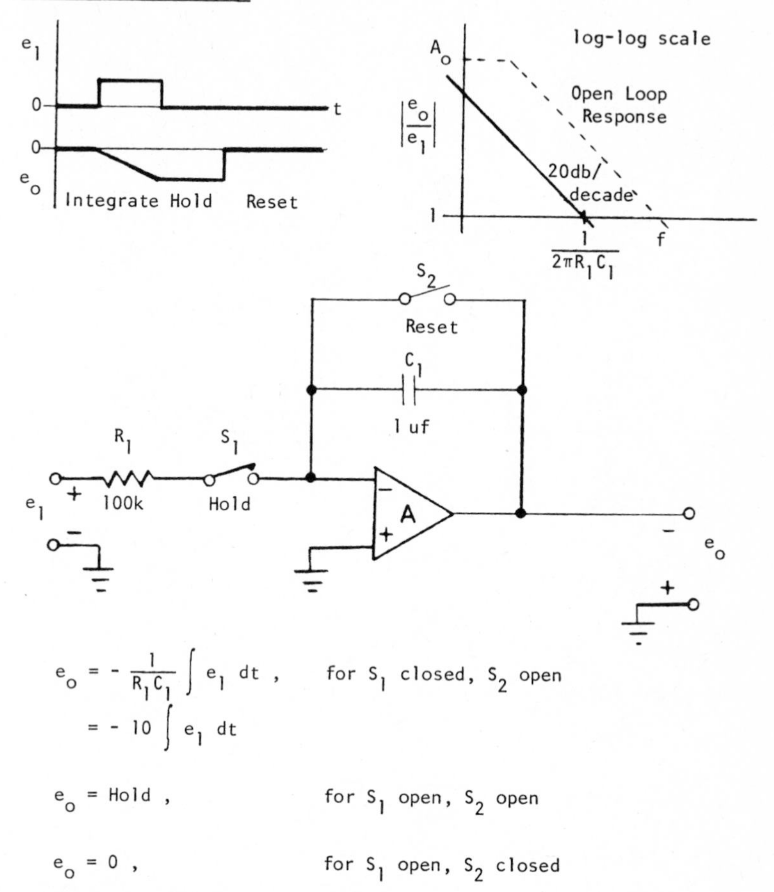

$$e_o = -\frac{1}{R_1 C_1}\int e_1\, dt, \qquad \text{for } S_1 \text{ closed, } S_2 \text{ open}$$

$$= -10\int e_1\, dt$$

$$e_o = \text{Hold}, \qquad \text{for } S_1 \text{ open, } S_2 \text{ open}$$

$$e_o = 0, \qquad \text{for } S_1 \text{ open, } S_2 \text{ closed}$$

Using a polystyrene capacitor and a good quality low leakage FET, or Chopper Stabilized Op Amp, this network can "hold" for hours to within 1% accuracy.

Simple Peak Detector

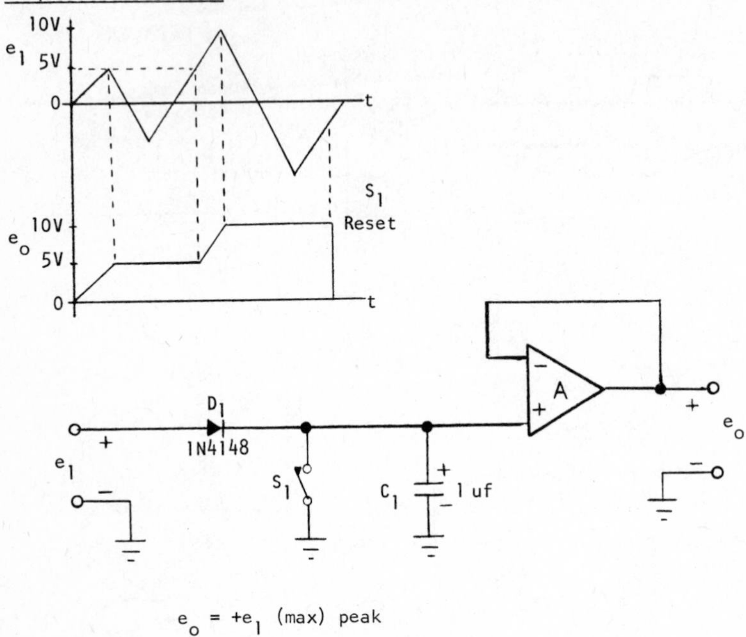

$$e_o = +e_1 \text{ (max) peak}$$

The Peak Detector concept is essentially a simple variation of the basic sample/hold network technique.

Capacitor C_1 will charge to the most positive value of the input voltage e_1. When e_1 drops below the voltage existing capacitor voltage, the diode D_1 becomes reversed biased, thus allowing the capacitor to retain its charge.

Improved Peak Detector

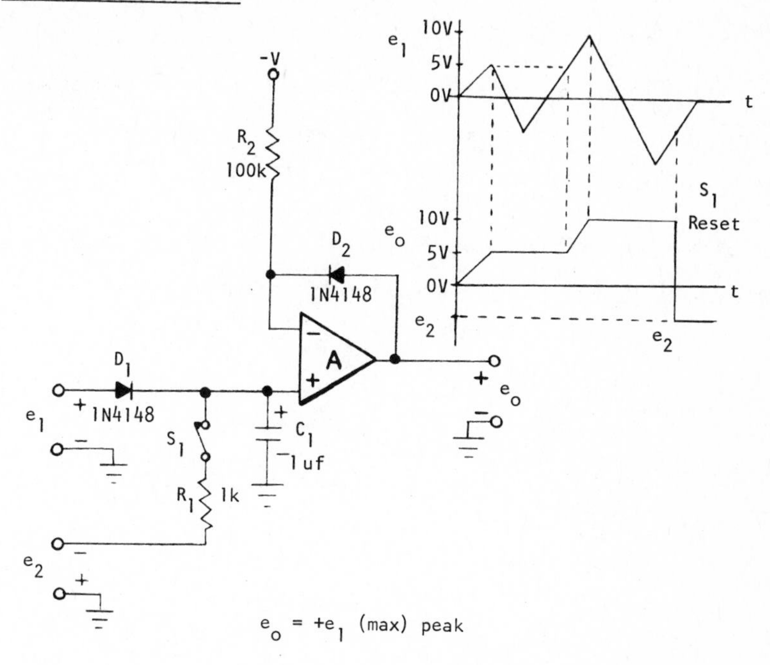

$e_o = +e_1$ (max) peak

The capacitor will store the most positive signal coming from the input e_1. Diode D_1 will forward bias to charge the capacitor C_1 to the peak input value. If the input e_1 drops below the charge on C_1, the diode D_1 will reverse bias, thus "storing" the former peak value. Diode D_2 serves to cancel the voltage drop error caused by diode D_1. The value of R_2 should be chosen to give a zero net offset for the most probable amplitude and duration of peak input voltages.

For high accuracy and long storage duration, diode D_1 should have low leakage. The leakage of D_2 is not critical because it is isolated from C_1. However, the temperature drift characteristics (voltage) of

Improved Peak Conditions (continued)

both diodes should be matched in order to optimize accuracy over a given temperature range. Therefore, it is advisable to use the same type of device for both D_1 and D_2.

The voltage e_2 establishes the "initial" conditions for the network. The resistor R_1 limits the discharge current through the switch S_1. For an output initial condition of zero, e_2 should be grounded.

Precision Peak Detector

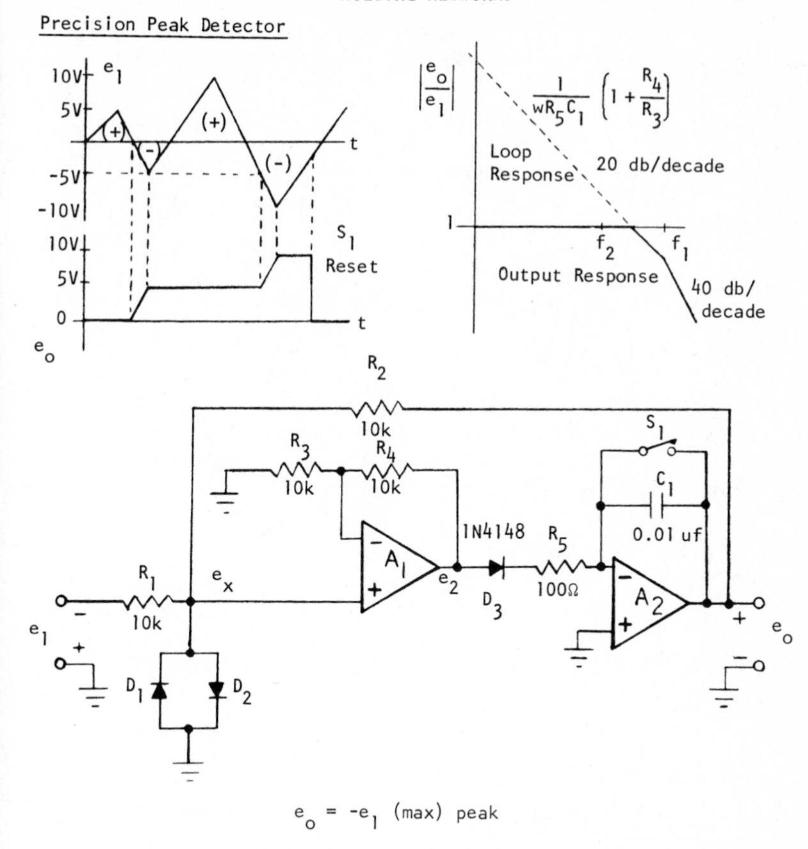

$$e_o = -e_1 \text{ (max) peak}$$

This network detects negative input peaks. Reversing diode D_3 will detect positive input peaks. The output is inverted.

It should be noted that this is a closed loop system. A negative going voltage e_x is amplified by the A_1 network which will, in turn, charge the integrator capacitor C_1.

Precision Peak Detector (continued)

At all time the output voltage e_o is equal to the charge on the capacitor C_1. When the output magnitude charges to the same level as the peak input voltage e_1 (but opposite polarity), the voltage e_x reduces to zero. This stops C_1 from charging. The capacitor will retain this charge even if e_1 reverses direction because of the reverse bias action of diode D_3. The diode D_1 and D_2 clamp e_x in order to prevent A_1 from saturating in the case of large input transients. The switch S_1 resets the output to zero.

Since this is a double Op Amp closed loop system, care must be taken in design to insure dynamic stability.

The composite response of the A_1 and A_2 networks must roll-off at a rate no steeper than 20 db/decade as it passes through unity gain (zero db).

From the response curves on the next page it can be seen that for overall stability, the frequency breakpoint f_1 should occur below unity gain. (For f_1 occurring exactly at unity gain, the phase margin is 45°. The network will tend to ring excessively for pulse inputs.) The farther below unity gain the f_1 breakpoint occurs, the better the frequency stability and greater the phase margin.

Therefore, lowering frequency f_2 will yield greater stability. The trade-off is, of course, the total system bandwidth will be sacrificed. Lowering frequency f_2 is accomplished by either decreasing the value of C_1 or increasing R_5. It is generally best to keep C_1 as large

Precision Peak Detector (continued)

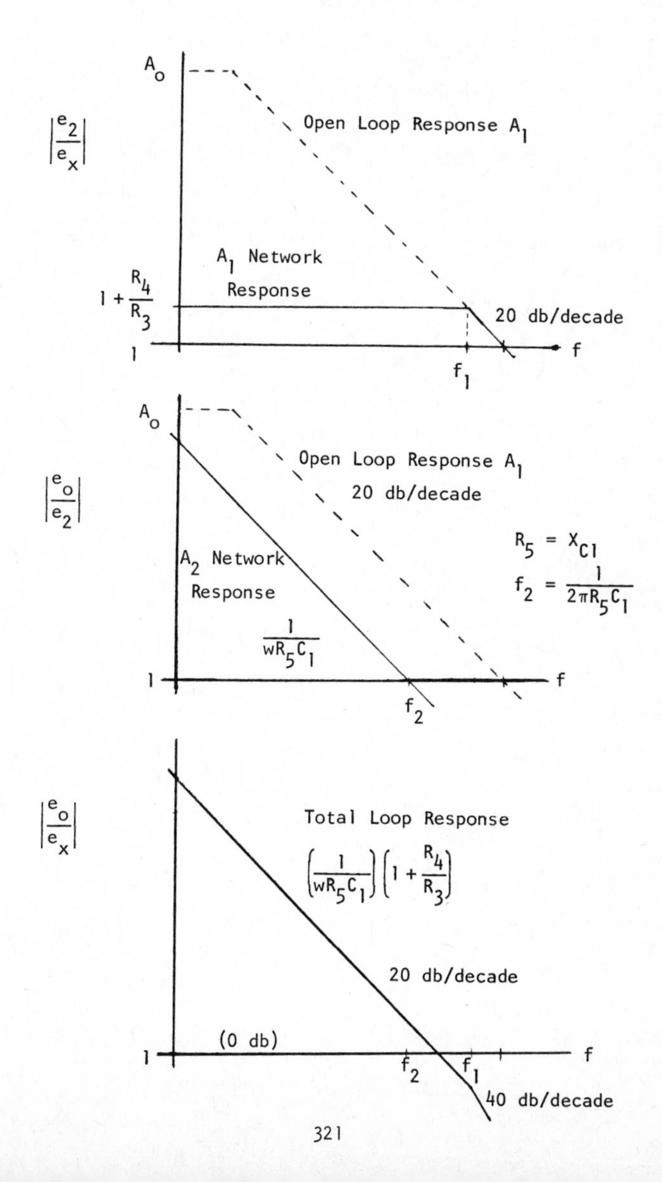

Precision Peak Detector (continued)

as possible in order to optimize the storage or "holding" of the peak detector. Therefore, resistor R_5 should be the "variable" component when adjusting for optimum dynamic stability.

The system has an inherent disadvantage. For high positive input diode D_3 will be heavily reversed biased. This can cause a relatively large leakage current to flow into C_1 during the "holding" mode.

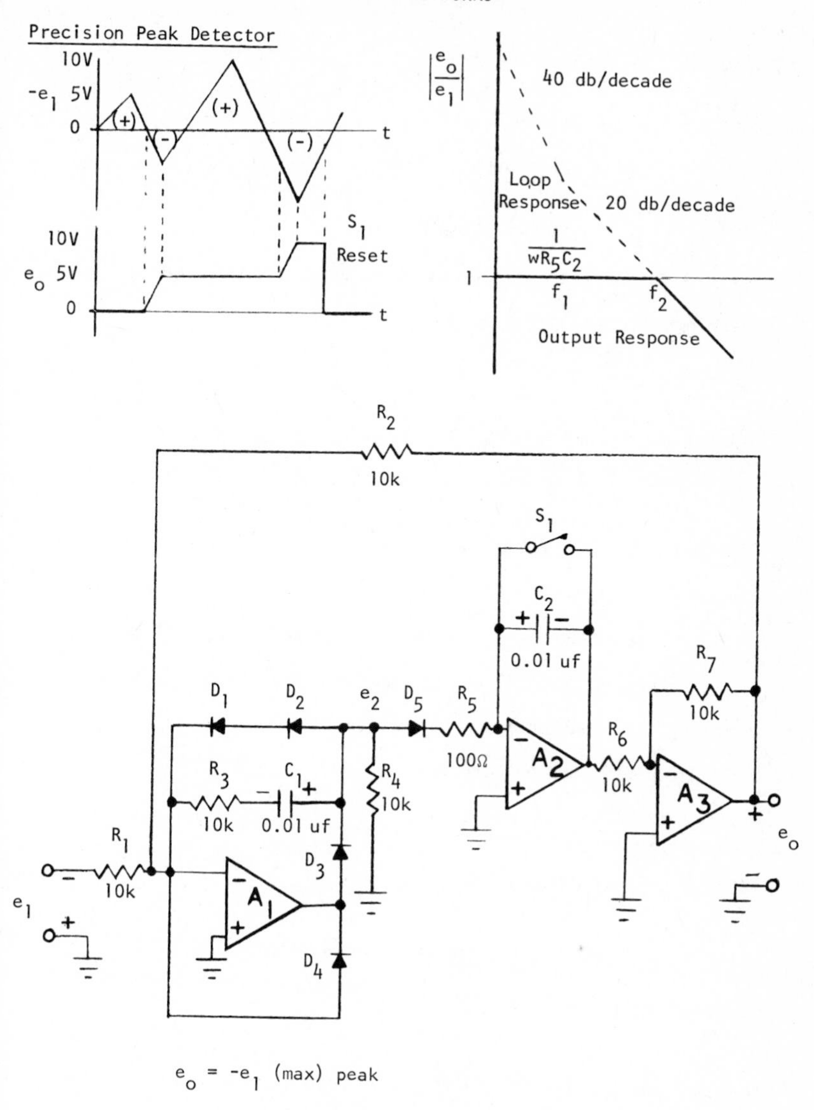

$e_o = -e_1$ (max) peak

This network detects negative input peaks. Reversing all diodes will detect positive input peaks. Note that the output is inverted.

Precision Peak Detector (continued)

The closed loop system offers the advantage of eliminating the error due to reverse bias leakage current of the diode D_5.

The A_1 network is a precision rectifier. The output e_2 goes to zero for positive inputs e_x. This protects D_5 from heavy reverse bias voltages, thus keeping leakage negligible.

Diodes D_1 and D_2 clamp the output of the A_1 network to protect it against saturation. Resistor R_4 provides a leakage path to ground for D_1, D_2 and D_5.

Since this is a multi-Op Amp closed loop system, dynamic stability must be carefully designed into the network. The frequency breakpoints should be located such that the loop roll-off passes through unity gain at a rate no greater than 20 db/decade. The loop response can be analyzed by "breaking" the feedback path at the junction of R_2 and e_o. The response of each Op Amp network can then be plotted. The total loop response G(total) is the composite of the separate responses.

$$G(\text{total}) = G_1 G_2 G_3$$

Since G_3 is unity gain, the plot can be simplified.

$$G(\text{total}) = G_1 G_2$$

In the above networks the open loop responses of all the Op Amps are identical. Therefore frequency f_x is considered the unity gain bandwidth (or gain-bandwidth product) of the total open loop response.

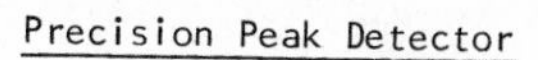

Precision Peak Detector

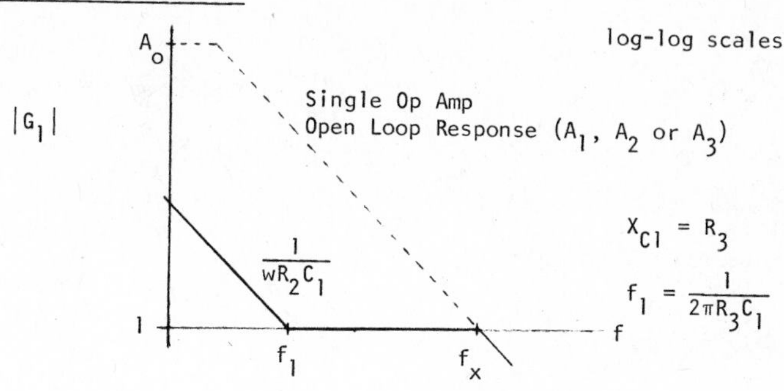

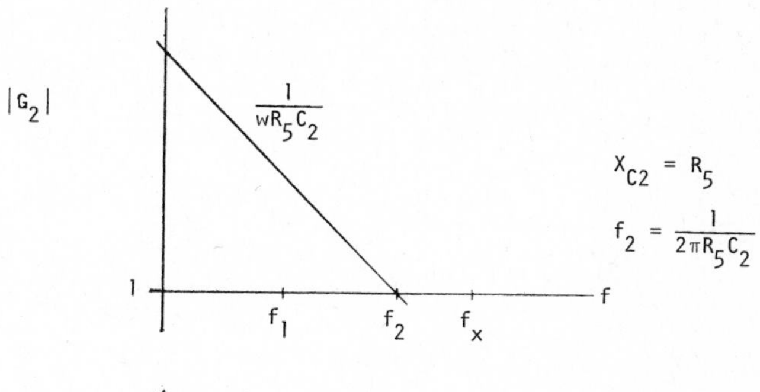

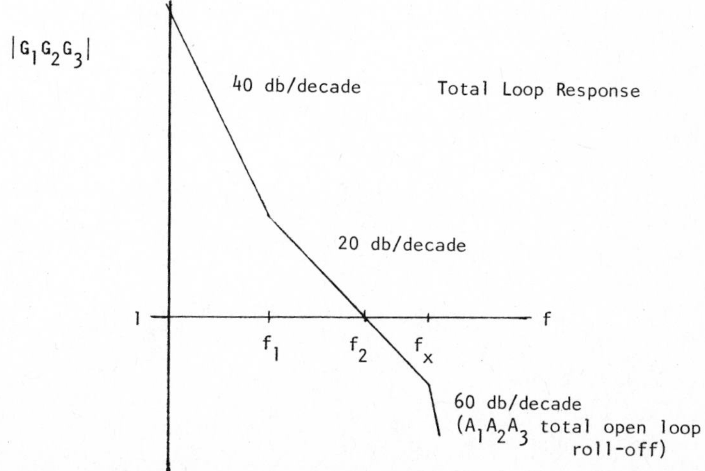

Precision Peak Detector (continued)

Basically, the A_2 network determines the unit gain bandwidth of the total system. However, the A_1 network breakpoint f_1 determines the relative stability. Frequency f_1 should break well in advance of f_2 to allow adequate phase margin at unity gain.

$$f_2 = \frac{1}{2\pi R_5 C_2}, \qquad \text{occurs where } \frac{X_{C2}}{R_5} = 1$$

$$= 160 \text{ kHz}$$

$$f_1 = \frac{1}{2\pi R_3 C_1}, \qquad \text{occurs where } R_3 = X_{C1}$$

$$= 1.6 \text{ kHz}$$

The basic disadvantage of this network is that three Op Amps are required.

Precision Peak Detector

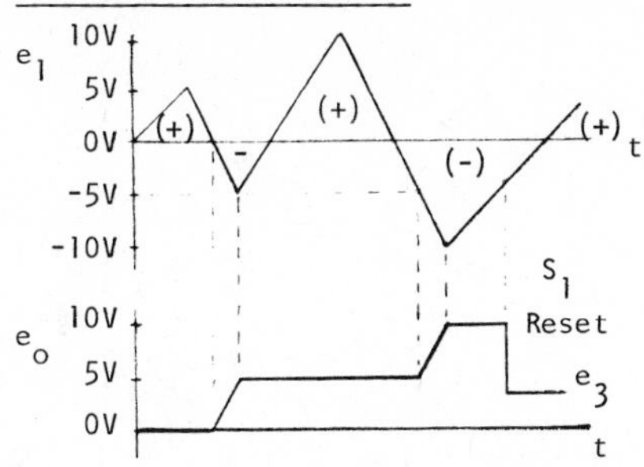

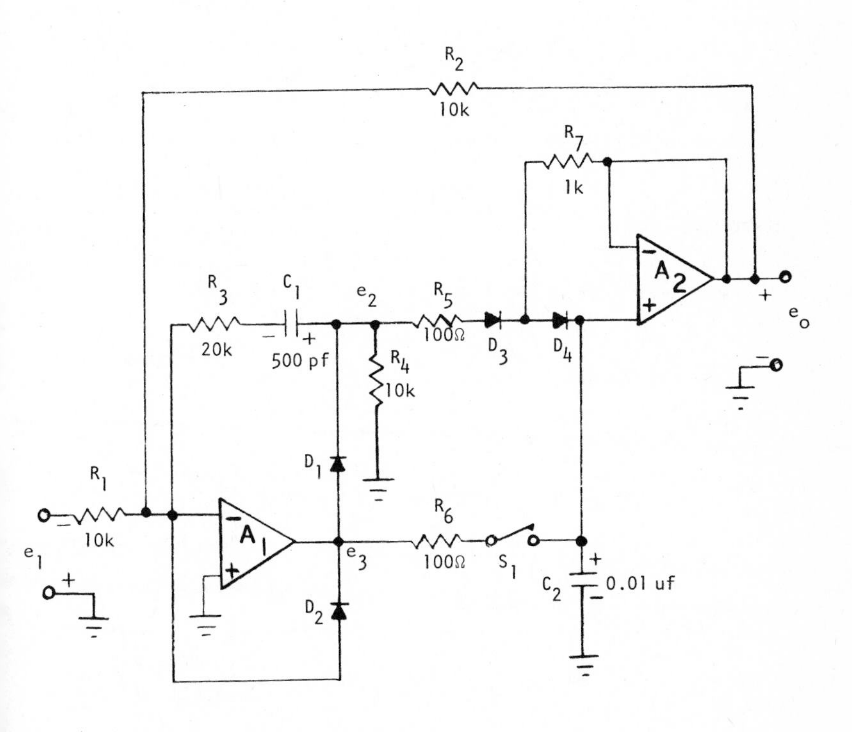

$e_o = -e_1$ (max) peak

Precision Peak Detector (continued)

The preceding network detects negative input peaks. The output is inverted. Reversing all diodes will change the network to a positive input peak detector.

This closed loop system offers the advantage of both eliminating the error due to reverse bias leakage current of the diodes (D_3, D_4) and requiring only two Op Amps.

The A_1 network is a precision rectifier used as an augmenting integrator. The breakpoint f_1 is needed to stabilize the network against the roll-off at f_2. When f_1 is designed to break well in advance of f_2, the total system response will be insured of dynamic stability.

The unique arrangement of diodes D_3, D_4 and R_7 protect capacitor C_2 from diode reverse bias leakage. When C_2 has charged to the peak value of e_1, the diodes D_3 and D_4 cease conduction and the output equals the peak value of the input e_1.

When e_1 moves away from the peak, the diodes D_3, D_4 reverse bias due to the voltage e_2 being less than the charge on the capacitor C_2. However, the reverse voltage across D_4 is forced to zero by feedback resistor R_7. This virtually eliminates D_4 leakage to the capacitor C_2. Leakage will occur from D_3, but it is passed through a low impedance R_7, resulting in a negligible output offset error.

Precision Peak Detector (continued)

In this particular network the reset circuit (S_1, R_6) is designed to acquire the "existing" value of the input at the moment of reset. The value of R_6 should be the same as R_5 to give accurate symmetry.

The loop response of the network is the composite of each Op Amp (A_1, A_2) network response.

$$|G_{total}| = |G_1 G_2|$$

As the preceding networks show, the open loop responses of both Op Amps are identical. Therefore, frequency f_x is considered the unity gain bandwidth (or gain-bandwidth product) of the total open loop response.

The A_1 network (G_1) is designed to "break" first at the frequency f_1, where the feedback components (R_3, C_1) impedances equal one another.

$$f_1 = \frac{1}{2\pi R_3 C_1}, \qquad \text{occurs where } X_{C1} = R_3$$

$$= 16 \text{ kHz}$$

The A_2 network (G_2) breaks next at f_2, where R_5 equals the reactance of C_2.

$$f_2 = \frac{1}{2\pi R_5 C_2}, \qquad \text{occurs where } X_{C2} = R_5$$

$$= 160 \text{ kHz}$$

Care should be taken when selecting the Op Amp A_1, that the frequency breakpoint f_3 is as far beyond the required unity gain frequency crossover point f_4 as possible. A small phase margin here will cause excessive ringing for step inputs.

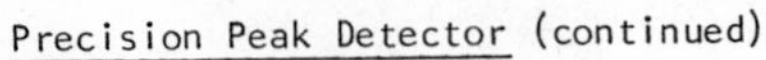

Precision Peak Detector (continued)

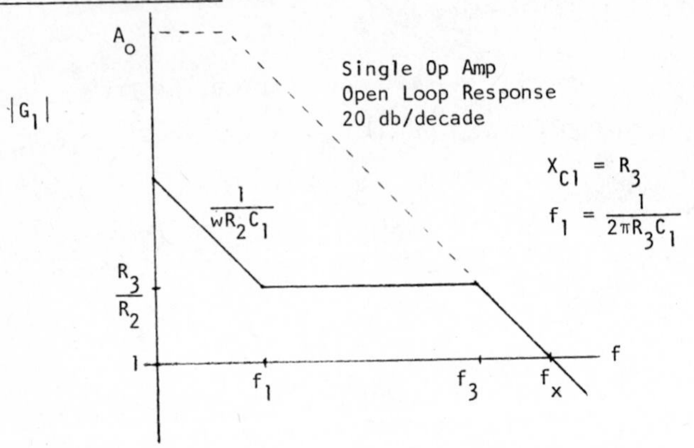

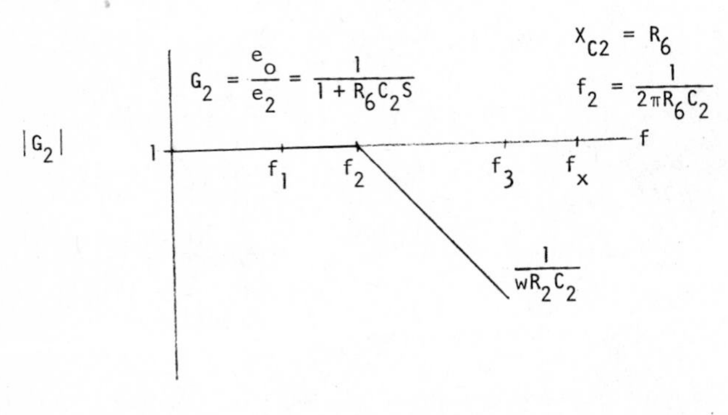

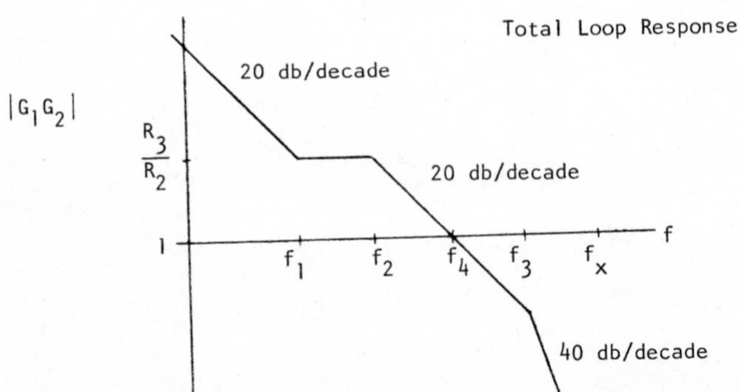

Peak-to-Peak Reader/Storage

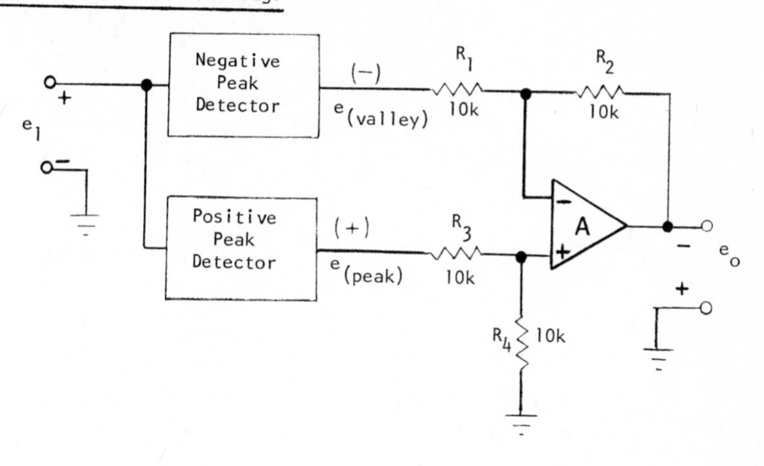

$$e_o = e_{(peak)} - e_{(valley)}, \qquad \text{for } R_1 = R_2 = R_3 = R_4$$

Combining the outputs of both positive and negative Peak Detectors with an Adder-Subtractor network results in a system that measures and stores peak-to-peak voltages. Refer to the section on DC Voltage Output networks for other Adder-Subtractor options.

Staircase Memory

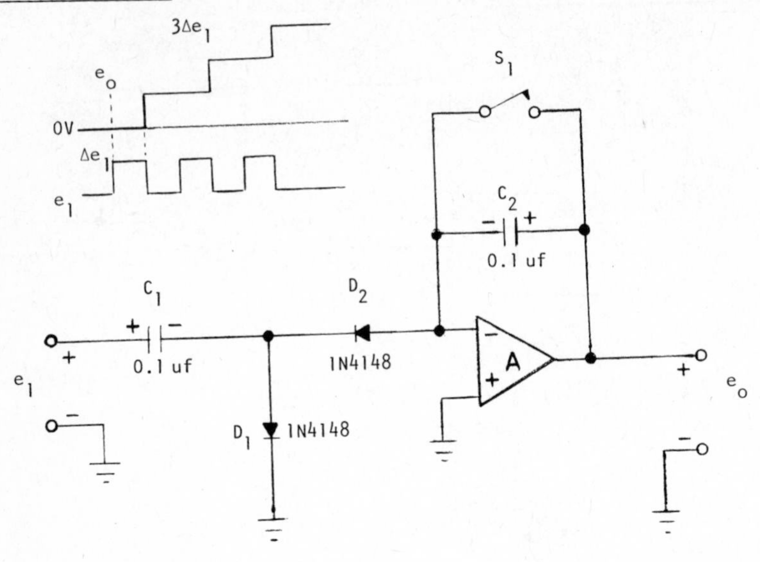

$$e_o = \sum_n \Delta e_{o_n} = \sum_n \Delta e_1 \frac{C_1}{C_2}\,,$$

where n is the number of input pulses applied between resets

$$= 3\Delta e_1\,,$$

for n = 3

The output climbs in incremental jumps resembling a staircase. Each input pulse generates an additional output step.

The first input pulse e_1 charges capacitor C_1 through the diode D_1. When e_1 drops back to zero (at the end of the first pulse), capacitor C_1 becomes grounded at the input and quickly transfers its charge (the negative voltage reverse biases D_1 and forward biases D_2) to the feedback capacitor C_2.

Staircase Memory (continued)

$$\Delta Q_1 = \Delta Q_2$$

$$C_1 \Delta e_1 = C_2 \Delta e_o$$

The output voltage is proportional to the gain of the network.

$$\Delta e_o = \frac{C_1}{C_2} \Delta e_o$$

At this point capacitor C_1 is discharged and ready for the next incoming pulse and the process repeats itself. Capacitor C_2 retains its accumulating charges due to diode D_2 being reversed biased.

The network is reset by closing the switch S_1.

Introduction

Oscillators have been broken into two categories in this section:

(1) Discontinuous output signals, such as square wave, triangular waves, or ramp functions.

(2) Continuous output signals, such as the fundamental sine wave.

The discontinuous output oscillators basically all work on the same general principle. They are positive feedback threshold switching circuits. The cyclic frequency is controlled by the corresponding RC time constants in the feedback circuits (See Multivibrators).

The basic idea in designing the continuous output (sine wave) oscillator is to maintain a loop phase margin of zero degrees at the unity gain cross-over frequency. It is the exact opposite of the unity gain stability criterion commonly referred to in all other Op Amp network designs.

This can be accomplished by either of two methods:

(1) Direct positive feedback of a portion of the output signal (180° phase shift) into the non-inverting input (+) terminal of the Op Amp (See Wien Bridge Oscillators).

(2) Indirect negative feedback, (-) input, using phase-shift network techniques for achieving an additional 180° phase shift from the output signal. This signal is then fed back to the (-) input of the Op Amp. This is usually done by connecting equivalent lag networks in tandem (see integrator Twin-Tee Oscillators).

Introduction (continued)

The major problems encountered when building sine wave oscillators are frequency stability and amplitude stability.

As a rule, the Wien Bridge approach offers excellent inherent frequency stability but relatively poor amplitude stability. However, this text will show methods to improve the amplitude stability.

The Integrator method (Twin-Tee networks) offers good amplitude control, but relatively poor frequency stability. However, these networks generally require fewer components and, therefore, may sometimes be a more desirable approach where a cost versus performance requirement situation exists.

It is advisable to use precision high quality resistors and capacitors in all oscillator networks for optimum performance. Metal film (non-inductive) resistors do well, and polystyrene or polycarbonate capacitors, respectively, are recommended.

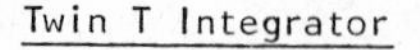

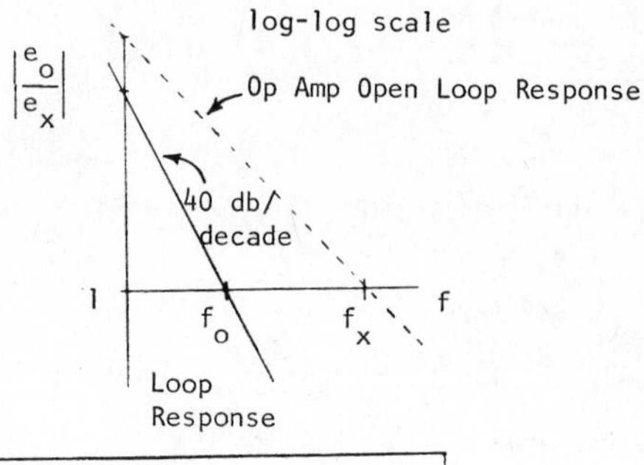

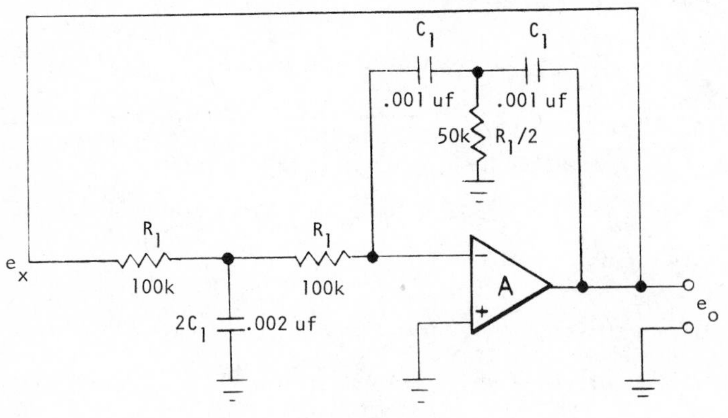

$f_o = \frac{1}{2\pi R_1 C_1}$, occurs where $\frac{e_o}{e_x} = 1$

$= 1.6$ kHz

$\frac{e_o}{e_x} = -\frac{1}{w^2 R_1^2 C_1^2}$, 40 db/decade roll-off with 180° phase shift (loop broken at e_x)

The frequency of oscillation occurs at the unity gain cross over point f_o. At this point, all the output is fed back to the input which results in a total phase shift of 360°.

Twin T Integrator (continued)

Oscillatory "buildup", or "damping out" will occur if the circuit components are not precisely matched. For this reason it is generally most practical to trim the Op Amp feedback resistor ($R_1/2$) until the oscillations are just barely sustained. The oscillation frequency f_o should be kept well within the Op Amp open loop response f_x, otherwise output impedance becomes large due to low loop gain. This will cause frequency instability with changing output loads. It is usually adviseable therefore to buffer the output with a high impedance non-inverter Op Amp network.

Improved Twin T Integrator

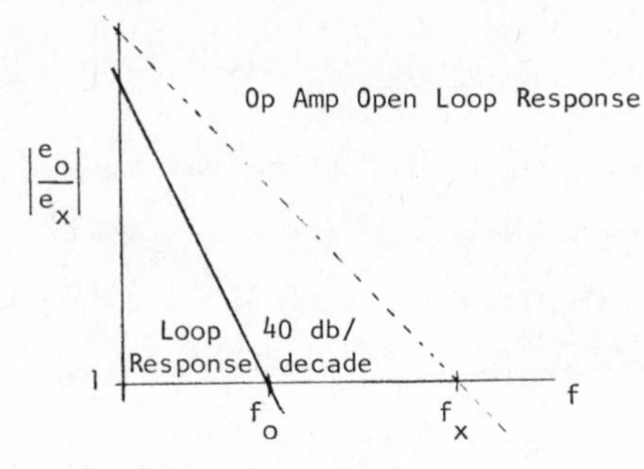

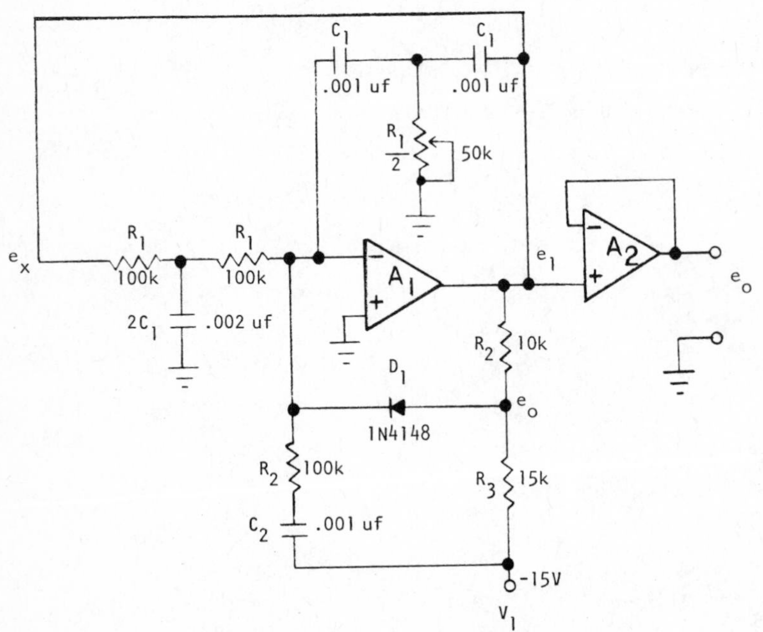

$$f_o = \frac{1}{2\pi R_1 C_1}, \qquad \text{occurs at } \frac{e_o}{e_x} = \frac{X_{C1}}{R_1} = 1$$

$$= 1.6 \text{ kHz}$$

$$e_1(\text{clamped}) = \frac{R_2}{R_3} V_1, \qquad \text{maximum peak amplitude of } e_o$$

$$= +10 \text{ volts}$$

Improved Twin T Integrator (continued)

This network is identical to the previous circuit with the exception that an output buffer has been added for frequency stability, and a clamping circuit (D_1, R_2, R_3) has been added for amplitude stability (see Clips and Clamps section).

The non-inverting Op Amp A_2 makes the network frequency f_o virtually independent of the output load.

The clamping circuit limits the output of A_1 from going above +10V, thus topping any oscillatory buildup on each cycle. The value of the feedback resistor ($R_1/2$) should be adjusted so that the circuit will have only slight oscillatory buildup. This will allow the clamping circuit to effectively stabilize the output amplitude while causing the minimum of waveform distortion.

It should be noted that the clamping voltage is affected both by the diode D_1 temperature drift and the supply voltage V_1.

The resistor capacitor arrangement R_2, C_2 is used as a quick self-starting technique. The optimum values are usually equal to R_1 and C_1, respectively. Since these two components actually shunt the Op Amp input to ground (through the supply), for high frequencies it may be necessary to deviate from this rule in order to increase the input impedance. As a rule of thumb

$$\frac{C_2}{C_1} \gtrsim 1 \qquad \text{and} \qquad \frac{R_1}{R_2} \gtrsim 1 \quad .$$

Care should be taken to by-pass the power supplies (V_1) to eliminate extraneous feedback paths.

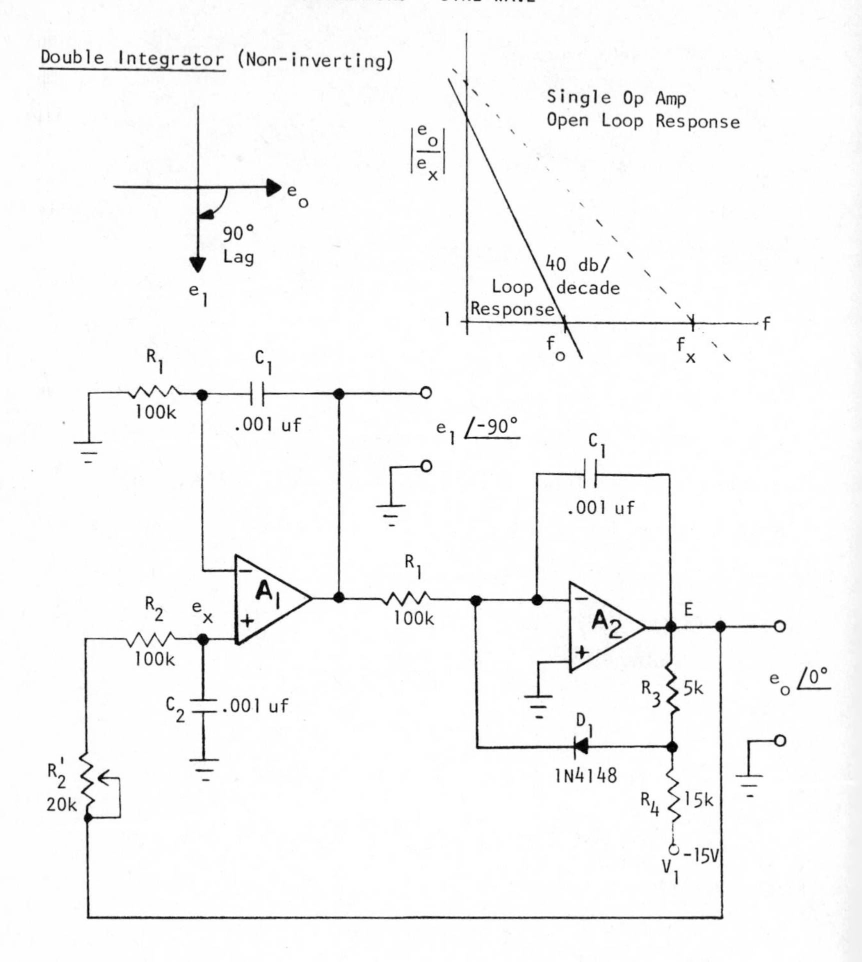

$$f_o = \frac{1}{2\pi R_1 C_1}, \qquad \text{occurs where } \frac{e_o}{e_x} = \frac{X_{C1}}{R_1} = 1$$

$$= 1.6 \text{ kHz}$$

$$E(\text{clamped}) = \frac{R_3}{R_4} V_1, \qquad \text{maximum peak amplitude of } e_o$$

$$= +5V$$

Double Integrator (continued)

This approach offers two separate output signals e_1, e_o with a 90 degree phase difference using only two Op Amps.

The clamping circuit D_1, R_3, R_4 serves to prevent oscillatory buildup. The ratio of R_3 and R_4 to the supply voltage limits the output amplitude to 5 volts peak.

The phase shift provided by the R_2C_2 breakpoint can be adjusted to an optimum value by R_2'. This will minimize the amplitude distortion that is caused by the clamping circuit while allowing a slight oscillatory buildup. Therefore, the oscillator stability is controlled by R_2' and the R_2C_2 time constant. This time constant will be slightly larger than the R_1C_1 to insure optimum operation.

Note that A_1 has a common mode input voltage, (+) input.

OSCILLATORS - SINE WAVE

Double Integrator (Inverting)

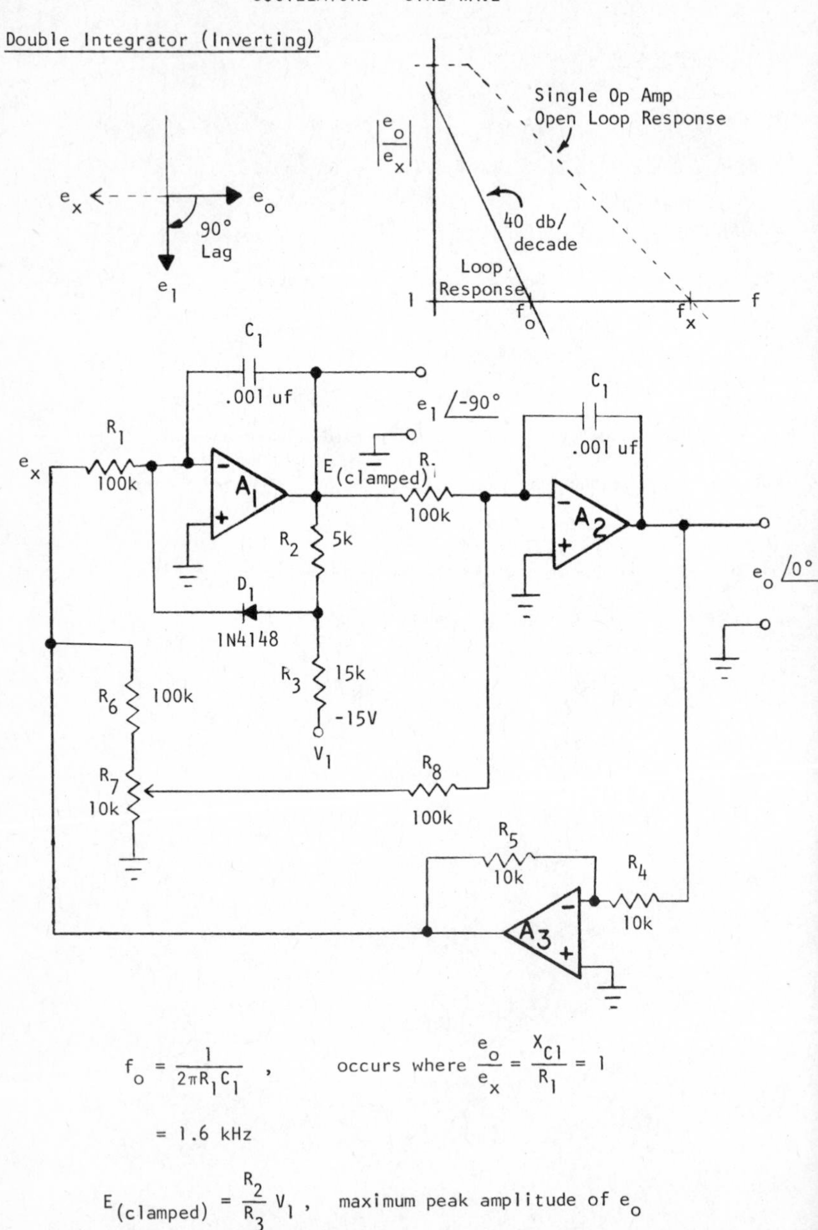

$$f_o = \frac{1}{2\pi R_1 C_1}, \qquad \text{occurs where } \frac{e_o}{e_x} = \frac{X_{C1}}{R_1} = 1$$

$$= 1.6\ \text{kHz}$$

$$E_{(clamped)} = \frac{R_2}{R_3} V_1, \quad \text{maximum peak amplitude of } e_o$$

$$= +5V$$

Double Integrator (continued)

This network offers a two phase output e_1, e_o using Op Amps in the inverting mode only. This makes it possible to use single ended ultra-low drift Chopper Stabilized Op Amps. This will result in high overall frequency stability versus temperature.

The clamping circuit D_1, R_2, R_3 stabilizes the output amplitude. The positive feedback circuit R_6, R_7 controls oscillatory build-up which is used to optimize the distortion trade-off of the clamping action.

OSCILLATORS - SINE WAVE

Three Phase Oscillator

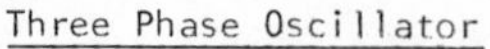

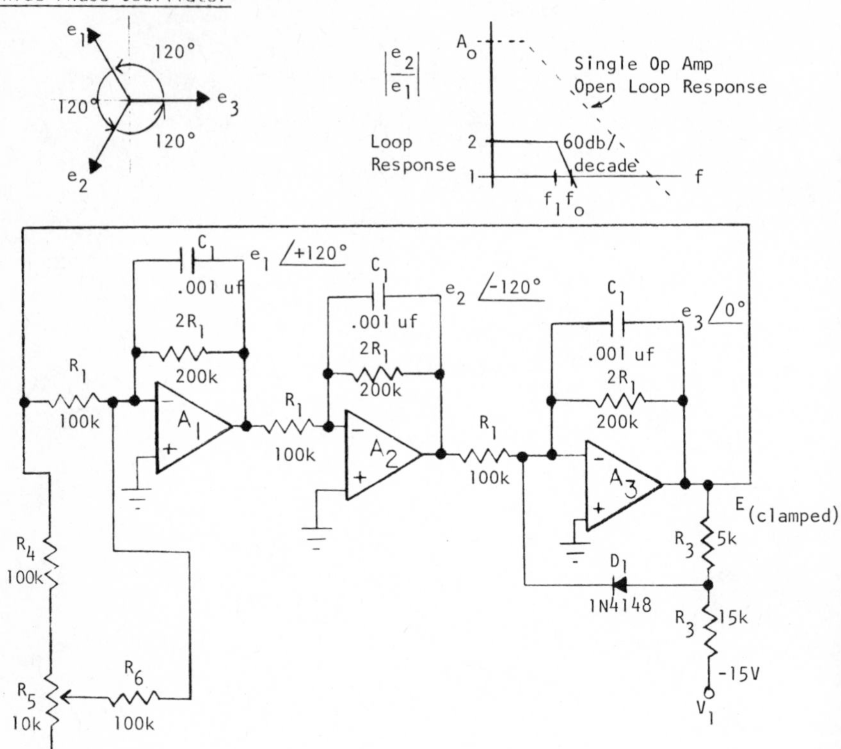

$$f_o = \frac{\sqrt{3}}{2}\left(\frac{1}{2\pi R_1 C_1}\right), \qquad \text{occurs where } \frac{e_1}{e_3} = \frac{e_2}{e_1} = \frac{e_3}{e_2} = -1\angle{-60^\circ}$$

$$= 1.4 \text{ kHz}$$

$$E_{(clamped)} = \frac{R_2}{R_3} V_1$$

$$= +5V$$

Three Phase Oscillator (continued)

The output of this network will generate three separate signals e_1, e_2, e_3 that are equal in amplitude and frequency. Each signal, however, will have a differential phase sequence of 120 degrees from the other resulting in a symmetrical three phase system.

The output amplitude is controlled by the clamping circuit D_1, R_2, R_3. The oscillatory build-up controller is located at the input of A_1 (R_4, R_5, R_6).

The frequency of oscillation occurs when the total loop gain becomes unity. At this point, the output phase shift of each amplifier (e_1, e_2, e_3) is 60 degrees lagging, plus the normal polarity inversion (180°). This gives a total combined phase shift of 120 degrees. The transfer equation for one amplifier section is

$$-\frac{e_o}{e_{in}} = \frac{1}{1 + 2R_1C_1S}$$

Therefore at the oscillation frequency f_o,

$$-\frac{e_o}{e_{in}} = 1\,\angle{-60°}$$

$$= \frac{1}{2} - j\,\frac{\sqrt{3}}{2}$$

Letting S = jw

$$-\frac{e_o}{e_{in}} = \frac{2}{1 + 2R_1C_1S}$$

$$= \frac{2}{1 + (2R_1C_1w)^2} - j^2\,\frac{(2R_1C_1w)}{1 + (2R_1C_1w)^2}$$

Three Phase Oscillator (continued)

Equating real (or imaginary) components and solving for w_o,

$$\frac{1}{2} = \frac{2}{1 + (2R_1C_1w_o)^2} \quad \text{, real component}$$

$$w_o = \frac{\sqrt{3}}{2R_1C_1}$$

$$\text{or} \quad f_o = \frac{\sqrt{3}}{2}\,\frac{1}{2\pi R_1C_1} \quad \text{, where } w = 2\pi f$$

Notice that the frequency breakpoint of the first pole f_1 of each amplifier occurs before the oscillation frequency f_o.

$$f_1 = \frac{1}{4\pi R_1C_1}, \qquad \text{occurs where } X_{C1} = 2R_1$$

This breakpoint, together with a DC closed loop gain of 2, makes the exact combination necessary for a 60 degree phase lag at unity gain (f_o).

From this conceptual approach, it can be seen that complex multi-phase networks can be easily designed by using additional amplifier circuits constructed with the proper gain-phase shift relationships.

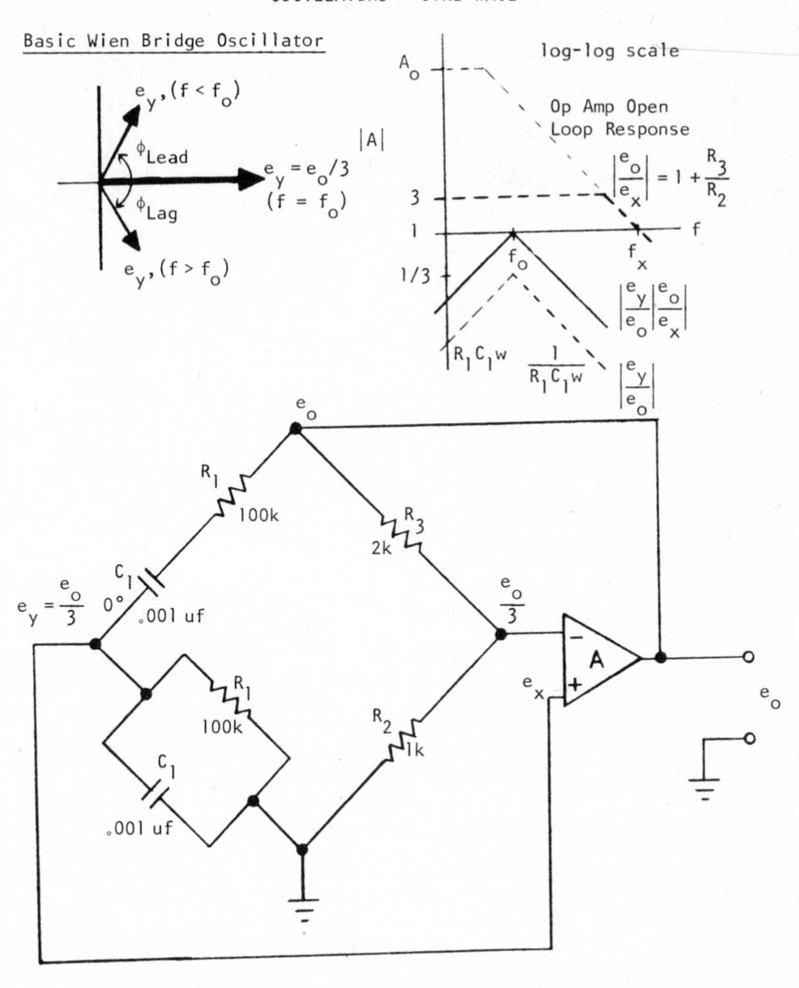

$$f_o = \frac{1}{2\ R_1C_1},$$ occurs where $X_{C1} = R_1$

$$= 1.6 \text{ kHz}$$

$$\frac{R_3}{R_2} = 2,$$ critical gain ratio to sustain oscillations

Basic Wien Bridge Oscillator (continued)

The preceding network offers good frequency stability, but relatively poor amplitude control. It is more easily understood if redrawn as follows.

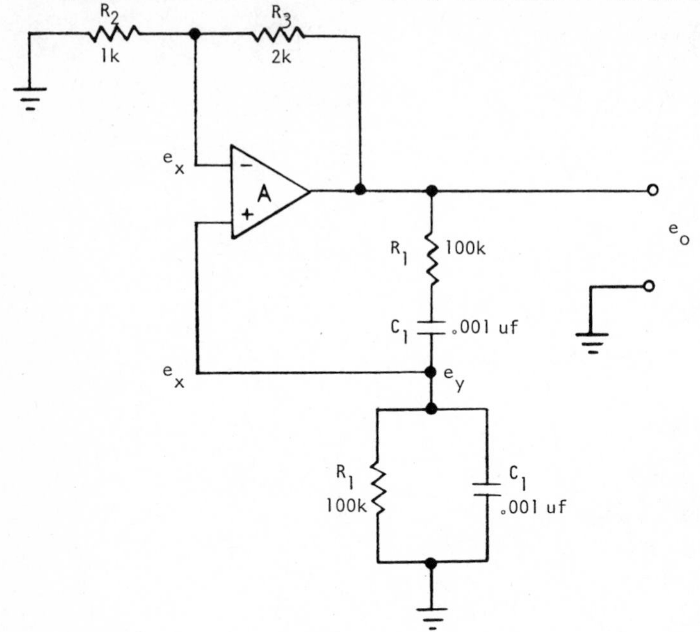

It can be seen that a feedback signal with zero phase shift at point e_y can cause sustained oscillations if provided with a sufficient gain by R_2 and R_3 to make the total loop gain unity.

$$\left|\frac{e_y}{e_o}\right| \times \left|\frac{e_o}{e_x}\right| = \left|\frac{e_y}{e_x}\right| = 1$$

By examining the reactive section separately, the response is

$$\frac{e_y(S)}{e_o} = \frac{1}{R_1C_1S + 3 + \frac{1}{R_1C_1S}}$$

Basic Wien Bridge Oscillator (continued)

At the oscillation frequency f_o where the phase shift is zero, R_1 is equal to X_{C1}. At precisely this point the transfer function becomes one-third of the output e_o.

$$\frac{e_y}{e_o} = \frac{1}{3}, \qquad \text{at } R_1 = X_{C1} \text{ (without amplifiers)}$$

Therefore, in order to regain the output signal that is necessary to sustain an oscillation, the amplifier network must have a gain of three (3).

$$\frac{e_o}{e_x} = 3, \qquad \text{amplifier gain for sustained oscillations}$$

Since

$$\frac{e_o}{e_x} = 1 + \frac{R_3}{R_2}$$

Then

$$3 = 1 + \frac{R_3}{R_2}$$

and

$$\frac{R_3}{R_2} = 2, \qquad \text{critical resistor ratio}$$

At the critical gain ratio, oscillations will just barely be sustained. If the gain is too high, the output will start to climb to the Op Amp saturation limits, causing excessive distortion. On the other hand, if the gain ratio is too low, the oscillations will dampen out to a zero output.

Basic Wien Bridge Oscillator (continued)

For this reason additional circuit techniques are usually required in order to control, or modulate, the gain to achieve the necessary amplitude stability. These techniques will be illustrated in the networks that follow.

Improved Wien Bridge Oscillator

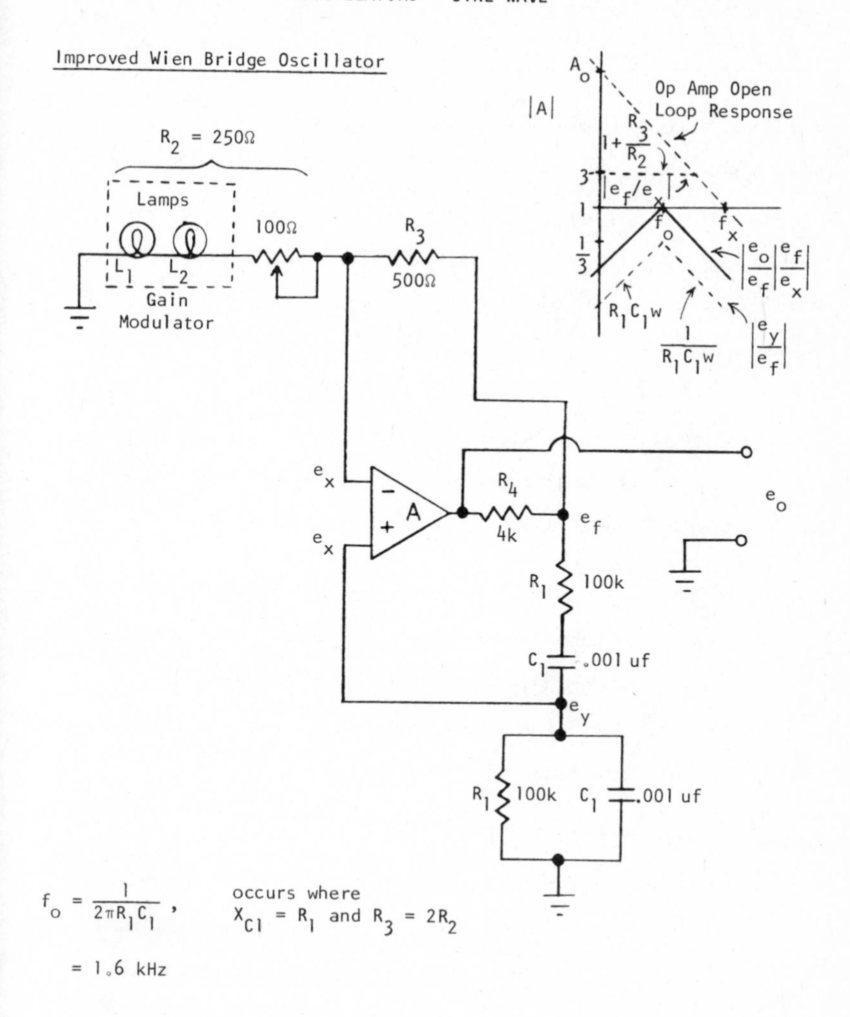

$$f_o = \frac{1}{2\pi R_1 C_1}, \quad \text{occurs where } X_{C1} = R_1 \text{ and } R_3 = 2R_2$$

$$= 1.6 \text{ kHz}$$

This network offers good frequency stability and improved amplitude stability.

Improved Wien Bridge Oscillator (continued)

The variable hot/cold negative resistance characteristics of the common low power incandescent lamp offers a very economical method to modulate gain for the purpose of achieving output amplitude stability. The lamps are operated below their visible illumination current level, or what is known as the "cold resistance" region (low resistance). For low amplitude e_x signals, the resistance of L_1 and L_2 is minimum; this makes the gain maximum. Conversely, if the amplitude of e_x becomes large, the lamp current increases, causing the device to gradually move into the "hot resistance" region (high resistance). The gain decreases correspondingly, bringing the network into equilibrium.

Since the resistance of these lamps is determined by thermal characteristics, the corresponding network output amplitude will, therefore, be temperature dependent. This effect can be controlled either by external temperature compensation circuitry, or by protecting the environment of the lamps themselves using oven-stabilization techniques.

The additional resistor R_4 is used to protect the Op Amp A from current overload. This becomes necessary because the operation of the lamps is usually less than 100 ohms, thus forcing R_3 to be quite low. By taking the output signal from the Op Amp A_1 itself, and only using e_f as a (correctly "proportioned") low level feedback signal ($e_f = 3e_x$), the oscillator network will function in a very linear mode.

Improved Wien Bridge Oscillator (continued)

$$\frac{e_y}{e_f}(S) = \frac{1}{R_1C_1S + 3 + \frac{1}{R_1C_1S}}, \qquad S = jw$$

$$= \frac{1}{3}\angle 0^\circ, \qquad \text{at } f = f_o$$

$$\frac{e_f}{e_x} = 1 + \frac{R_3}{R_2} = 3\angle 0^\circ$$

The value of R_4 is determined by relating the common mode voltage at e_x to the maximum feedback current I_2.

Designing the feedback current to a maximum value of 2 ma will establish the common mode input voltage level.

$$e_x = I_2R_2, \qquad \text{Op Amp common mode input voltage}$$

$$= (2\text{ ma})(250\Omega)$$

$$= 0.5V$$

The corresponding voltage at e_f will be

$$e_f = \left(1 + \frac{R_3}{R_2}\right) e_x = 3e_x, \qquad \text{for } R_3 = 2R_2$$

$$= 3(0.5V)$$

$$= 1.5V$$

Similarly, the voltage at e_o will be

$$e_o = \left(1 + \frac{R_3+R_4}{R_2}\right) e_x, \qquad \text{assuming the } R_1C_1 \text{ network draws negligible current}$$

Improved Wien Bridge Oscillator (continued)

or, $$e_o = \left(3 + \frac{R_4}{R_2}\right) e_x \,, \qquad \text{for } R_3 = 2R_2$$

$$= \left(3 + \frac{4k\Omega}{250\Omega}\right) (0.5V)$$

$$= 9.5V$$

Increasing or decreasing the value of R_4 will increase or decrease the output voltage e_o, respectively.

It is generally advisable to buffer the output e_o with an additional non-inverting amplifier. This netowrk will serve to isolate the sensitive oscillator circuit from external output loading frequency stability effects.

Precision Wien Bridge Oscillator (DC Coupled Modulator)

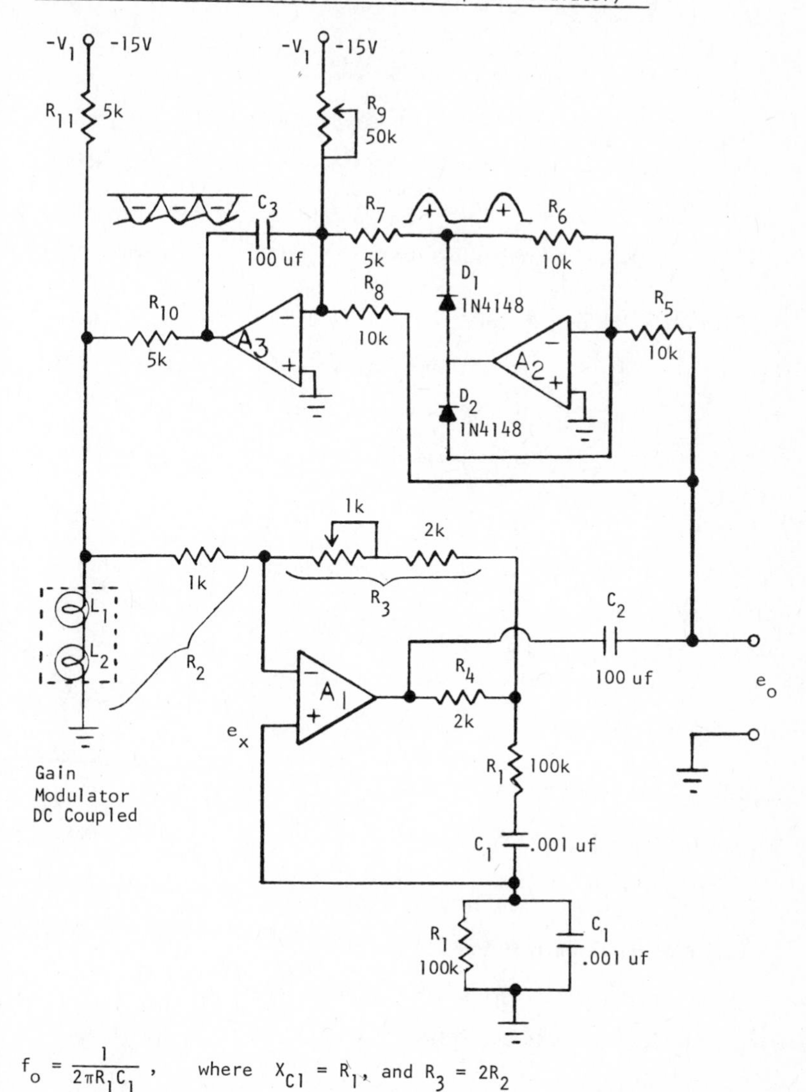

$$f_o = \frac{1}{2\pi R_1 C_1}, \quad \text{where } X_{C1} = R_1, \text{ and } R_3 = 2R_2$$

$$= 1.6 \text{ kHz}$$

Precision Wien Bridge Oscillator (continued)

This network utilizes a DC feedback loop (A_2, A_3) that serves as a bias current controller to the gain modulator lamps L_1, L_2. (Some lamps formerly used by Op Amp manufacturers have been GE1869 and CAL-GLO Type 80). The lamp DC quiescent current is supplied through R_{11}. At equilibrium, no net DC current flows through R_{10} (this is trimmed by R_9).

However, if the output amplitude e_o should try to increase, the rectifier output A_2 will also increase, causing the ouptut of the integrator to add a net DC current into the modulator lamps. This effect will result in decreasing the output amplitude back to its equilibrium state.

Notice the output is AC coupled by capacitor C_2. This is necessary in order to eliminate the DC offset voltage caused by the gain modulator DC biasing circuitry (V_1, R_{11}).

The addition of resistor R_8 makes the net rectification full wave rather than half wave, as seen by A_3.

Although the circuit concept is relatively straightforward, this network is difficult to design because of the interaction between lamp modulator DC offset voltage and the critical AC voltage gain.

It should be noted that the RC time constant of the integrator circuit (A_3) should be larger than the temperature characteristic time constant of the lamps. This consideration will eliminate the output amplitude from "hunting" due to overall closed loop instability.

Precision Wien Bridge Oscillator (AC Coupled Modulator)

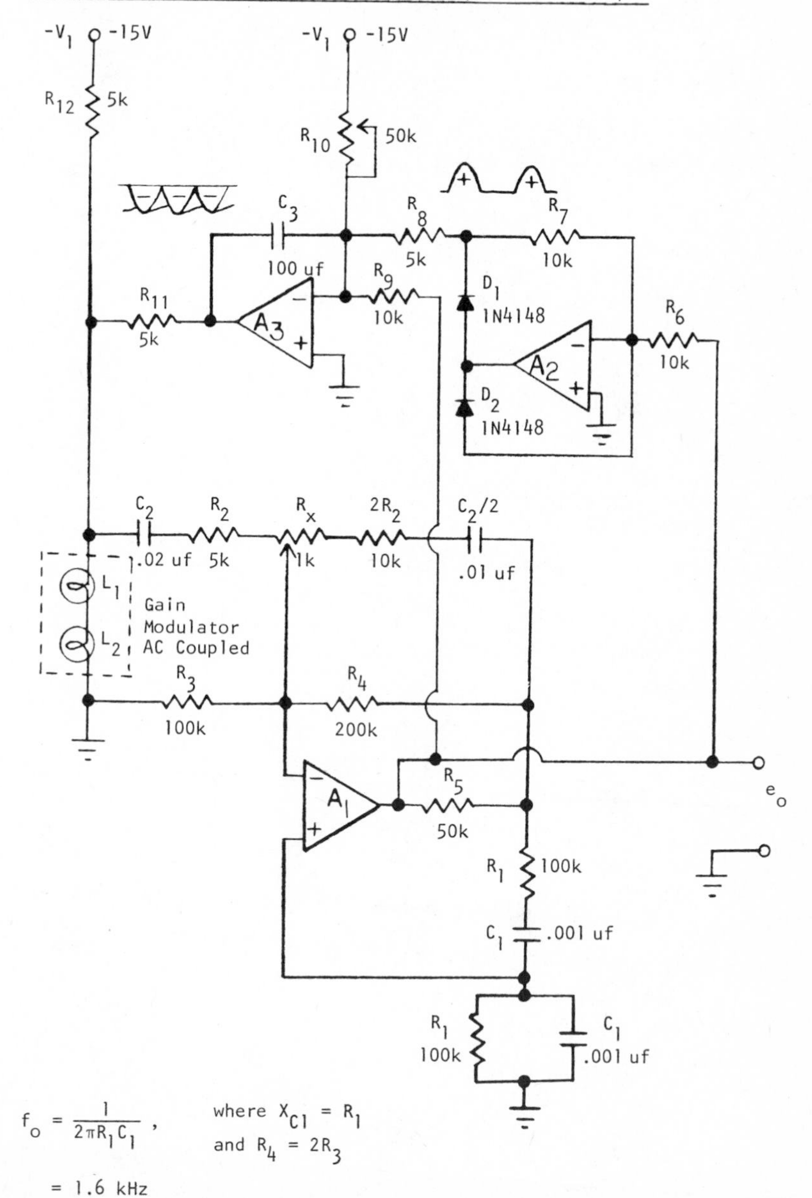

$$f_o = \frac{1}{2\pi R_1 C_1}, \quad \text{where } X_{C1} = R_1 \text{ and } R_4 = 2R_3$$

$= 1.6$ kHz

Precision Wien Bridge Oscillator (continued)

The network employs a DC feedback loop A_2, A_3 to modulate the AC coupled dynamic resistance of the lamps L_1, L_2 in order to cancel output amplitude variations due to temperature changes. The circuit offers the advantage of eliminating the DC offset voltage caused by the R_{12}, L_1, L_2 quiescent current biasing circuit. The L_1, L_2 dynamic resistance is AC coupled through the R_2, C_2 feedback combinations.

The output e_o is symmetrical about zero volts, and therefore can be direct-coupled to a non-inverting buffer for excellent load isolation characteristics.

The output amplitude is virtually independent of temperature variations due to the action of the integrator A_3 error correction current supplied to L_1, L_2 through R_{11}.

Precision Wien Bridge Oscillator (Solid-State Modulator)

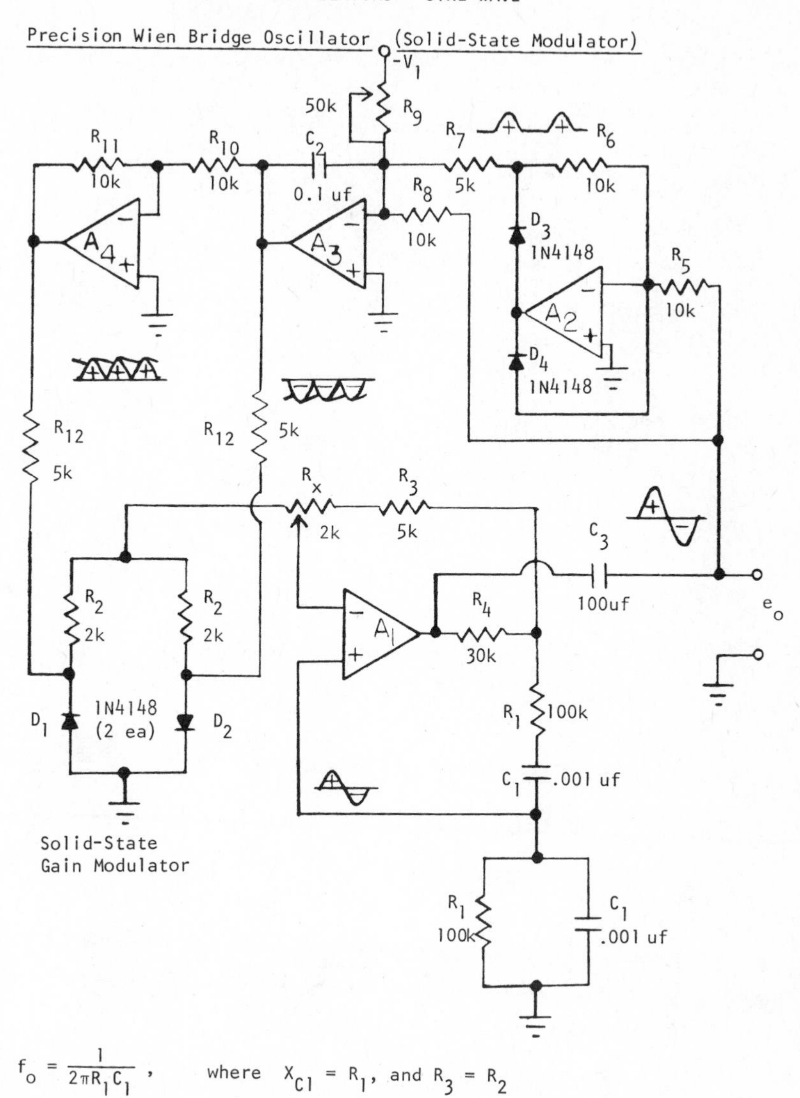

$$f_o = \frac{1}{2\pi R_1 C_1}, \qquad \text{where } X_{C1} = R_1, \text{ and } R_3 = R_2$$

$$= 1.6 \text{ kHz}$$

Precision Wien Bridge Oscillator (continued)

This network utilizes a solid-state gain modulator (D_1, D_2) instead of lamps. The output feedback signal e_o is rectified (A_2 network), filtered (A_3 network) and inverted (A_4). The outputs of A_3 and A_4 provide push-pull, or complement DC voltages, that are used to control the gain modulator circuit.

Understanding the modulator action is more easily studied by slightly rearranging the circuit schematic and replacing A_3 and A_4 outputs with equivalent variable voltage sources $-V_m$ and $+V_m$, respectively.

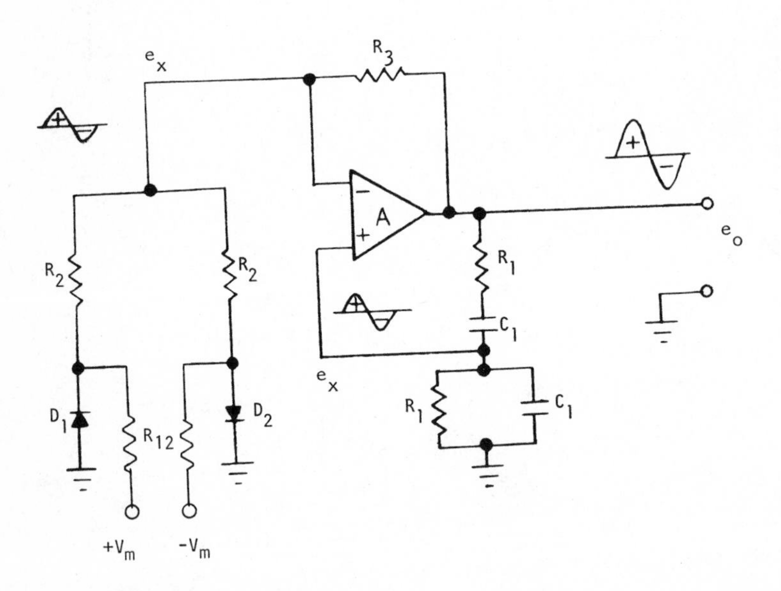

Precision Wien Bridge Oscillator (continued)

During the positive half cycle of e_x, diode D_2 conducts, and D_1 is reverse biased. Conversely, during the negative half cycle D_1 conducts and D_2 is reverse biased.

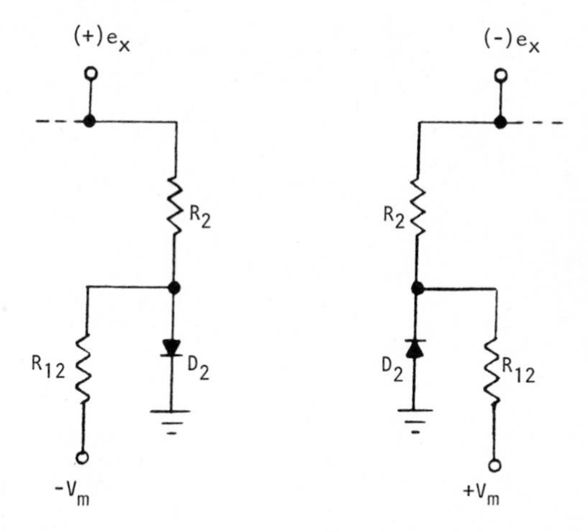

Modulation is accomplished by the voltage V_m generating an opposing bias current through R_{12} into the diode circuit. By changing this bias, the amount of current in R_2 can be raised or lowered.

As the absolute value of V_m decreases, the current through R_2 increases, which increases the current in the feedback resistor R_3. This, in turn, increases the output voltage of the A_1 non-inverting network. The gain is essentially modulated by the value of V_m. The value of V_m is controlled by the output voltage of A_1 which is e_o.

Precision Wien Bridge Oscillator (continued)

Should the output signal e_o increase, however, then V_m will increase. This increases the bias current in R_{12} which, in turn, lowers the gain. The output signal will now drop back to equilibrium. The network time response (error recovery time) is controlled by capacitor C_2. If the capacitor is too small the output will "hunt"; if the value is too large the output will correct too slowly. An optimum value can be determined empirically for the particular frequency of oscillation f_o.

Resistor R_9 sets the desired peak value of e_o for the equilibrium condition.

The diodes D_1 and D_2 should be matched in order to minimize distortion. The R_2 resistor pair, as well as the R_{12} resistor pair, should also be matched.

This network approach overcomes the basic disadvantages of the previous networks that used the thermal characteristics of lamps.

The output stability of this network can be affected by the external load that interfaces with e_o. An additional amplifier can be inserted in the loop just before the output capacitor C_3 in order to "buffer" any of these external loading effects.

Symmetrical Free Running Multivibrator

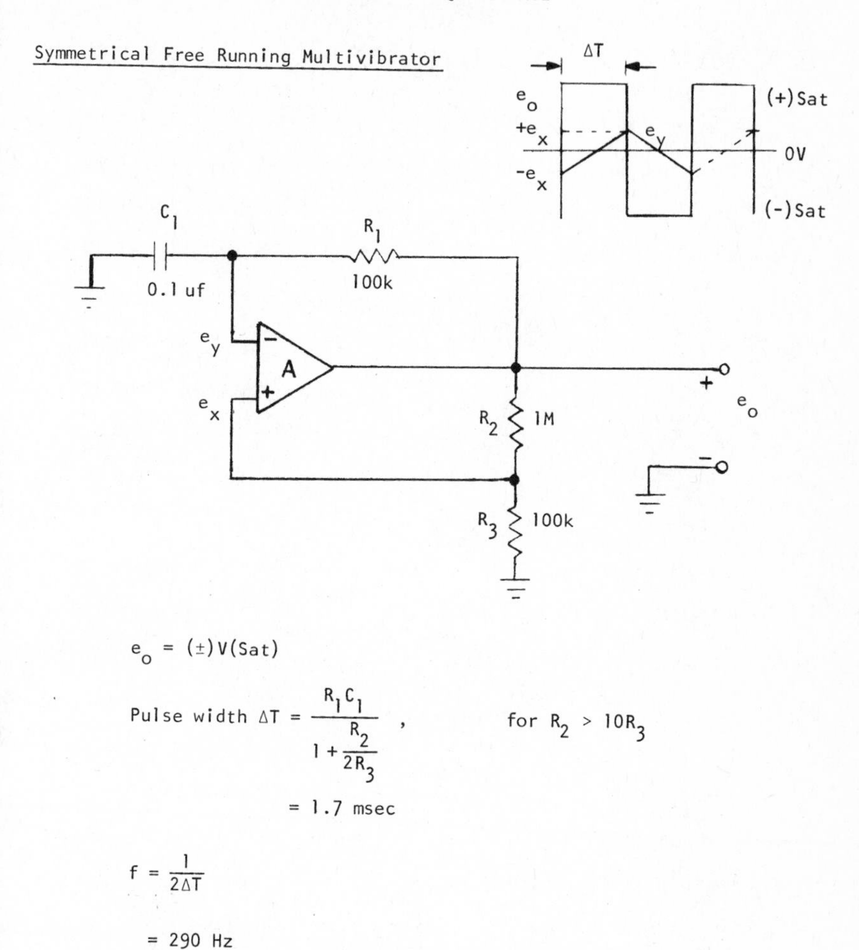

$e_o = (\pm)V(Sat)$

$$\text{Pulse width } \Delta T = \frac{R_1 C_1}{1 + \frac{R_2}{2R_3}}, \qquad \text{for } R_2 > 10R_3$$

$$= 1.7 \text{ msec}$$

$$f = \frac{1}{2\Delta T}$$

$$= 290 \text{ Hz}$$

The output signal of this network alternately swings from positive saturation to negative saturation to form a symmetrical square wave.

Symmetrical Free Running Multivibrator (continued)

The pulse width (ΔT) of the square wave is controlled by the charging rate of the capacitor C_1. For instance, assume the capacitor has no charge and the output is at positive saturation voltage. Assuming the Op Amp saturation voltage is 11 volts, then

$$e_x = V(\text{Sat}) \frac{R_3}{R_2+R_3}$$

$$= +1 \text{ volt} \quad , \qquad \text{for } V(\text{Sat}) = +11 \text{ volts}$$

$$e_y = 0 \; , \qquad \text{for no charge on } C_1$$

The output will remain in (+) saturation until the capacitor has charged just beyond the value of e_x (+1 volt). At this point the output will switch to negative saturation, and e_x will change to -1 volt. The residual charge left on C_1 is +1 volt, but the output polarity has reversed; therefore, the capacitor will start charging toward the negative output potential. When e_y reaches -1 volt, the output will again switch. This cycle repeats itself, alternately switching at the (+) or (-) voltage of e_x. As long as the value of e_x is small compared to V(sat), the charging rate of the capacitor can be assumed to be linear (rather than exponential) for simplicity.

$$I_1 = C_1 \frac{\Delta e}{\Delta T} = C_1 \frac{(2e_x)}{\Delta T} = \frac{V(\text{sat}) + e_x}{R_1}$$

$$\frac{V(\text{sat}) + e_x}{R_1} = C_1 \frac{(2e_x)}{\Delta T}$$

Symmetrical Free Running Multivibrator (continued)

Solving for pulse width

$$\Delta T = \frac{2R_1C_1e_x}{V(sat) + e_x}$$

Since e_x is directly related to the saturation voltage

$$e_x = \frac{R_3}{R_2+R_3} V(sat)$$

Substituting for e_x yields

$$\Delta T = \frac{R_1C_1}{1 + \frac{R_2}{2R_3}}$$

It is generally advisable to make e_x as small as possible to keep the capacitor C_1 charging linearly. However, making e_x too small will amplify the Op Amp temperature drift effects if e_x approaches the voltage offset specification range. The frequency will change. If e_x is made too large the triggering point will "jitter" due to the "sagging" slope of the capacitor charging exponentially. The network becomes very susceptible to random triggering on varying noise signals. The value of e_x = 1 volt is generally the optimum.

A more precise exponential equation for the pulse duration is derived in Appendix O.

$$\Delta T = R_1C_1 \ln \left(1 + \frac{2R_3}{R_2}\right)$$

Unsymmetrical Free Running Multivibrator

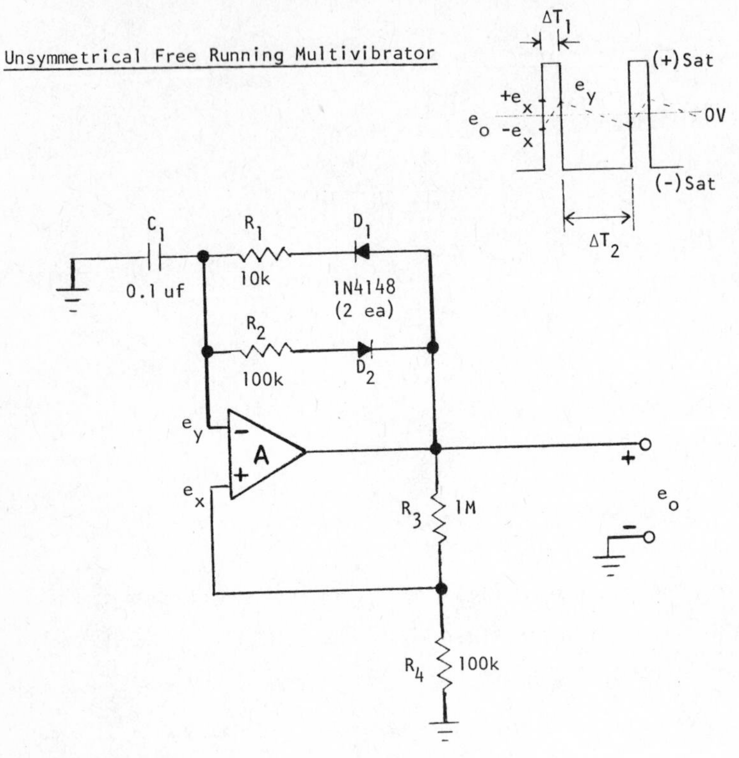

$$e_o = (\pm)\ V(\text{Sat})$$

$$\Delta T_1 = \frac{R_1 C_1}{1 + \frac{R_3}{2R_4}}\ , \qquad \text{for } R_3 > 10R_4$$

$$= 0.17 \text{ msec}$$

$$\Delta T_2 = \frac{R_2 C_1}{1 + \frac{R_3}{2R_4}}$$

$$= 1.7 \text{ msec}$$

Unsymmetrical Free Running Multivibrator (continued)

Changing the value of the feedback resistor (R_1, R_2) during each half cycle produces an unsymmetrical pulse train. During (+) saturation the changing time constant is R_1C_1. The (-) saturation time constant is R_2C_2. The threshold switching point occurs alternately at $(+)e_x$ and $(-)e_x$.

As in the case with the Symmetrical Multivibrator, the more precise exponential equation is the same as that derived in Appendix 0.

$$\Delta T_1 = R_1C_1 \ln \left(1 + \frac{2R_4}{R_3}\right)$$

$$\Delta T_2 = R_2C_1 \ln \left(1 + \frac{2R_4}{R_3}\right)$$

Synchronized Free Running Multivibrator

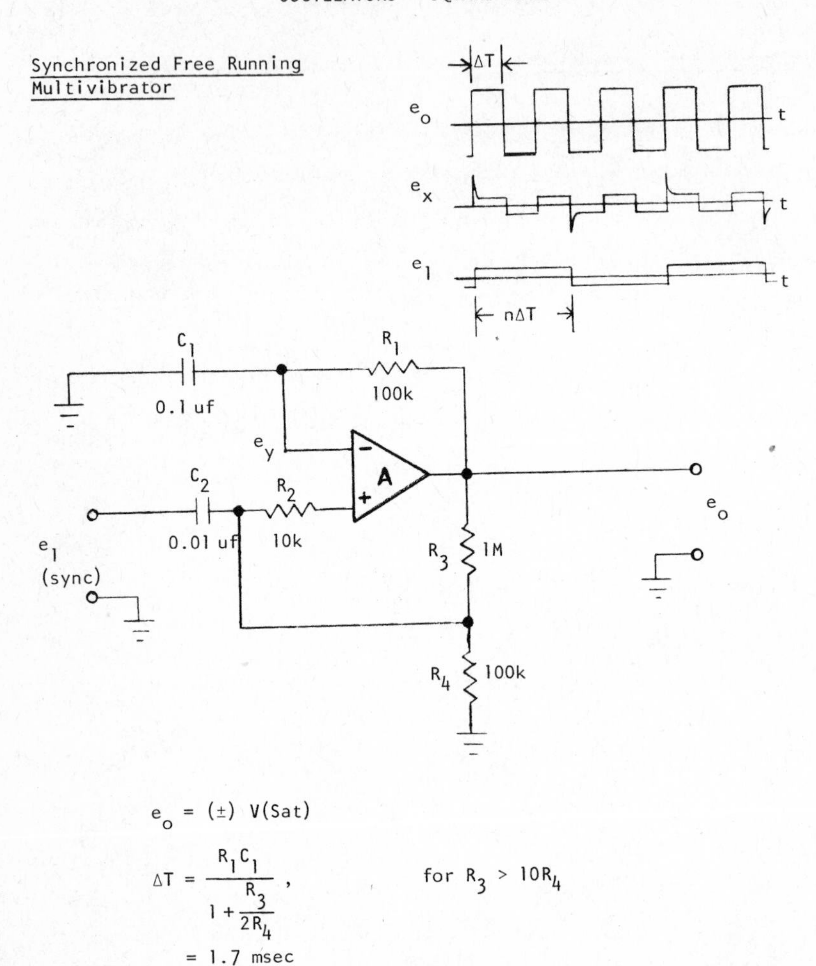

$$e_o = (\pm)\ V(Sat)$$

$$\Delta T = \frac{R_1 C_1}{1 + \frac{R_3}{2R_4}}, \qquad \text{for } R_3 > 10R_4$$

$$= 1.7 \text{ msec}$$

$$f = \frac{1}{2\Delta t}$$

$$= 290 \text{ Hz}$$

Synchronized Free Running Multivibrator (continued)

Utilizing a synchronization pulse e_1 stabilizes the output frequency. The sync pulse width is equal to, or a direct multiple of, ΔT.

Resistor R_2 is used to buffer the input.

Clamped Output Free Running Multivibrator

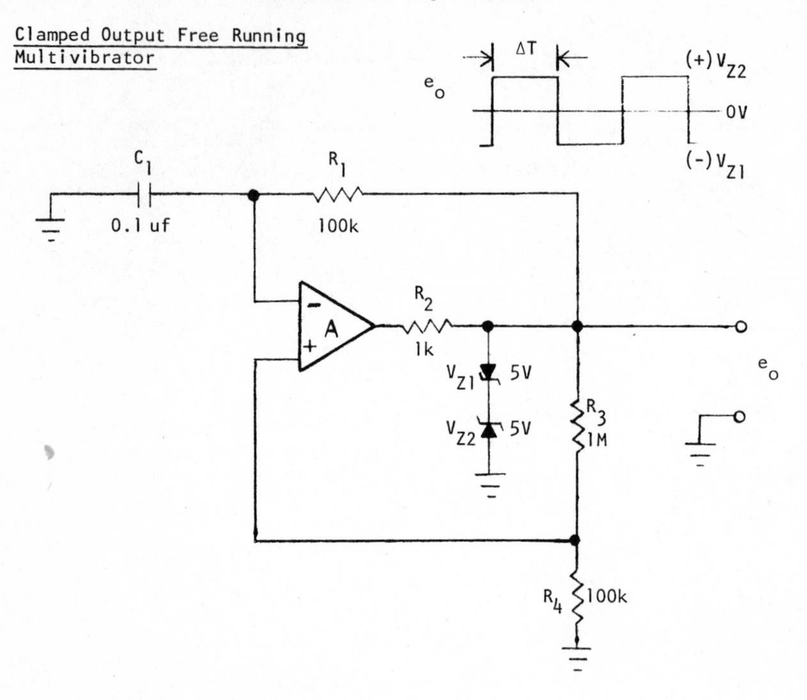

$$(+)e_o = V_{Z2} + 0.6V \quad , \qquad \text{for } (+)\text{Sat}$$
$$= (+)5.6V$$

$$(-)e_o = V_{Z1} + 0.6V \quad , \qquad \text{for } (-)\text{Sat}$$
$$= (-)5.6V$$

$$\Delta T = \frac{R_1 C_1}{1 + \frac{R_3}{2R_4}} \quad , \qquad \text{for } R_3 > 10R_4$$

Clamped Output Free Running Multivibrator (continued)

The output of the oscillator is clamped by the zener diodes V_{Z1}, V_{Z2}. Resistor R_2 serves as the current limiting resistor for the Op Amp.

$$I_{max} = \frac{V(Sat) - V(Clamped)}{R_2}$$

$$= \frac{11V - 5.6V}{1k}, \qquad \text{for } V(Sat) = \pm 11V$$

$$= 5.4 \text{ ma}$$

The basic advantage of this technique is that it presents a "clean" square wave to the output. Overshoot and ringing are virtually eliminated.

See Clips and Clamps section for additional information on other possible clamping circuit techniques.

Monostable Multivibrator

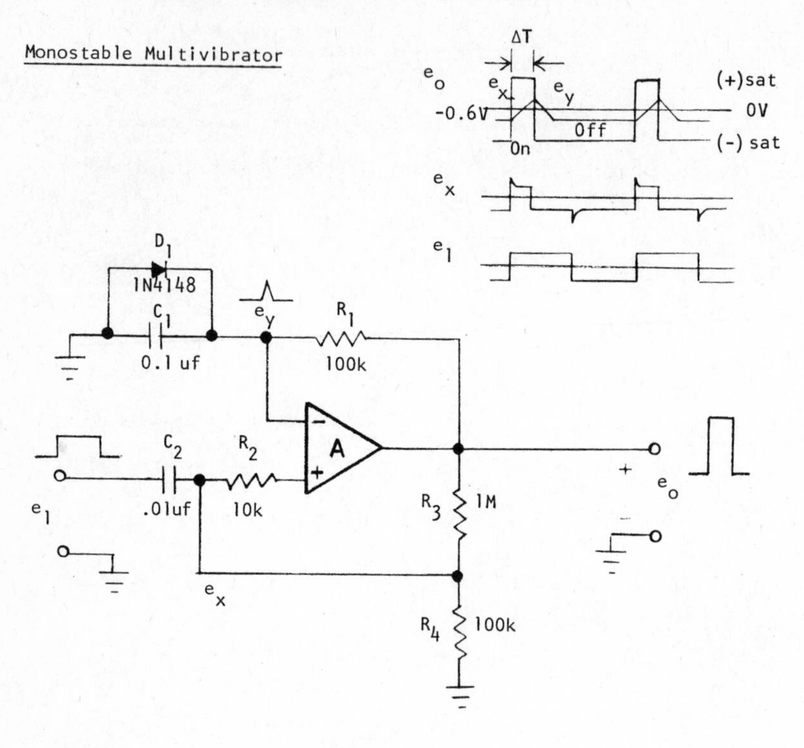

$$e_o = (\pm)\ V(sat)$$

(approximate) $$\Delta T = \frac{R_1 C_1}{1 + \frac{R_2}{2R_4}},$$

$$= 1.7 \text{ msec},$$

$$e_x = V(sat)\frac{R_4}{R_3 + R_4}$$

$$= 1V$$

for $V(sat) = 11V$
and $0.6V < e_x < 1V$

actual value will always be slightly less (1.5 msec)

Monostable Multivibrator (continued)

This circuit functions in a similar manner to the Free Running Multivibrator except an external start pulse e_1 is used to trigger the network.

Diode D_1 only allows the capacitor C_1 to charge in the positive direction. The time required for C_1 to charge to $(+)e_x$ determines the "On" pulse width ΔT of the Monostable Multivibrator. When C_1 reaches $(+)e_x$ the output switches "Off" to negative saturation $(-)V(sat)$. At this point the capacitor C_1 discharges toward $(-)e_x$, but is stopped at -0.6V, the D_1 forward voltage drop. This clamping action inhibits the Multivibrator from returning itself to the "On" position. Only the next external trigger pulse e_1 will cause the cycle to repeat itself.

When designing this network, care should be taken to insure that e_x be somewhat larger than the forward diode voltage drop of D_1.

$$e_x = V(sat) \frac{R_4}{R_3+R_4} > 0.6V$$

Below this critical value the Multivibrator will "free run."

A more accurate, but somewhat more tedious, method for calculating the "On" pulse duration time is the time constant exponential equation.

$$\Delta e_y = [V(Sat) + 0.6V] \, [1 - \varepsilon^{-T/R_1C_1}]$$

Monostable Multivibrator (continued)

The capacitor starts charging from -0.6V ("Off" state) and goes positive to $(+)e_x$. Therefore, the value of Δe_y in the preceding equation is the sum of the diode drop plus e_x.

$$\Delta e_y = e_x + 0.6V \text{ ,} \qquad \text{where } e_x = V(sat) \frac{R_4}{R_3+R_4}$$

Substituting and rearranging terms,

$$\varepsilon^{-\Delta T/R_1 C_1} = 1 - \frac{e_x - 0.6V}{V(Sat) + 0.6V}$$

Solving for ΔT,

$$\Delta T = R_1 C_1 \ln \left[\left(1 + \frac{0.6V}{V(Sat)} \right) \left(1 + \frac{R_4}{R_3} \right) \right]$$

$$= 1.5 \text{ msec ,} \qquad \text{for } V(Sat) = 11V$$

Capacitor C_2 AC couples the input trigger pulse and resistor R_2 serves as an input buffer. Note that the total pulse width of e_1 can exceed ΔT.

Reversing the diode D_1 will invert the polarity of the "On" state. The triggering polarity is also reversed.

Improved Monostable Multivibrator

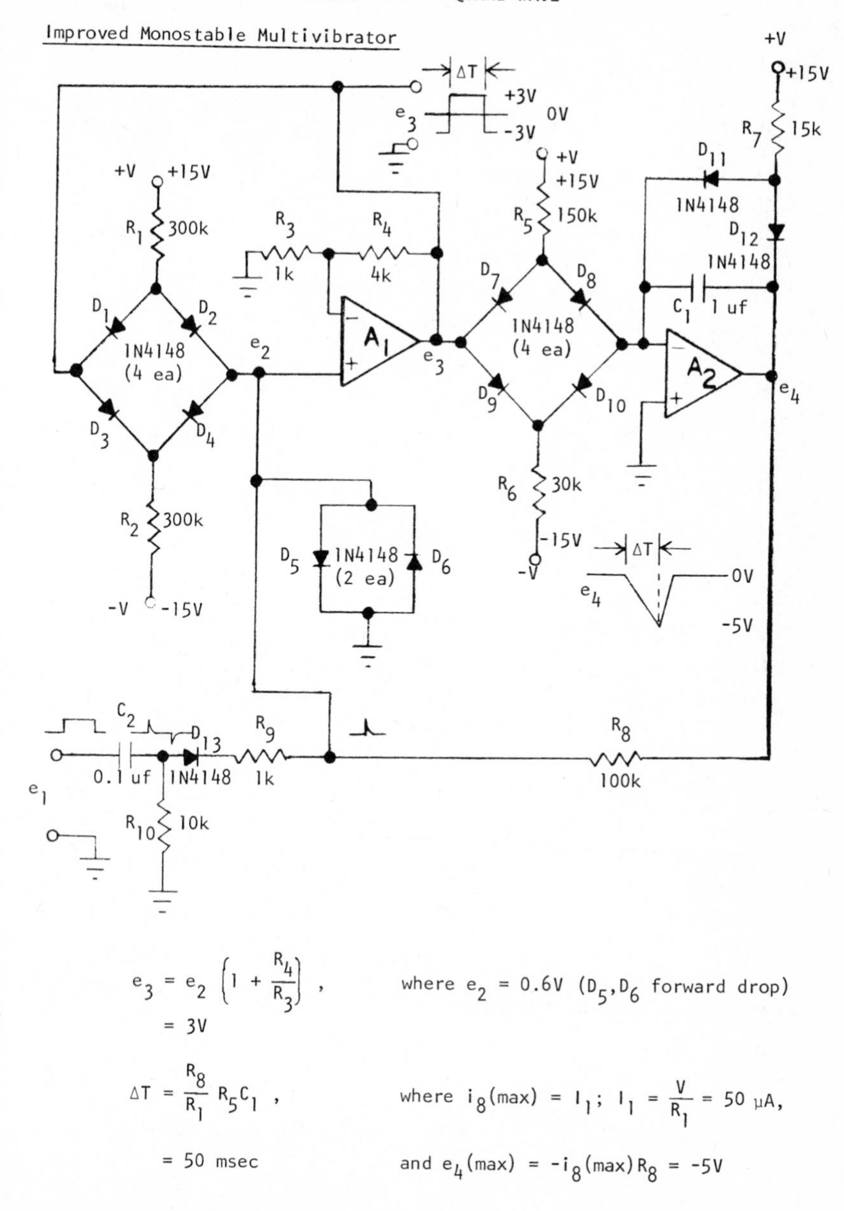

$$e_3 = e_2\left(1 + \frac{R_4}{R_3}\right), \qquad \text{where } e_2 = 0.6\text{V } (D_5, D_6 \text{ forward drop})$$

$$= 3\text{V}$$

$$\Delta T = \frac{R_8}{R_1} R_5 C_1, \qquad \text{where } i_8(\text{max}) = I_1;\ I_1 = \frac{V}{R_1} = 50\ \mu\text{A},$$

$$= 50 \text{ msec} \qquad \text{and } e_4(\text{max}) = -i_8(\text{max}) R_8 = -5\text{V}$$

Improved Monostable Multivibrator (continued)

This network offers the advantage of a precise linear charge rate for the capacitor (C_1) which determines the pulse width ΔT.

The two diode bridge circuits (D_1 to D_4 and D_7 to D_{10}) are current switches both driven by Op Amp A_1. When e_3 is negative ("Off" state), both current switches will transfer a negative input signal to its corresponding Op Amp.

In the "Off" state e_2 is clamped by D_6 ($e_2 = 0.6V$); current flow goes through D_4 and R_2 while D_2 (and D_3) is reversed biased. Also, at the same time, the output of A_2 is held at zero volts by the clamping action of R_7, C_{11}, and D_{12}. (See Clips and Clamps section.) In this case, the input current flow goes through R_6 and D_{10}; diode D_8 (and D_9) is reversed biased.

A positive going pulse at the trigger input e_1 will cause the output of Op Amp A_1 to quickly switch positive (e_3). This reverses the action of both the diode bridges. The input to A_1 is changed to a positive signal coming from R_1 and D_2; diodes D_4 and D_1 become reversed biased. Similarly, the input to A_2 becomes a positive current coming from R_5 and D_8 (diodes D_{10}, D_7 become reversed biased). A "positive" input to A_1, however, does not clamp e_4 (diode D_{11} becomes reversed biased), and therefore, capacitor C_1 charges linearly in a "negative" direction (e_4). The capacitor (C_1) will continue to charge (e_4) until a current through R_8 equals the current supplied by the diode bridge (R_1, D_2).

Improved Monostable Multivibrator (continued)

At this point, the input to A_1 will reverse polarity again and the Op Amp output e_3 will return to the negative "Off" state.

Meanwhile, capacitor C_1 will discharge back to zero at a rate determined by the current through R_6 and D_{10}. Notice that the capacitor (C_1) recovery rate (discharge) is considerably faster than the charge rate in order that the network may "reset" itself as quickly as possible. It should be noted, however, that the "reset" current from R_6 should never exceed the clamping current from R_7 or the clamping action will be lost.

$$\frac{V}{R_6} < \frac{V}{R_7}$$

A good rule of thumb is to keep R_6 at least twice as large as R_7. This, of course, will limit the minimum recovery time of the network.

The Monostable pulse width (ΔT) output is taken from e_3. The pulse is controlled by the leading edge, only, of the triangular wave output e_4.

$$I_5 = C_1 \frac{de_4}{dt}, \qquad \text{where } I_5 = I_{C1}$$

Rearranging,

$$e_4 = \frac{1}{C_1} \int_{\Delta T} I_5 \, dt$$

At the maximum output of the triangular wave e_4(max), Op Amp A_1 output will switch "Off". Therefore,

Improved Monostable Multivibrator (continued)

$$e_4(\text{max}) = \frac{I_5 \Delta T}{C_1}$$

and,

$$\Delta T = \frac{e_4(\text{max})\ C_1}{I_5}$$

Since,

$$I_5 = \frac{V}{R_5}$$

and,

$$e_4(\text{max}) = i_8(\text{max}) R_8 = I_1 R_8 \;, \qquad \text{where } I_1 = \frac{V}{R_1}$$

$$= \frac{R_8}{R_1} V$$

Substituting,

$$\Delta T = \frac{R_8}{R_1} R_5 C_1$$

This equation only holds true if positive and negative supply voltages are equal (±15V).

A similar equation can be derived for the recovery time Δt_r.

$$I_6 = C_1 \frac{de_4}{dt} \;, \qquad \text{where } I_6 = \frac{V}{R_6} \quad , \; e_4(\text{max}) = \frac{R_8}{R_1} V$$

$$= C_1 \frac{e_4(\text{max})}{\Delta t_r}$$

Substituting and rearranging,

$$\Delta t_r = \frac{R_8}{R_1} R_6 C_1$$

$$= 10 \text{ msec}$$

Improved Monostable Multivibrator (continued)

This network may also be used as a Ramp Generator by taking the output from e_4. The polarity of the signals may be reversed by reversing diodes D_{11}, D_{12}, and D_{13}. Resistors R_5 and R_6 must be interchanged and R_7 must now go to the negative supply voltage -15V.

OSCILLATORS - TRIANGULAR WAVE

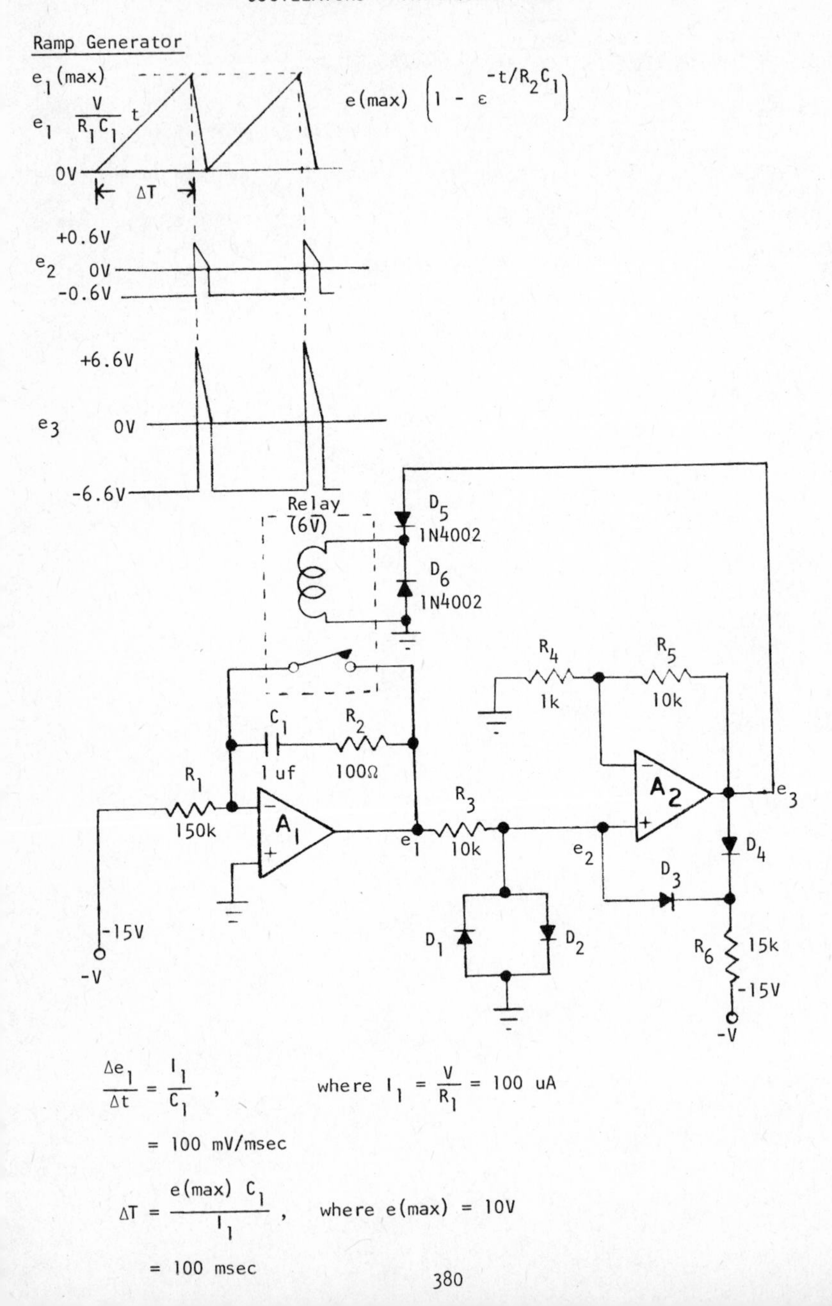

$$\frac{\Delta e_1}{\Delta t} = \frac{I_1}{C_1}, \qquad \text{where } I_1 = \frac{V}{R_1} = 100 \text{ uA}$$

$$= 100 \text{ mV/msec}$$

$$\Delta T = \frac{e(\max)\ C_1}{I_1}, \qquad \text{where } e(\max) = 10\text{V}$$

$$= 100 \text{ msec}$$

Ramp Generator (continued)

$$e_3 = e_2 \left(1 + \frac{R_5}{R_4}\right), \qquad \text{where } e_2 = 0.6V \text{ (forward drop of } D_1/D_2)$$

$$= 6.6V$$

The output ramp voltage is taken at output e_1. The waveform is generated from the integrator circuit A_1. Capacitor C_1 charges at rate determined by the current through R_1.

$$I_1 = C_1 \frac{\Delta e_1}{\Delta t}, \qquad \text{where } I_1 = \frac{V}{R_1}$$

The output e_1 will continue to linearly charge positive until the current through R_3 reaches the value set by R_6.

$$i_3(\text{max}) = I_6, \qquad \text{where } I_6 = \frac{V}{R_6} = 1 \text{ ma}$$

$$\frac{e_1(\text{max})}{R_3} = \frac{V}{R_6}$$

At this point, the output e_3 switches from negative to positive. The diode D_5 is forward biased causing the relay to energize. The close relay switch discharges the capacitor C_1 resetting the output e_1 to zero. Resistor R_2 is used to protect the relay contacts by limiting the discharge current. Making this resistor too large, however, will adversely affect (increase) the reset time.

The diode steering circuit D_3, D_4, R_6 reverse biases D_3 during the reset time. The diode D_4 forward biases during reset. Once e_1 is reset back to zero, diode D_3 again becomes forward biased and the cycle repeats itself.

Ramp Generator (continued)

The output amplitude of the A_2 network e_3 can be adjusted to fit the specifications of the particular relay chosen (reed relays are recommended).

$$e_3 = e_2 \left(1 + \frac{R_5}{R_4}\right), \qquad \text{where } e_2 = 0.6V$$

$$= 6.6V, \qquad \text{output for use with a 6V relay.}$$

Depending on the Op Amp and relay chosen, an output current booster may be required. The booster should be connected in series with the relay (rather than in the loop) for obtaining the best frequency stability. Diode D_6 is used to clamp the "inductive kickback" of the relay coil.

It should be noted that the output voltage e_3 is directly affected by the temperature variations of the diodes D_1, D_2.

Triangle and Square Wave Generator

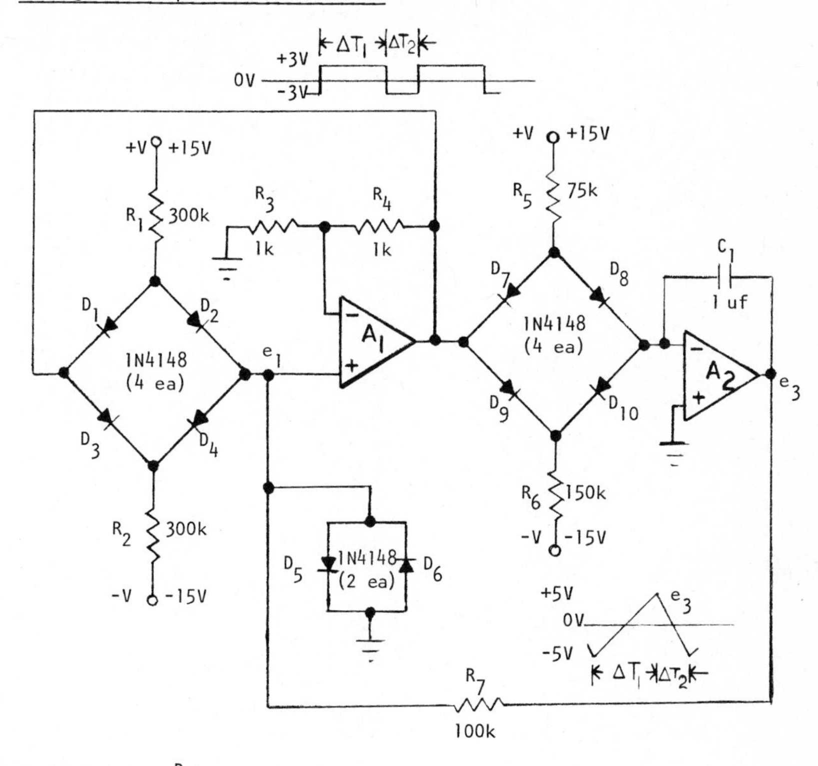

$$e_2 = \left(1 + \frac{R_4}{R_3}\right) e_1 = \pm 3V \ , \qquad \text{square wave output}$$

$$e_3 = \pm 5V \ , \qquad \text{triangular wave output}$$

$$\Delta T_1 = \frac{C_1 \ \Delta e_3}{I_6} \ ,$$

where Δe_3 is the peak-to-peak value of e_3, or $\Delta e_3 = 10V$ and $I_6 = V/R_6 = 100 \ \mu A$

$$= 100 \text{ msec}$$

$$\Delta T_2 = \frac{C_1 \ \Delta e_3}{I_5} \ ,$$

where $I_5 = V/R_5 = 200 \ \mu A$ and $\Delta e_3 = 10V$

$$= 50 \text{ msec}$$

Triangle and Square Wave Generator (continued)

This network can be utilized either as a square wave generator e_2 or a triangular wave generator e_3.

The integrator circuit (A_2) generates a ramp output e_3. The linear charge rates of the capacitor C_1 determine the precise pulse widths ΔT_1 and ΔT_2.

The two diode bridge circuits (D_1 to D_4 and D_7 to D_{10}) are current switches both driven by Op Amp A_1. When e_1 is negative, diodes D_9, D_8 and D_3, D_2 are reversed biased while D_7, D_{10} and D_1, D_4 are forward biased. Resistor R_6 sets the current I_6 that charges the capacitor C_1 at the linear rate.

$$I_6 = C_1 \frac{\Delta e_3}{\Delta t}, \qquad \text{where } I_6 = \frac{V}{R_6} = 100\ \mu A$$

$$\frac{\Delta e_3}{\Delta t} = \frac{I_6}{C_1}, \qquad \text{charge rate during time } \Delta T_1$$

$$= 100\ V/sec$$

$$= 0.1\ V/msec$$

The ramp output e_3 will continue to increase until the current through resistor R_7 becomes equal to the bridge current set by the negative supply voltage (-)V through R_2, D_4.

$$I_2 = \frac{V}{R_2}$$

$$= 50\ \mu A$$

Triangle and Square Wave Generator (continued)

Therefore, when current i_7 equals I_2, the input voltage to the Op Amp A_1 will reverse polarity causing e_2 to switch to the "positive" state.

$$i_7(\max) = I_2$$

$$\frac{e_3(\max)}{R_7} = 50\ \mu A$$

$$e_3(\max) = +5V\ , \qquad \text{positive peak}$$

When e_2 reaches the positive state both current bridge circuits switch polarity (D_9, D_8 conduct while D_7, D_{10} reverse bias and D_3, D_2 conduct while D_1, D_4 reverse bias). Capacitor C_1 now begins to charge in the opposite direction (negative) at a rate determined by R_5, C_1.

$$I_5 = C_1 \frac{\Delta e_3}{\Delta T_2}\ , \qquad \text{where } I_5 = \frac{V}{R_5} = 200\ \mu A$$

$$\frac{\Delta e_3}{\Delta T_2} = \frac{I_5}{C_1}$$

$$= 200\ V/sec$$

$$= 0.2\ V/msec$$

The capacitor will continue to go negative until e_3 reaches the negative threshold. The negative ramp switching point is established by the current set in R_1.

$$I_1 = \frac{V}{R_1}$$

$$= 50\ \mu A$$

Triangle and Square Wave Generator (continued)

As in the case with the positive ramp, the voltage e_1 will switch polarity again when i_7 reaches the value of I_1.

$$-i_7(\max) = I_1$$

$$-\frac{e_3(\max)}{R_7} = 50\ uA$$

$$e_3(\max) = -5V, \qquad \text{negative peak}$$

It can be seen that the interplay of resistor values within the bridge circuits can control both the output peak amplitudes of e_3 as well as the pulse durations.

The output amplitude of the square wave e_2 is controlled by the gain of the Op Amp A_1 network R_3, R_4.

$$e_2 = e_1\left(1 + \frac{R_4}{R_3}\right), \qquad \text{where } e_1 = 0.6V \ (D_5, D_6 \text{ forward drop})$$

$$= 3V$$

The pulse width equations (ΔT_1, ΔT_2) were derived on the bases of symmetrical voltage peaks about zero. However, this criterion was selected arbitrarily for the purpose of clarity in presenting the circuit concepts.

Bistable Multivibrator (Inverting Trigger)

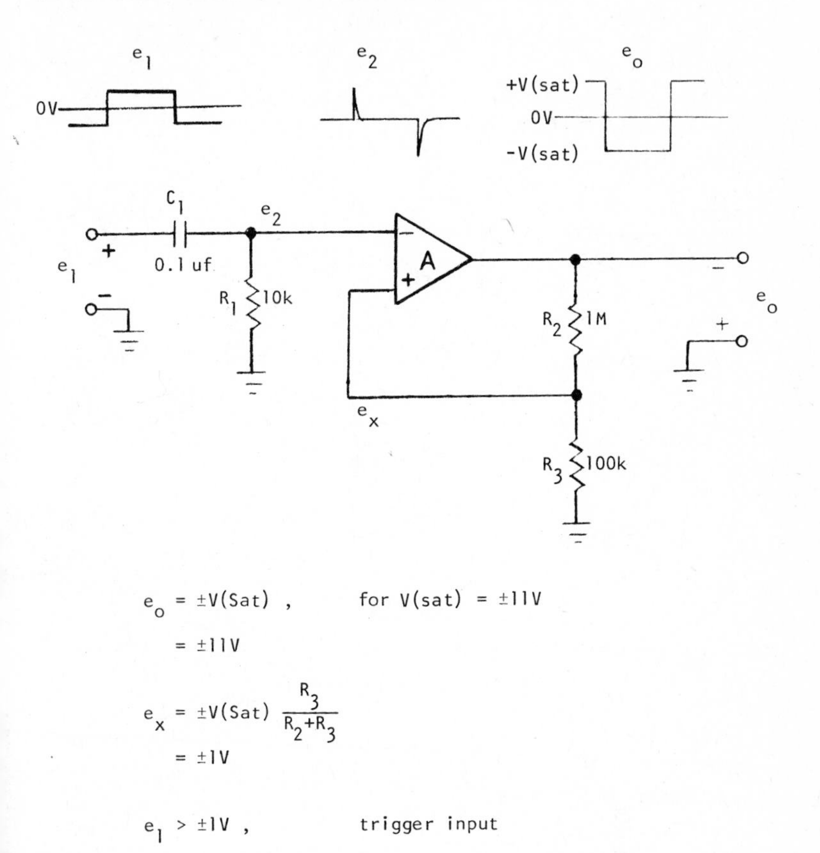

$$e_o = \pm V(\text{Sat}) \,, \qquad \text{for } V(\text{sat}) = \pm 11V$$

$$= \pm 11V$$

$$e_x = \pm V(\text{Sat}) \frac{R_3}{R_2+R_3}$$

$$= \pm 1V$$

$$e_1 > \pm 1V \,, \qquad \text{trigger input}$$

The network has two stable states: positive saturation +V(sat) and negative saturation -V(sat). The output will change states when a trigger pulse e_1 of opposite polarity and sufficient amplitude (must exceed e_x) is applied to the input e_2 of the Op Amp. Resistors R_2 and R_3 provide a positive feedback that "latches" the network for either stable state.

Bistable Multivibrator (Non-inverting Trigger)

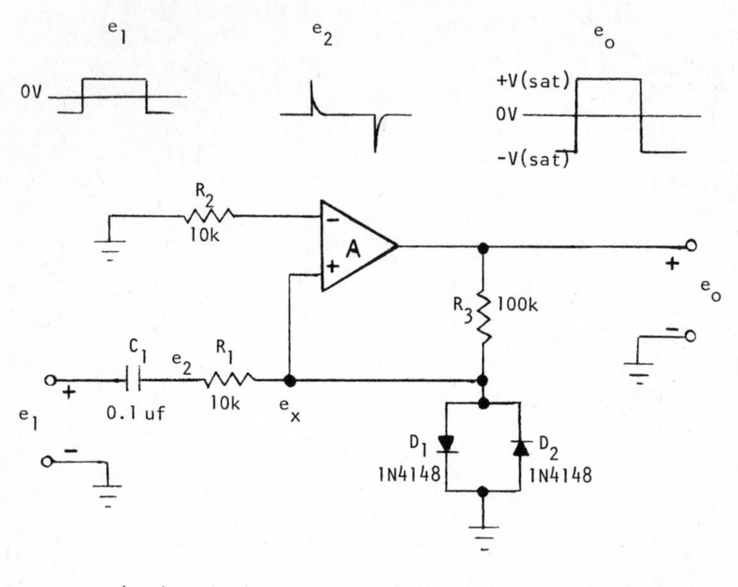

$e_o = \pm V(sat)$

$e_1 > \pm 1V$, trigger input

$e_x = \pm 0.6V$, forward diode drop of D_1, D_2

The network has two stable states, positive saturation +V(sat) and negative saturation -V(sat). The output changes state only when a trigger pulse e_1 is applied. The output will switch in the same direction as the leading edge of the trigger signals (non-inverting).

Positive feedback "latching" is provided by the R_3, D_1, D_2 circuit. Resistors R_1 and R_2 are used to reduce input loading to the trigger signal e_1.

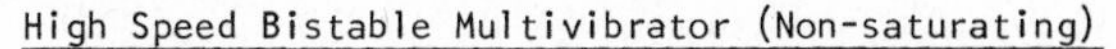

High Speed Bistable Multivibrator (Non-saturating)

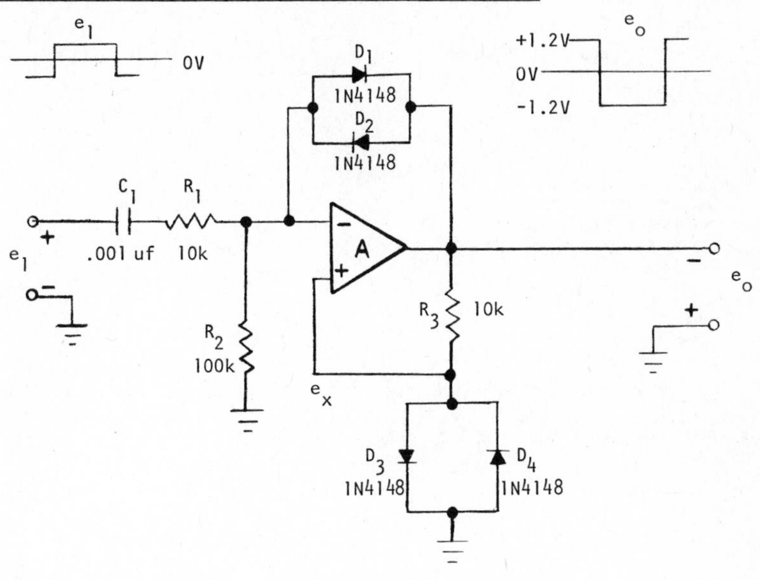

$e_o = \pm 1.2V$, forward diode drops of D_1, D_2 and D_3, D_4

$e_x = \pm 0.6V$

$e_1 > I_2 R_1$ minimum trigger input where $I_2 = I_1$, $I_2 = \frac{e_x}{R_2}$

$= 60 \ mV_{p-p}$

The speed of this Multivibrator is fast due to the low voltage clamping circuit D_1, D_2. The Op Amp is kept from saturating; this increases the switching response.

Since the Op Amp operates linearly, the input trigger level becomes correspondingly sensitive. The input triggering current generated by e_1, R_1 must simply overcome the current through R_2.

$$I_2 = \frac{e_x}{R_2} = 6 \ \mu A$$

High Speed Bistable Multivibrator (continued)

Therefore, resistor R_1 can be adjusted to minimize input loading with respect to the input voltage level.

$$R_1 = \frac{e_1}{I_1}, \qquad \text{where } I_1 \geq I_2$$

For an input trigger level of $0.6V_{p-p}$, the maximum limit for R_1 is

$$R_1 = \frac{0.6V}{6\ \mu A} = 100k$$

The value of C_1 should be decreased correspondingly to keep the input time constant within a safe recovery range.

Bistable Counter

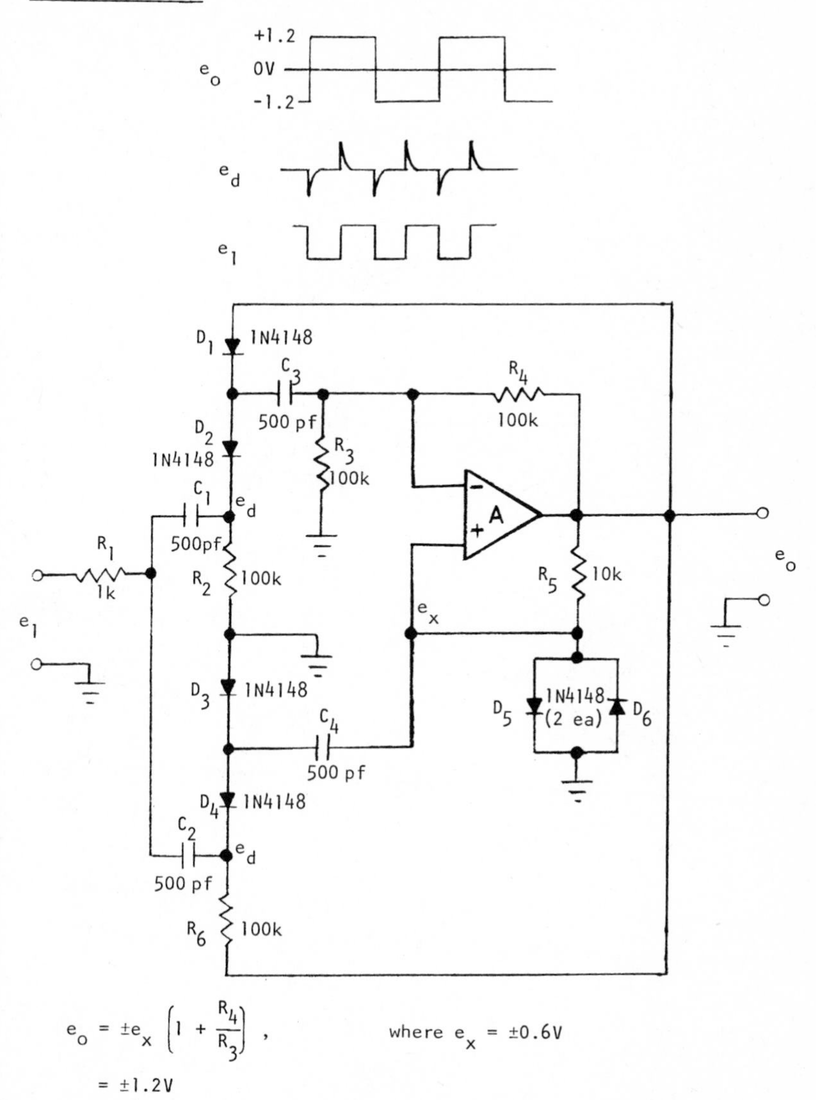

$$e_o = \pm e_x \left(1 + \frac{R_4}{R_3}\right), \qquad \text{where } e_x = \pm 0.6V$$

$$= \pm 1.2V$$

$$e_1 = I_5 R_1,$$ minimum trigger input. Networks trigger only on negative going edge

$$= 120 \text{ mV}_{p-p}$$ where $I_5 = e_o/R_5$

Bistable Counter (continued)

This network constitutes a modulus two counter. The multivibrator switches only on the negative edges of the input trigger pulses e_1.

The diode steering circuitry D_1, D_2, R_2 and D_3, D_4, R_6 controls the direction of signal flow to the input of the Op Amp. When the output e_o is negative, the capacitor C_4 is clamped by D_3, which inhibits the input signal from passing. However, capacitor C_3 input is open for a negative signal through D_2 (D_1 is reversed biased). A negative signal will switch the Op Amp to positive e_o. The positive output e_o now forward biases D_1, D_2 blocking signals to C_3. Conversely, the path to C_4 is opened for a negative input signal through D_4.

Therefore, it can be seen that the diode steering circuitry functions in a manner that alternately directs a negative input signal to both the inverting and non-inverting inputs of the Op Amp.

These networks may be cascaded to achieve frequency division or counters. The output voltage is symmetrical around zero, and the amplitude may be changed by changing the gain resistor R_3 and R_4.

APPENDIX A

Bandwidth and Phase Shift Error

$$\text{Output gain} = \overset{\text{gain}}{\frac{R_2}{R_1}} \overset{\text{error factor}}{\left(\frac{1}{1 + \frac{1}{A_L}}\right)}$$

The phase shift of loop gain can be expressed mathematically.

$$A_L = A_L \angle\phi = \frac{A_L}{1 + j\frac{f}{f_c}},$$

f_c = open loop pole cutoff frequency

A_L = DC loop gain

$$A_L = A\beta, \quad \text{where } \beta = \frac{R_1}{R_1+R_2}$$

$$A = A_{OL}\left(\frac{1}{1 + j\frac{f}{f_c}}\right)$$

β = feedback factor

A = Open Loop Gain

A_{OL} = DC Open Loop Gain

$$\text{Error factor} = \frac{1}{1 + \frac{1}{A_L \angle\phi}}$$

$$= \frac{1}{1 + \frac{1}{A_L\left(\frac{1}{1 + j\frac{f}{f_2}}\right)}} = \frac{1}{1 + \frac{1 + j\frac{f}{f_c}}{A_L}}$$

$$= \frac{A_L}{(A_L+1) + j\frac{f}{f_c}} = \frac{A_L\left(A_L' - j\frac{f}{f_c}\right)}{(A_L')^2 + (\frac{f}{f_c})^2}$$

Letting $A_L' = A_L + 1$

$$\text{Error factor} = \frac{A_L A_L' \left(1 - j\frac{f}{A_L' f_c}\right)}{(A_L')^2 \left[1 + \left(\frac{f}{A_L' f_c}\right)^2\right]} = \frac{A_L}{A_L'} \left[\frac{1 - j\frac{f}{A_L' f_c}}{1 + \left(\frac{f}{A_L' f_c}\right)^2}\right]$$

Since $\frac{A_L}{A_L'} = \frac{1}{1 + \frac{1}{A_L}}$

$$= \left[\frac{1}{1 + \frac{1}{A_L}}\right] \left[\frac{1}{1 + \left(\frac{f}{A_L' f_c}\right)^2}\right] \left[1 - j\frac{f}{A_L' f_c}\right]$$

$$= \left[\frac{1}{1 + \frac{1}{A_L}}\right] \left[\frac{\sqrt{1 + \left(\frac{f}{A_L' f_c}\right)^2}}{1 + \left(\frac{f}{A_L' f_c}\right)^2}\right] \angle \tan^{-1} \frac{f}{A_L' f_c}$$

For $A_L >> 1$, the $A_L' = A_L + 1 = A_L$

$$\text{Error factor} = \underbrace{\underbrace{\left[\frac{1}{1 + \frac{1}{A_L}}\right]}_{\text{DC error factor}} \underbrace{\left[\frac{1}{\sqrt{1 + \left(\frac{f}{A_L f_c}\right)^2}}\right]}_{\text{AC error factor}}}_{\text{magnitude}} \underbrace{\angle \tan^{-1} \frac{f}{A_L f_c}}_{\text{phase angle}}$$

$$\text{Error phase angle } \phi = \tan^{-1} \frac{f}{A_L f_c}$$

APPENDIX B

Slew Rate and Bandwidth

Understanding the slew rate and how it relates to bandwidth is best accomplished by reviewing the fundamentals of capacitance and charge currents.

The slew rate limitation of an Op Amp can easily be simulated by a basic integrator network.

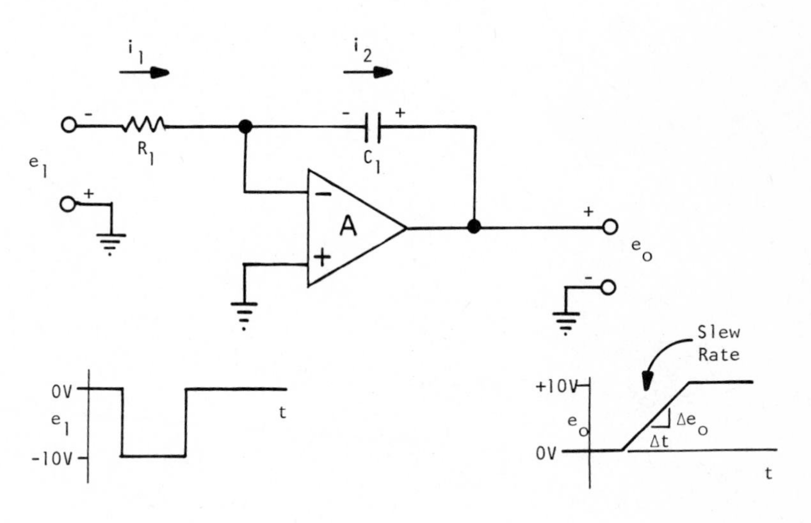

$$\text{Slew rate} = \frac{\Delta e_o}{\Delta t} = \frac{i_1}{C}$$

Figure 1. Slew Rate Simulating Circuit

The fundamental charge equation for a capacitor determines the rate of voltage increase across the capacitor.

$$i = C \frac{\Delta e}{\Delta t}$$

In Figure 1 one side of the capacitor is connected to the output, and the other side is connected to the Op Amp input, which is essentially at ground potential. This means the voltage across the capacitor is equal to the output voltage.

$$e_o = e_c$$

Therefore, from Figure 1, $i_2 = i_c$

$$i_2 = C \frac{\Delta e_o}{\Delta t}$$

The slew rate is $\frac{\Delta e_o}{\Delta t}$ volts/sec. So,

$$\text{Slew rate} = \frac{\Delta e_o}{\Delta t} = \frac{i_2}{C}$$

And from basic Op Amp fundamentals,

$$i_2 = i_1$$

Therefore,

$$\text{Slew rate} = \frac{i_1}{C}$$

This is exactly what happens within the Op Amp. The phase compensation capacitor is the major factor in determining the Op Amp slew rate specification. The size of this capacitor is designed to override all internal roll-off networks in order to achieve frequency stability for all external gain configurations. This means only one single pole is allowed to show itself externally on a correctly compensated general purpose Op Amp. This insures that the output phase shift will never exceed 90°, therefore eliminating the possibility of any ringing or oscillations.

To achieve this requirement, the phase compensation capacitor (usually located in the first stage of amplification in the Op Amp) must be relatively large. Conversely, the quiescent current or operating current of the first stage is usually relatively low in order to minimize input drift and bias errors and maximize output gain.

The combination of these two factors determines the design tradeoffs necessary for making fast-slew, high-speed Op Amps.

General purpose Op Amp slew rates range from 1.0 to 100 V/μsec without excessive engineering trade-offs. However, beyond this point the input bias currents, and voltage drifts, will begin to increase by orders of magnitude and in some cases open loop gain may be sacrificed by several orders of magnitude.

Practically in every case, the quiescent current (standby current consumption) will increase from 1 to 10 ma (as in the general purpose Op Amp) to 10 to 100 ma for the fast-slew, high-speed Op Amps.

Slew rates have been attained in the 1000 to 4000 V/μsec range with the preceding combinations of trade-offs. For a 10 volt output, the full power frequency range here is from 16 MHz to 64 MHz. These DC Op Amps actually can be operated in the video frequency range. The small signal range for these devices usually extends over 100 MHz.

The next question that may be asked is, "how does the slew rate physically and conceptually relate to the small signal bandwidth?"

The best way to answer this is to go back to the fundamental RC Time constant.

The open loop small signal response of a general purpose Op Amp is similar to that of an ordinary RC network. The only difference is the output of the RC network has no gain. For purposes of illustration let us ignore output gain since it has no value in developing the concept.

The frequency response of the RC network is determined solely by the value of R and C, where the cutoff frequency occurs when the reactance of the capacitor is equal to the resistance of the resistor (see Figure 2)

$$R = X_C = \frac{1}{2\pi f_c C}$$

Therefore

$$f_c = \frac{1}{2\pi RC}$$

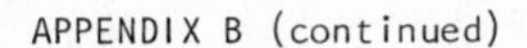

APPENDIX B (continued)

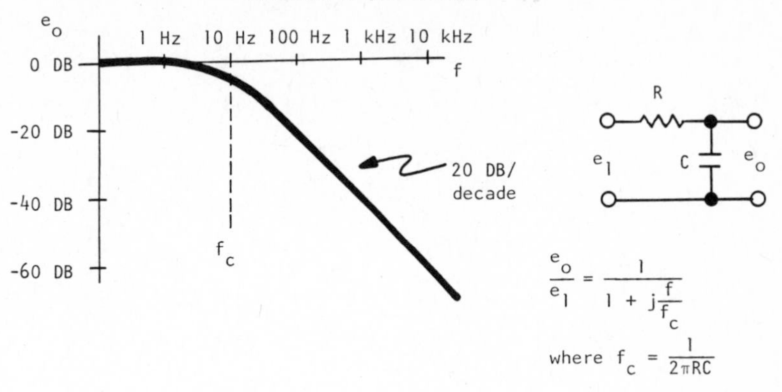

Figure 2. Frequency Response

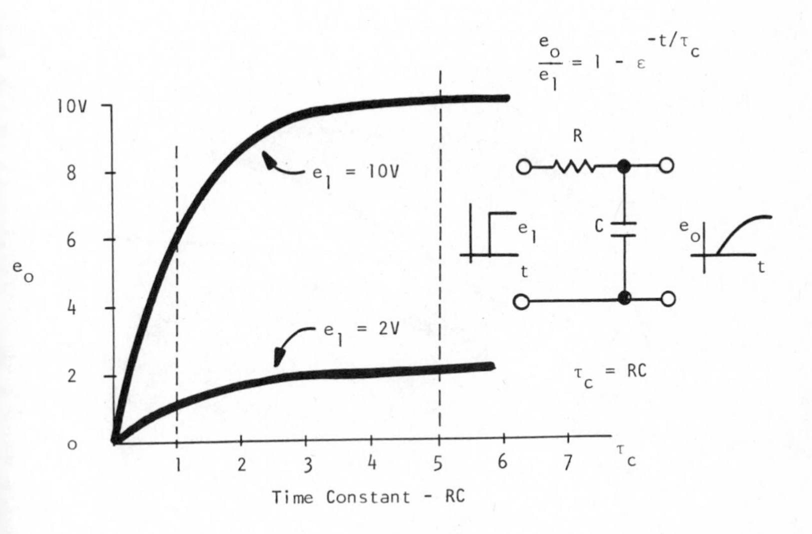

Figure 3. Square Wave Response

The pulse or square wave response of this network is shown in Figure 3.

Notice that this response is completely independent of current or voltage. It is a function of the RC Time Constant only.

$$T_c = RC$$

$$f_c = \frac{1}{2\pi T_c}$$

$$\frac{e_o}{e_1} = \frac{1}{1 + j2\pi T_c f}$$

This is the key difference between frequency response and slew rate full power response. Frequency response is governed by the RC time constant of the basic Op Amp response, where as full power response is governed by voltage magnitude as well as the RC component values.

To demonstrate this relationship refer to Figure 3. This is the square wave response of the RC network.

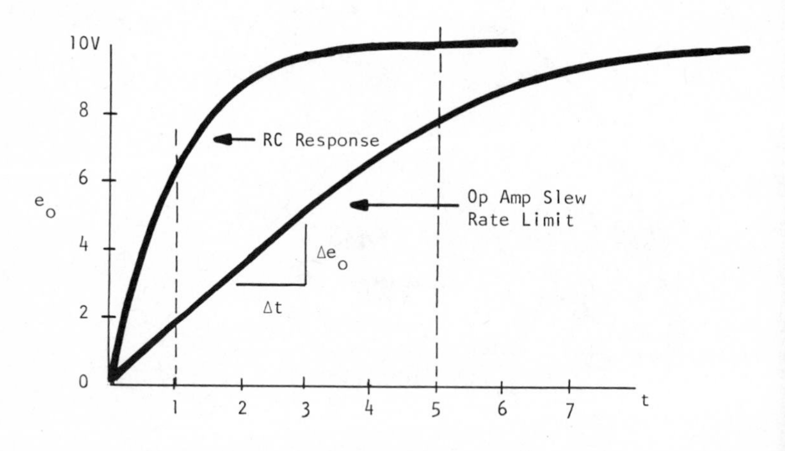

Figure 4. Slew Rate Limiting Slows Down Frequency Response for Large Output Signals

Fundamental physics shows that in one time constant the output voltage will rise to about 63% of its final value, or input level (it takes about five time constants to reach 95% of the final value).

Notice that the output reaches the 63% point always at the same time regardless of the amplitude of the signal. It is the slew rate that must increase in order to meet this criterion.

Therefore, it can be seen that if the Op Amp has a slew rate limitation less than the slew rate required by the RC time constant response, the frequency response of the output will be jeopardized for large voltage swings (see Figure 4).

Finally, to convert this square response to the Full Power Frequency requires the following mathematics.

Deriving the Full Power Frequency Response requires a closer look at an output sine wave.

In Figure 5 it can be seen that the point of maximum slew rate for an output sine wave occurs as the signal passes through zero. Deriving this value and setting it equal to the maximum slew rate of the Op Amp will result in the maximum frequency sine wave that the Op Amp will respond to before slew rate limiting.

$$\text{slew rate} = \frac{\Delta e_o}{\Delta t} = \frac{de_o}{dt}$$

$$\frac{de_o}{dt} = 2\pi f\, V_o \cos 2\pi ft$$

$$e_o = V_o \sin 2\pi ft$$

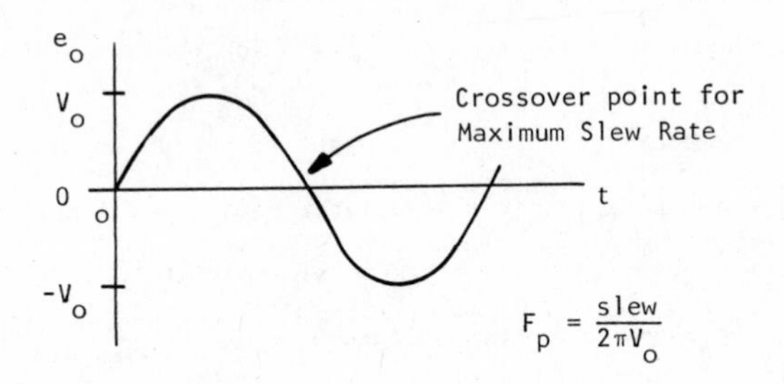

Figure 5. Output Full Power Frequency

At the maximum slew rate point where t = 0 (Figure 5)

$$\left.\frac{de_o}{dt}\right|_{t=0} = \text{Max. Slew Rate} = 2\pi F_p V_o$$

Solving for full power frequency,

$$F_p = \frac{\text{slew rate}}{2\pi V_o}$$

APPENDIX C

Low Leakage Considerations

When dealing with ultra-low bias Op Amps of less than 10 pA, care should be taken with the circuit layout.

Fundamental concepts concerning voltage gradients and material resistivities must be more closely analyzed.

Material resistivities must be watched closely especially where printed circuit boards are utilized. Surface cleanliness is also a prime consideration. When surfaces are cleaned it should be recognized that some cleaners leave a surface film which is essentially a conductor relative to these applications. Ethyl alcohol is an excellent cleaner for this purpose. Teflon terminals make excellent resistivity isolation in critical areas.

Substantial error currents will flow where voltage gradients are high. Voltage gradients are reduced two ways: (1) increasing the physical distance between the two voltage points and (2) inserting a low impedance electrical potential between the two points to "shield" one from the other. The second method is the most effective and most commonly used technique. Figures 1 (a) and 2 (a) show how this technique is applied to both inverting and non-inverting Op Amp configurations. Figure 1 (b) and 2 (b) illustrate the printed circuit board layout with respect to the physical pins, connections and terminals.

APPENDIX C (continued)

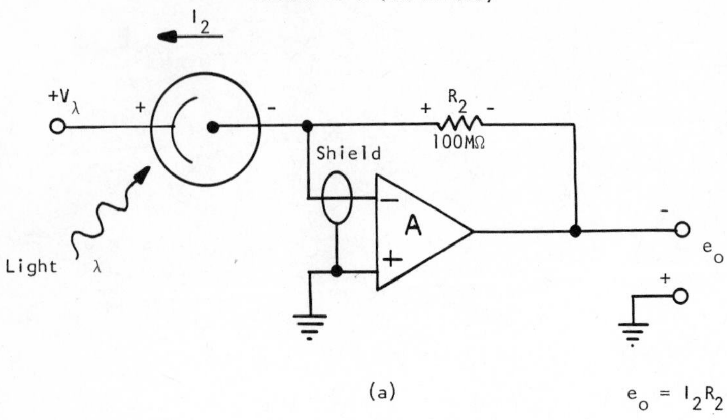

(a) $e_o = I_2R_2$

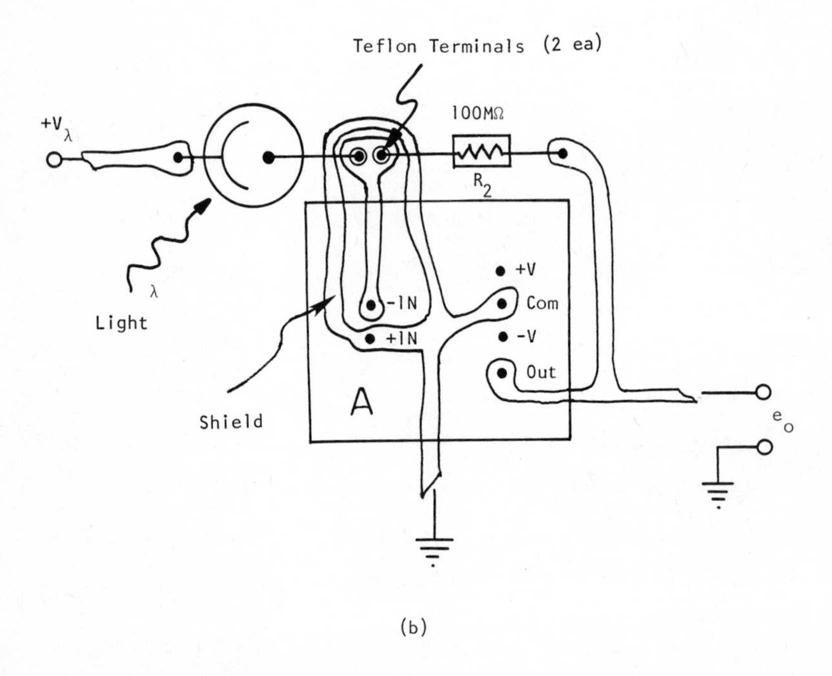

(b)

Figure 1. Inverting Shielding Technique

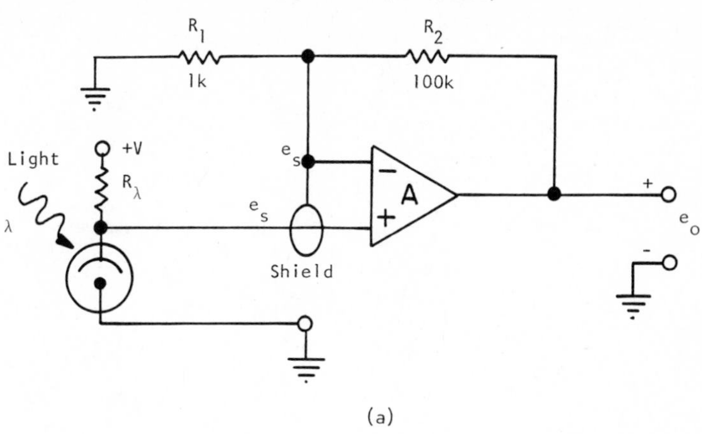

(a)

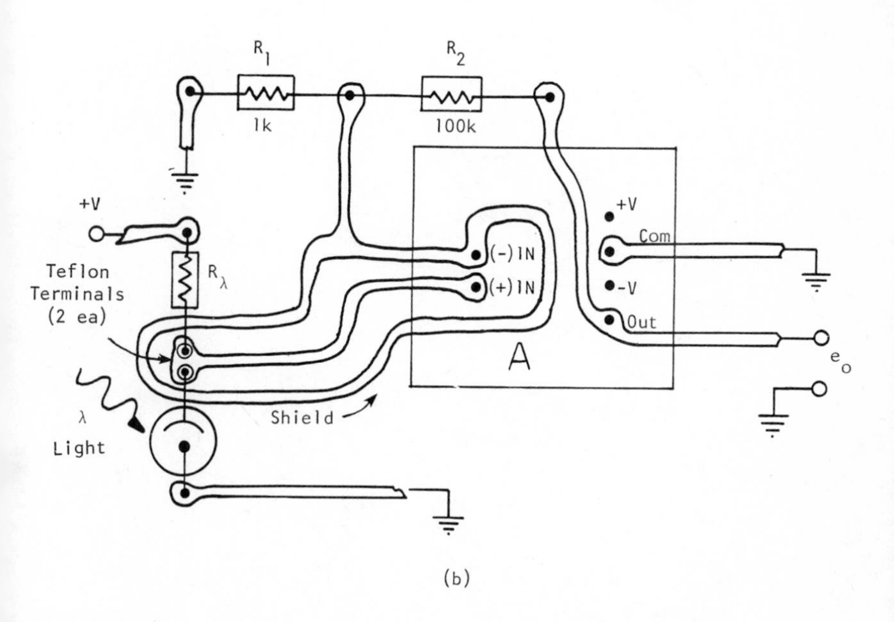

(b)

Figure 2. Non-inverting Shielding Technique

APPENDIX D

Common Mode Rejection

It should be understood that the parameter Common Mode Rejection Ratio (CMRR) the specification that enables us to transfer common mode voltage error into an equivalent input error.

$$\text{Common Mode Input Error} = V_{error(CM)} = \frac{e_1}{CMRR}$$

The output error can then be determined by multiplying this error by the closed loop gain.

$$e_o\ error(CM) = \left(1 + \frac{R_2}{R_1}\right) V_{error(CM)} = \left(1 + \frac{R_2}{R_1}\right) \frac{e_1}{CMRR}$$

To derive the general expression for output error factor caused by common mode voltage swing, take the ratio of the output error voltage to the actual output voltage

$$\frac{e_o\ error(CM)}{e_o} = \frac{\left(1 + \frac{R_2}{R_1}\right) \frac{e_1}{CMRR}}{e_o} = \left(1 + \frac{R_2}{R_1}\right) \left(\frac{e_1}{e_o}\right) \frac{1}{CMRR}$$

and since $\frac{e_o}{e_1} = 1 + \frac{R_2}{R_1}$,

$$\frac{e_o\ error(CM)}{e_o} = \frac{1}{CMRR}$$

Therefore, the percent error at the output due to common mode input is

$$\%\ error(CM) = \frac{1}{CMRR} \times 100\%$$

APPENDIX E

Definition of Effective Noise Bandwidth

The effective noise bandwidth is the power bandwidth of a rectangular equivalent of the actual bandwidth response. The area enclosed by each of these geometries is equal (see Figure 1).

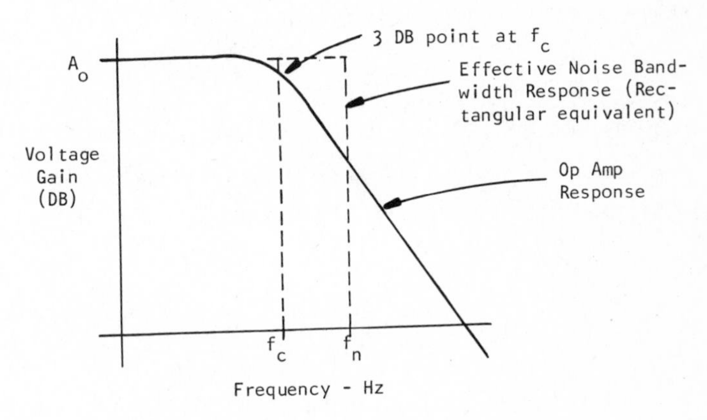

Figure 1. Effective Noise Bandwidth

The mathematical expression for this definition can be expressed by the calculus

$$f_n = \frac{1}{A_o^2} \int_o^{\infty} |A(jw)|^2 \, df$$

where f_n = Effective Noise Bandwidth

A_o = DC Open Loop Op Amp Gain

$|A(jw)|$ = absolute magnitude of Op Amp gain versus frequency

When the Op Amp response is 20 DB/decade, which is the normal Op Amp response, the voltage gain versus frequency is as shown.

$$A(jw) = \frac{A_o}{1 + j\frac{w}{w_c}} \qquad w = 2\pi f \qquad w_c = 3 \text{ DB point}$$

The Op Amp voltage gain is down 3 DB at frequency f_c as shown in Figure 1. This frequency corresponds to the half-power point (-6 DB) of the power pass-band.

Squaring the voltage gain gives an expression proportional to the power gain as a function of symmetry. Substituting into the integral equation,

$$f_n = \frac{1}{A_o^2} \int \left| \frac{A_o}{1 + j\frac{w}{w_c}} \right|^2 df$$

$$f_n = \frac{\pi}{2} f_c, \qquad \text{effective noise bandwidth}$$

This is the effective noise bandwidth for an amplifier that rolls off at 20 DB/decade.

It should be mentioned that this is the same as the effective noise bandwidth of a simple single pole RC filter.

APPENDIX F

Resistor T Network

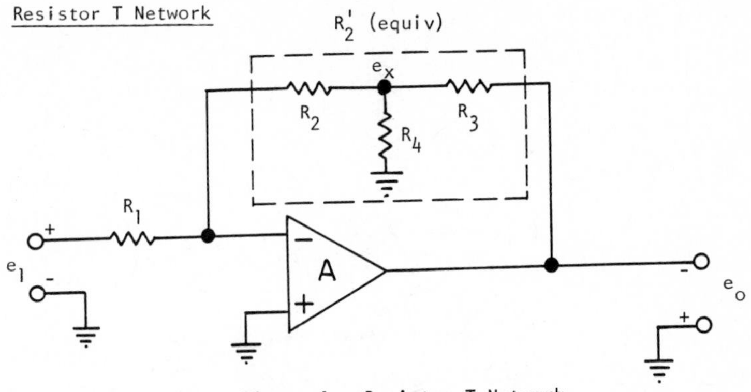

Figure 1. Resistor T Network

Rearranging the network in Figure 1 clarifies the derivation. Figure 2 shows the same circuit arranged in a manner that looks somewhat like a non-inventing configuration.

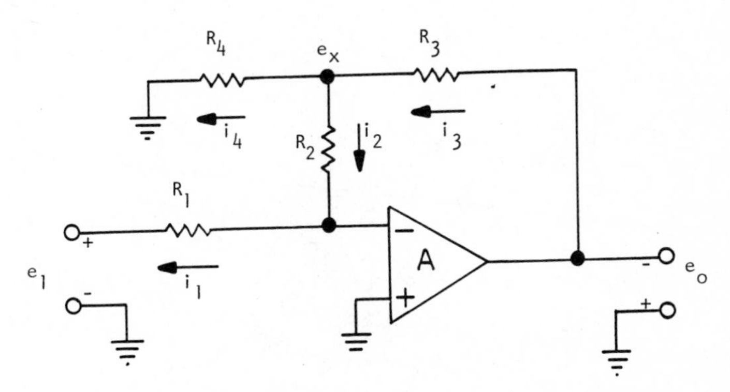

Figure 2. Rearranged T Network.

$$e_x = -\frac{R_2}{R_1} e_1 , \qquad \text{since } -i_1 = i_2$$

Assuming that i_2 is small compared to i_4, then $i_3 \approx i_4$, and

$$\frac{e_o - e_x}{R_3} = \frac{e_x}{R_4}$$

Therefore,

$$e_o = e_x \left(1 + \frac{R_3}{R_4}\right) \quad ,$$

Substituting for e_x,

$$e_o = -\frac{R_2}{R_1}\left(1 + \frac{R_3}{R_4}\right) e_1 \quad .$$

Referring back to the original Resistor T Network circuit diagram, an equivalent feedback resistor R_2' can be established.

Since $\quad -i_1 = i_2'$

then $\quad -\frac{e_1}{R_1} = \frac{e_o}{R_2'}$

and $\quad e_o = -\frac{R_2'}{R_1} e_1 \quad .$

It can be seen by comparing this equation to the network equation that the equivalent resistor is

$$R_2' = R_2 \left(1 + \frac{R_3}{R_4}\right) \quad .$$

If the assumption is not made that i_4 is equal to i_3, then the network equation becomes slightly more complicated.

$$i_3 = i_2 + i_4$$

$$\frac{e_o - e_x}{R_3} = \frac{e_x}{R_2} + \frac{e_x}{R_4} \quad .$$

Substituting for e_x,

$$e_o = -\frac{R_2}{R_1}\left(1 + \frac{R_3}{R_2} + \frac{R_3}{R_4}\right) e_1$$

$$e_o = -\frac{R_2''}{R_1} e_1$$

where

$$R_2'' = R_2\left(1 + \frac{R_3}{R_2} + \frac{R_3}{R_4}\right)$$

APPENDIX G

Current Sources

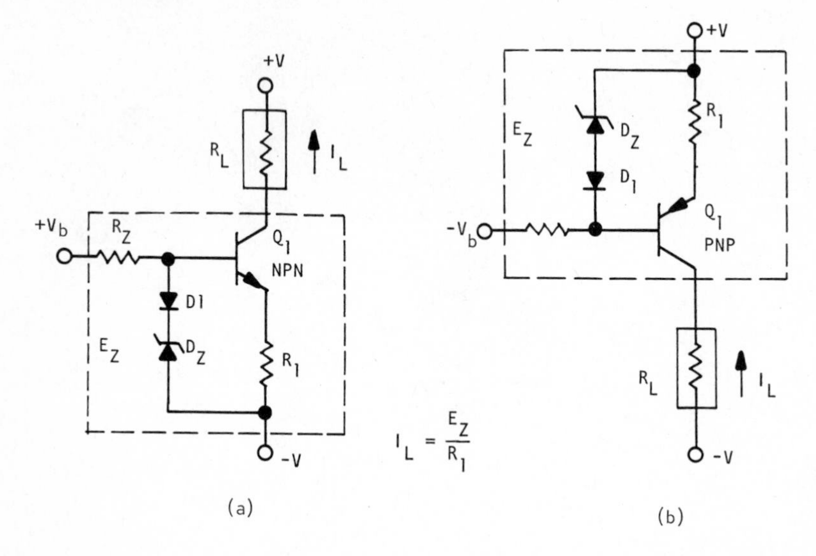

Figure 1. Bipolar Transistor Current Sources

Both circuits of Figure 1 operate essentially the same. The voltage from the Zener diode established a fixed voltage across R_1. This, in turn, generates a fixed current that will flow into the load R_L, regardless of its size. Diode D_1 is used for temperature compensation of the base-emitter junction of Q_1. (It is assumed here that the Zener diode is temperature compensated.)

Often the Zener diode can be replaced by a simple resistor, thus letting a resistor divider network establish the reference voltage E_Z.

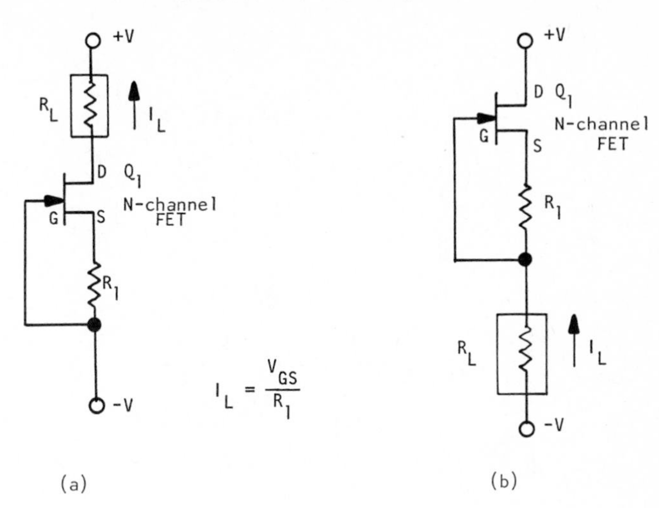

Figure 2. FET Current Sources

Figure 2 shows the simplest type of transistor current source. Its main disadvantage is that the precise value of R_1 must be determined somewhat empirically because the value of V_{GS} can vary over a relatively large range for a given type of FET. Also, temperature variations can cause significant errors unless the current is chosen near the zero temperature compensation point (Zero TC point) of the particular FET type. This value can be obtained from the data sheet or directly from the factory. Most general purpose N-channel junction FETS have a zero TC current around 100 μa - 500 μa (typically 300 μa). The exact value can be determined by using an oscilloscope curve tracer and spot heating the device while closely observing the crossover point on the I_D vs. V_{GS} curve where drain current does not change with temperature changes (the "pivot" point, see Figure 3).

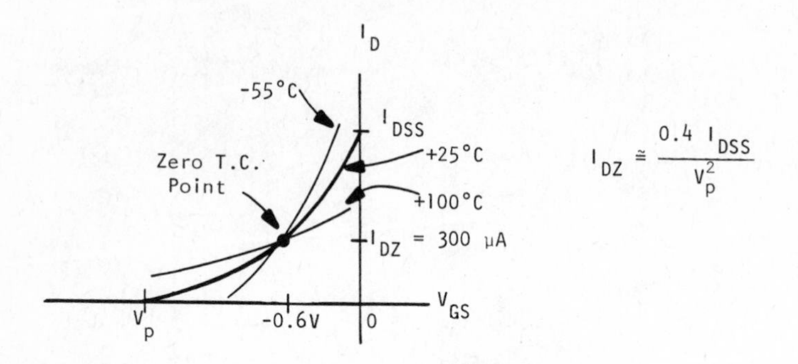

Figure 3. I_D vs. I_{GS} Curve

When designing the FET current source it is best to use the zero TC point for optimum temperature characteristics. In the circuit shown in Figures 2a and 2b, the value of R_1 using the FET characteristics in Figure 3 is

$$R_1 = \frac{V_{GS}}{I_D} = \frac{V_{GSZ}}{I_{DZ}} = \frac{0.6\text{V}}{300\ \mu\text{A}} = 2\ \text{k}\Omega$$

Figure 4 shows networks that are logical extensions of the same concepts shown in Figures 1 and 2. The additional components are used to increase output current or improve performance without being constrained to the zero TC point.

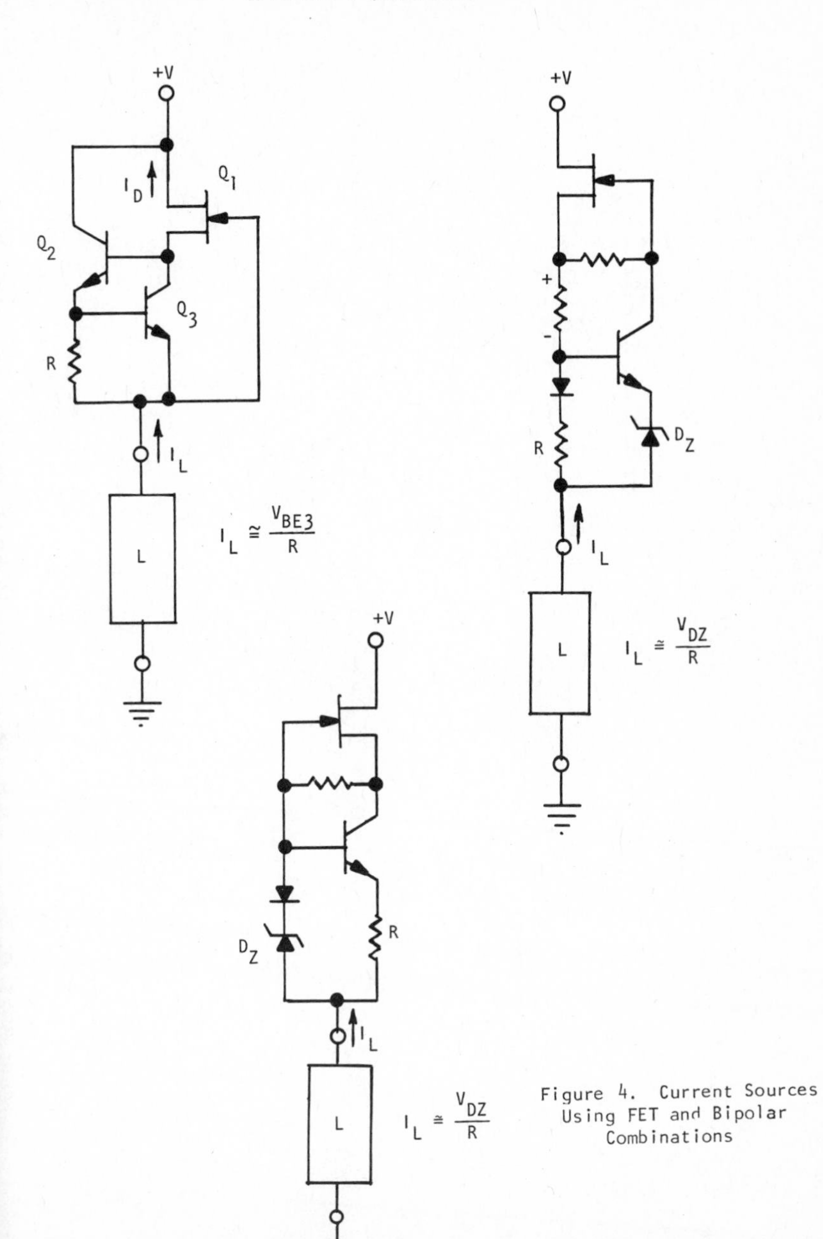

Figure 4. Current Sources Using FET and Bipolar Combinations

APPENDIX H

Level-Shift Circuit

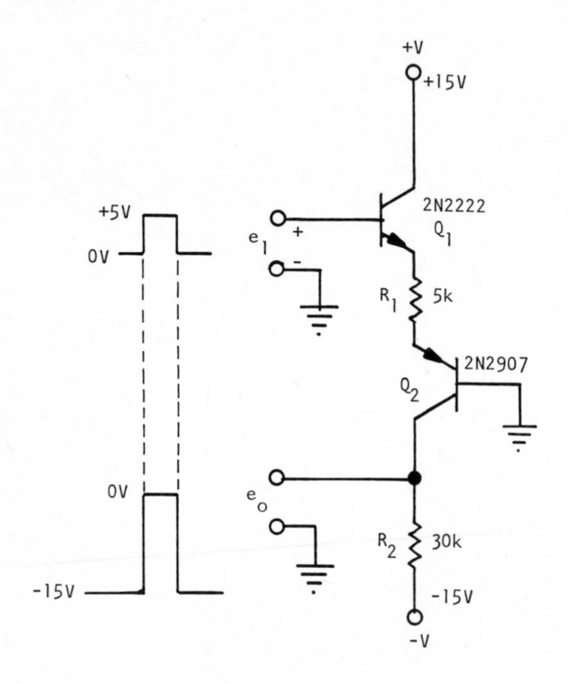

This is a general purpose circuit that converts a standard TTL +5V logic signal to a large 0 to -15V switching signal. The output e_o is used for "on-off" gating of standard N-channel FET transistors.

APPENDIX I

Double Integrator Derivations

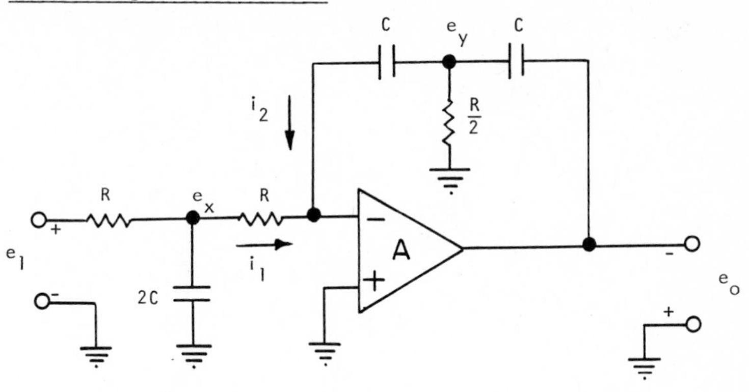

For the sake of simplicity and clarity, the derivation will be done in the frequency domain.

The basis of the derivation comes from the fact that i_1 must equal i_2.

$$i_1 = i_2$$

and

$$i_1 = \frac{e_x}{R}, \qquad i_2 = \frac{e_y}{X_C} = CS\, e_y, \qquad \text{where } X_C = \frac{1}{CS}$$

The voltages e_x and e_y can be expressed in terms of the input and output voltages, e_1 and e_o.

$$e_x = e_1 \frac{R \,||\, \frac{1}{2CS}}{R + R \,||\, \frac{1}{2CS}}$$

$$= e_1 \frac{1}{2(1 + RCS)}$$

where $$R||\frac{1}{2CS} = \frac{R}{1+2RCS}$$

which is the parallel combination of R and 2C. This is a valid assumption since i_1 flows into a virtual ground at the summing junction.

Since $$i_1 = \frac{e_x}{R}$$

then $$i_1 = \frac{e_1}{2R(1+RCS)}$$

Similarly for e_y,

$$e_y = -e_o \frac{X_C||\frac{R}{2}}{X_C + X_C||\frac{R}{2}}$$

$$= -e_o \frac{RCS}{2(1+RCS)}$$

where $$X_C||\frac{R}{2} = \frac{R}{2+RCS}$$

which is the parallel combination of C and R/2. This is a valid assumption since i_2 flows into a virtual ground at the summing junction.

Since $$i_2 = \frac{e_y}{X_C} = CS\ e_y$$

then $$i_2 = -\frac{RC^2S^2}{2(1+RCS)} e_o$$

Equating i_1 and i_2,

$$\frac{e_1}{2R(1+RCS)} = -\frac{RC^2S^2}{2(1+RCS)} e_o$$

Solving for e_o,

$$e_o = - \frac{e_1}{R^2C^2S^2}, \qquad \text{frequency domain}$$

From the Laplace transform, the output can be expressed in the time domain,

$$e_o(t) = - \frac{1}{R^2C^2} \iint e_1(t)\, dt^2 , \quad \text{time domain}$$

APPENDIX J

AC Buffer

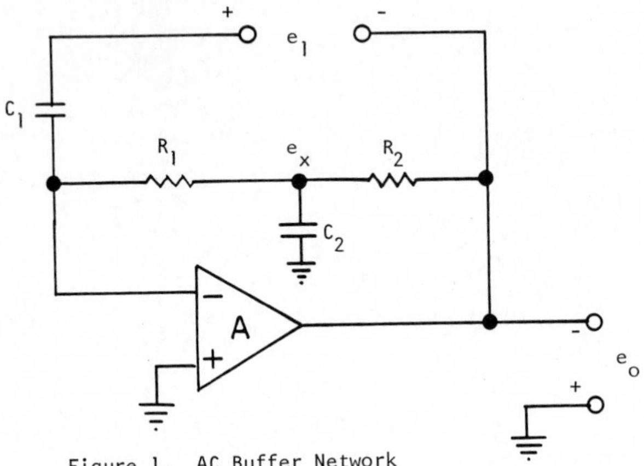

Figure 1. AC Buffer Network

Rearranging the circuit will more clearly show the derivation technique. Figure 2 shows the same circuit drawn differently.

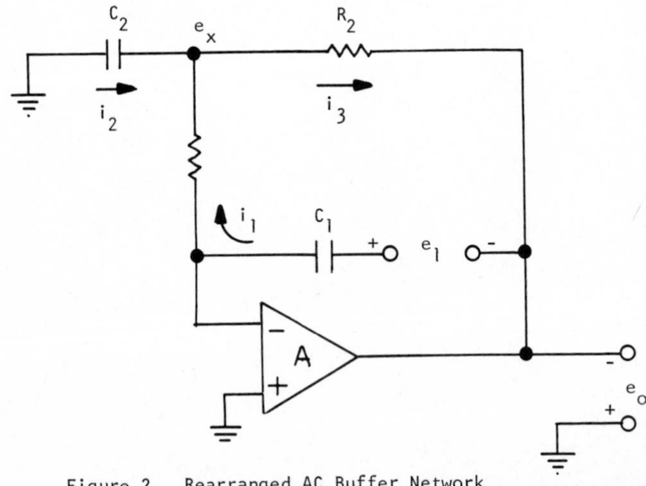

Figure 2. Rearranged AC Buffer Network

The basic assumption is that i_2 and i_3 are equal. This is true for a practical purpose because i_1 is extremely small due to the bootstrap effect of e_1 and e_o.

$$i_2 = i_3$$

$$\frac{e_x}{XC_2} = \frac{e_o - e_x}{R_2}$$

$$e_o = e_x \left(1 + \frac{R_2}{X_{C1}}\right)$$

The value of e_x is

$$e_x = i_1 R_i \,, \qquad \text{where } i_1 = \frac{e_1 - e_o}{X_{C1}}$$

$$e_x = (e_1 - e_o) \frac{R_1}{X_{C1}}$$

Substituting e_x into the output equation

$$e_o = (e_1 - e_o) \frac{R_1}{X_{C1}} \left(1 + \frac{R_2}{X_{C2}}\right)$$

In terms of the "S" transform,

$$X_C = \frac{1}{CS} \,, \qquad \text{where } S = jw$$

and solving for network gain,

$$\frac{e_o(S)}{e_1(S)} = \frac{1 + R_2C_2S}{\frac{1}{R_1C_1S} + 1 + R_2C_2S}$$

The input impedance is also very high due to the bootstrapping effect.

$$Z_{in} = \frac{e_1}{i_1} , \qquad i_1 = \frac{e_1 - e_o}{X_{C1}}$$

Therefore,

$$Z_{in} = X_{C1} \frac{1}{1 - \frac{e_o}{e_1}}$$

Substituting in the general gain equation in place of $\frac{e_o}{e_1}$ yields

$$Z_{in} = R_1 \left(\frac{1}{R_1C_1S} + 1 + R_2C_2S\right)$$

APPENDIX K

Difference Voltage to Current Converter (Grounded Load)

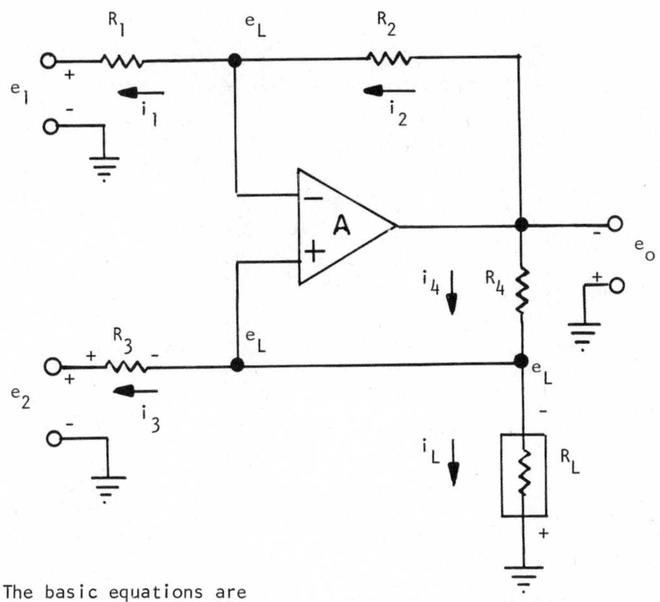

The basic equations are

$$i_L + i_3 = i_4 \qquad\qquad i_1 = i_2$$

$$i_L + \frac{e_L - e_2}{R_3} = \frac{e_o - e_L}{R_4} \qquad\qquad \frac{e_1 - e_L}{R_1} = \frac{e_L - e_o}{R_2}$$

Combining the above equations and solving for i_L,

$$i_L = \frac{1}{R_3}(e_2 - e_L) - \frac{R_2}{R_1 R_4}(e_1 - e_L)$$

Letting $\quad \frac{1}{R_3} = \frac{R_2}{R_1 R_4}$, or $\frac{R_4}{R_3} = \frac{R_2}{R_1}$,

then $\quad i_L = \frac{e_2 - e_1}{R_3}$.

APPENDIX L

Difference Voltage to Current Converter (Higher Input Impedance)

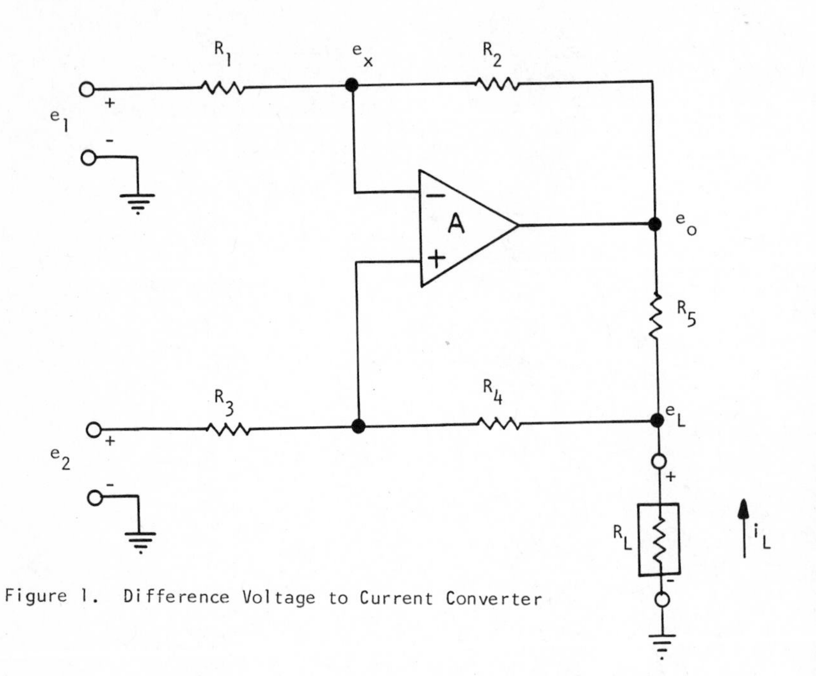

Figure 1. Difference Voltage to Current Converter

The basic concept used in developing this network is that the positive feedback voltage taken at the load point e_L through R_4 will cause the Op Amp output e_o to change an equal amount. The result will be that the voltage across R_5 will always remain constant regardless of the value of load resistance R_L. Likewise the current through R_5 will always remain constant. Therefore, if the current through R_4 is negligible compared to i_5, then the load current I_L, for all practical purposes, is equal to the current through R_5.

$$i_5 \approx I_L \qquad \text{for } i_4 << i_5$$

This is easily accomplished in practice by using large value resistors for R_3 and R_4.

In deriving the equations it is best to disconnect the positive feedback loop at e_L for clarity. The network then becomes a standard adder-subtractor circuit as shown in Figure 2.

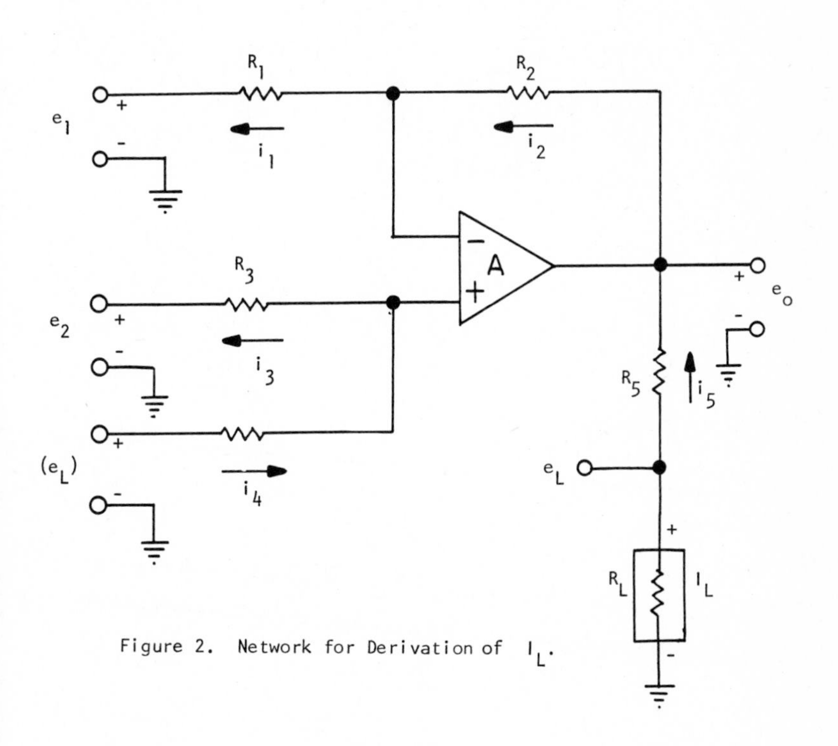

Figure 2. Network for Derivation of I_L.

The basic equations from Figure 2 are

$$i_1 = i_2 \qquad\qquad i_3 = i_4$$

$$\frac{e_1 - e_x}{R_1} = \frac{e_x - e_o}{R_2} \qquad\qquad \frac{e_2 - e_x}{R_3} = \frac{e_x - e_L}{R_4}$$

$$e_o = -\frac{R_2}{R_1} e_1 + \left(1 + \frac{R_2}{R_1}\right) e_x \qquad\qquad e_x = \frac{e_2}{\left(1 + \frac{R_3}{R_4}\right)} + \frac{e_L}{\left(1 + \frac{R_4}{R_3}\right)}$$

Substituting e_x and solving for the Op Amp output voltage,

$$e_o = -\frac{R_2}{R_1} e_1 + \frac{R_4}{R_3} \frac{\left(1 + \frac{R_2}{R_1}\right)}{\left(1 + \frac{R_4}{R_3}\right)} e_2 + \frac{\left(1 + \frac{R_2}{R_1}\right)}{\left(1 + \frac{R_4}{R_3}\right)} e_L$$

By letting the feedback ratios be equal,

$$\frac{R_2}{R_1} = \frac{R_4}{R_3}$$

the Op Amp output simplifies,

$$e_o = \frac{R_2}{R_1} (e_2 - e_1) + e_L$$

At this point it can be seen that when e_L changes, the Op Amp output e_o will change an equal amount. This is the key requirement necessary.

Since the current through R_4 is negligible compared to the load current, then

$$I_L = i_5 = \frac{e_o - e_L}{R_5}$$

Substituting in the output e_o reduces the final equation to

$$I_L = \frac{R_2}{R_1 R_5} (e_2 - e_1)$$

APPENDIX M

Buffer Power Booster

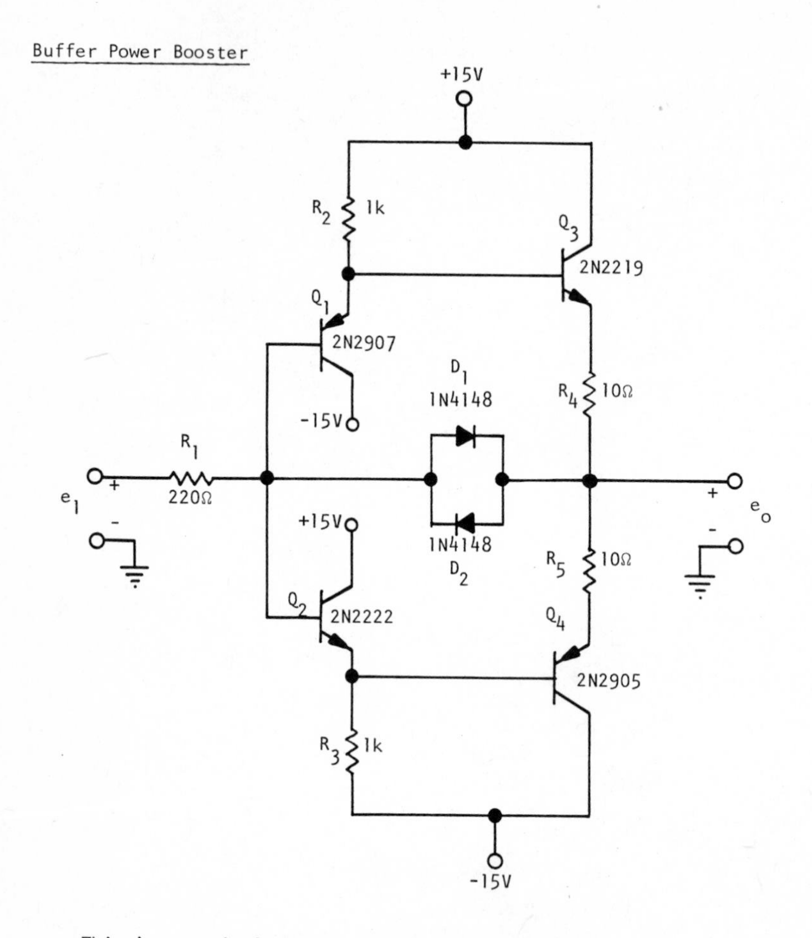

This is a typical current booster, or buffer amplifier. The network is a symmetrical combination of basic emitter follower circuits. Diodes D_1 and D_2 in conjunction with resistors R_4 and R_5 serve as the short circuit current limits. When the voltage drop across R_4 (or R_5) equals the diode forward drop D_1 (or D_2), the output load current will no longer increase. This technique protects the output transistors against inadvertent shorts to ground.

$$I_{L(max)} = \frac{V_{D1}}{R_4} = \frac{V_{D2}}{R_5}$$

$$= \frac{0.6}{10} = 60 \text{ ma}$$

Care should be taken to heatsink the output transistors properly.

Note that this design approach eliminates any amplitude cross-over distortion at zero. This advantage is due to Q_1 and Q_2 slightly biasing Q_3 and Q_4 with a "keep-alive" quiescent current. Therefore, when the input passes through zero volts, the output will follow without any discontinuity of the waveform.

APPENDIX N

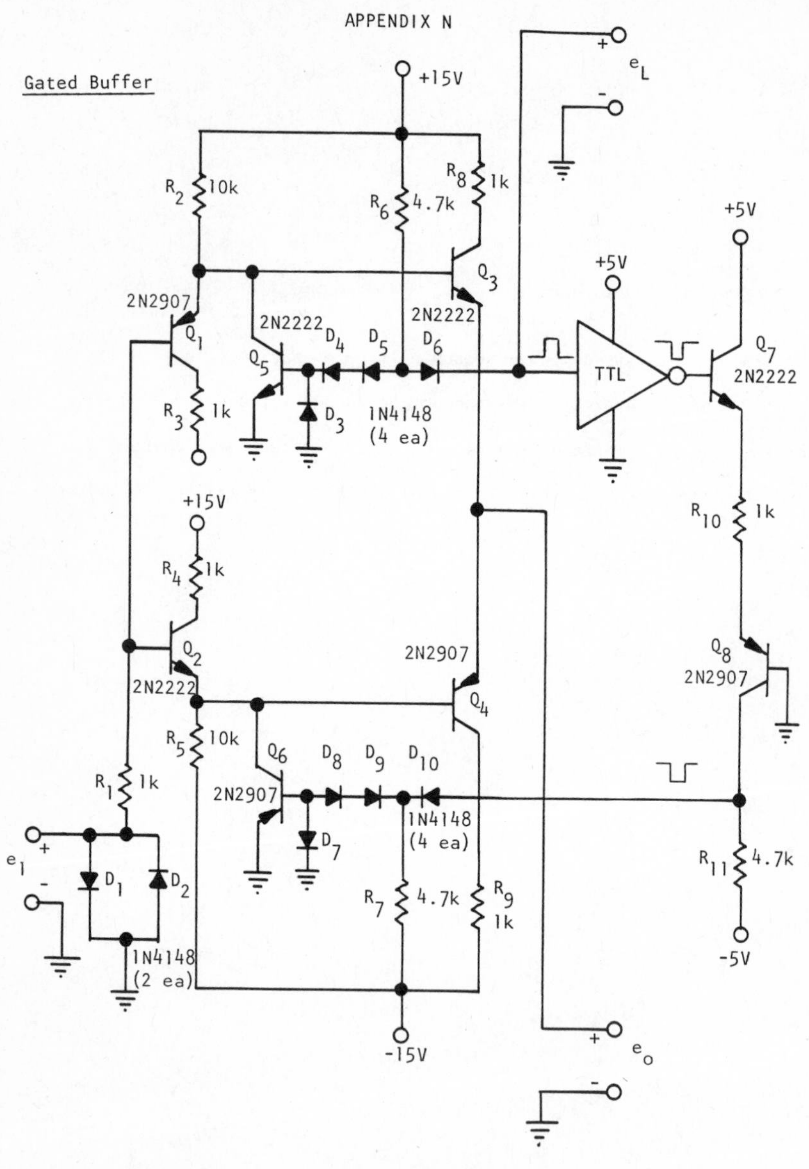

e_o = open (hold) , for e_L = +5V(TTL)

e_o = closed (track) , for e_L = 0V (TTL)

This circuit is fundamentally the same as the typical buffer shown in Appendix M except that the output transistors Q_3 and Q_4 are switched on and off by gating (Q_5, Q_6) circuitry.

The TTL gate drives the level shifting circuit Q_7, Q_8 to control the Q_4 output. This circuit is not optimized for any specific applications. However, it does demonstrate a common technique used in accomplishing the design objective.

Free Running Multivibrator

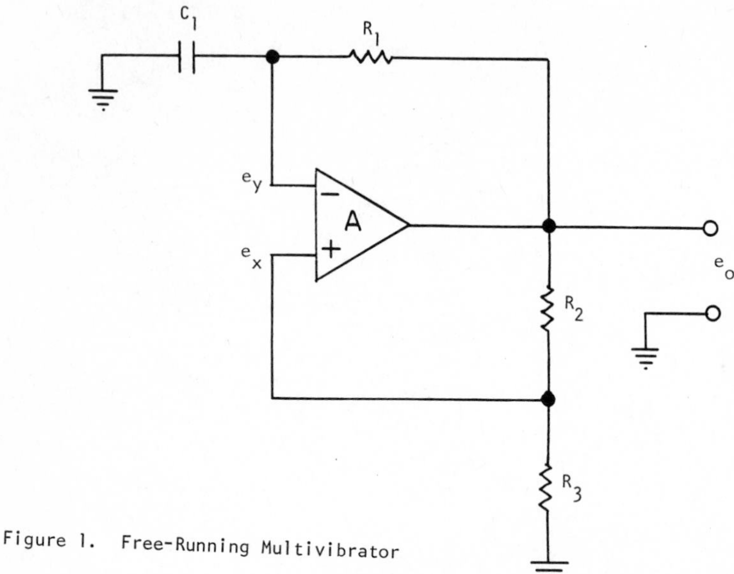

Figure 1. Free-Running Multivibrator

In the circuit shown in Figure 1 the output e_o switches when the capacitor voltage e_y reaches the value of e_x.

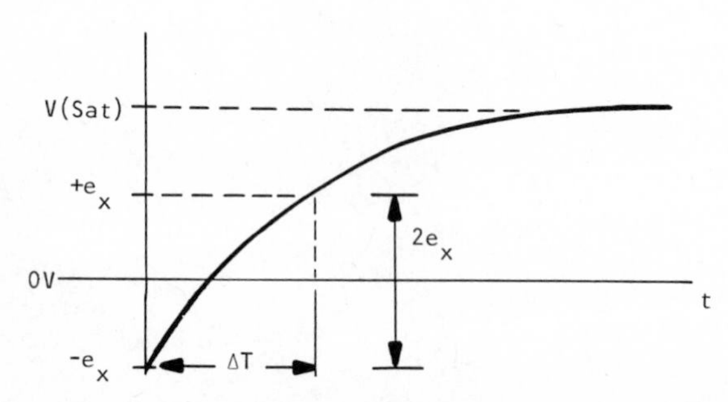

Figure 2. Charging Voltage of Capacitor C_1.

The exponential charging of the capacitor C_1 as shown in Figure 2 is

$$e = E_{max}\ (1 - \varepsilon^{-t/RC})$$

Starting from $(-)e_x$ and substituting at $(+)e_x$ gives a pulse duration ΔT.

$$e_o = (+)e_x - (-)e_x = 2e_x\ , \qquad E_{max} = V(Sat) + e_x$$

$$2e_x = \left(V(Sat) + e_x\right)\left(1 - \varepsilon^{-\Delta T/R_1C_1}\right) .$$

Rearranging,

$$\varepsilon^{-\Delta T/R_1C_1} = 1 - \frac{2e_x}{V(Sat) + e_x}$$

Since,

$$e_x = V(Sat)\ \frac{R_3}{R_2+R_3}$$

Substituting for e_x,

$$\varepsilon^{-\Delta T/R_1C_1} = \frac{1}{1 + \frac{2R_2}{R_3}}$$

and the final equation for determining ΔT is

$$\Delta T = R_1C_1\ \ln\left(1 + \frac{2R_2}{R_3}\right)$$

The result is a square-wave with a period of $2\Delta T$,

or a frequency of $\qquad f = \frac{1}{2\Delta T}$

APPENDIX P

Gain Error

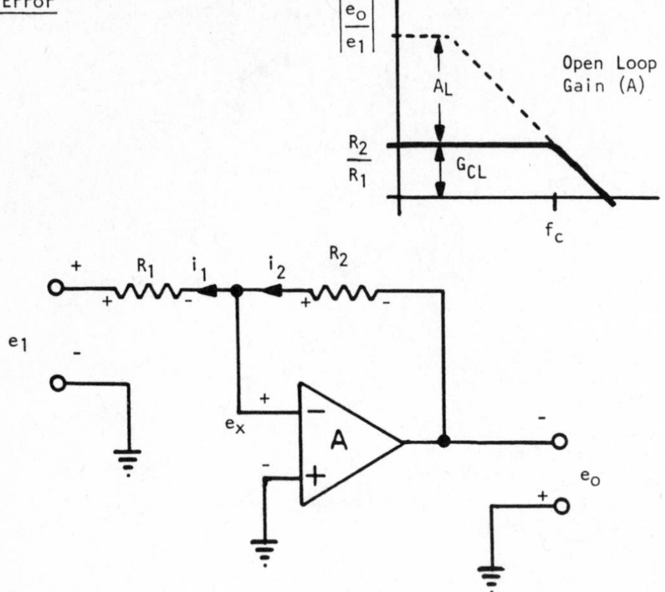

The error caused by the finite open loop gain (A) of the Op Amp can be derived from the above network.

$$i_1 = i_2 \quad , \quad e_x = \frac{e_o}{A}$$

$$i_1 = \frac{e_1 - e_x}{R_1} \quad , \quad i_2 = \frac{e_o + e_x}{R_2}$$

Substituting,

$$\frac{e_1 - \frac{e_o}{A}}{R_1} = \frac{e_o + \frac{e_o}{A}}{R_2}$$

solving for output gain,

$$\left|\frac{e_o}{e_1}\right| = \left(\frac{R_2}{R_1}\right) \left[\frac{1}{1 + \dfrac{1 + \dfrac{R_2}{R_1}}{A}}\right]$$

Ideal Gain (first factor); Gain Error (second factor)

From classical feedback theory, the term β is the feedback attenuation and is defined as the factor by which the output voltage (e_o) is reduced to produce the error voltage (e_x) with the forward gain open and with the input source replaced by its Thevenin equivalent.

$$\beta = \frac{e_x}{e_o} = \frac{R_1}{R_1 + R_2}$$

rearranging,

$$\frac{1}{\beta} = 1 + \frac{R_2}{R_1}$$

Substituting β into the output gain equation the error can now be expressed in terms of loop gain $A\beta$.

$$\frac{e_o}{e_1} = \frac{R_2}{R_1}\left(\frac{1}{1 + \frac{1}{A\beta}}\right)$$

For R_2 greater than R_1,

$$\frac{1}{\beta} \simeq \frac{R_2}{R_1} = \frac{e_o}{e_1} = \text{Closed Loop Gain}$$

Consequently, Loop Gain ($A\beta$) is approximately the ratio of Open Loop Gain and Closed Loop Gain (G_{CL}).

$$A\beta \simeq \frac{A}{G_{CL}} = A_L$$

The Op Amp network gain error can now be expressed in terms of the Op Amp network Loop Gain (A_L)

$$\frac{e_o}{e_1} \simeq \underbrace{\frac{R_2}{R_1}}_{\text{Ideal Gain}} \underbrace{\left(\frac{1}{1 + \frac{1}{A_L}} \right)}_{\text{Gain Error}}$$

INDEX